Lecture Notes in Artificial Intelligence 5729

Edited by R. Goebel, J. Siekmann, and W. Wahlster

Subseries of Lecture Notes in Computer Science

Václav Matoušek   Pavel Mautner (Eds.)

Text, Speech
and Dialogue

12th International Conference, TSD 2009
Pilsen, Czech Republic, September 13-17, 2009
Proceedings

 Springer

Series Editors

Randy Goebel, University of Alberta, Edmonton, Canada
Jörg Siekmann, University of Saarland, Saarbrücken, Germany
Wolfgang Wahlster, DFKI and University of Saarland, Saarbrücken, Germany

Volume Editors

Václav Matoušek
Pavel Mautner
University of West Bohemia at Pilsen
Department of Computer Science and Engineering
E-mail:{matousek, mautner}@kiv.zcu.cz

Library of Congress Control Number: 2009933673

CR Subject Classification (1998): I.2, J.5, H.5.2, I.2.7, I.2.6, H.3

LNCS Sublibrary: SL 7 – Artificial Intelligence

ISSN 0302-9743

ISBN 978-3-642-04207-2 Springer Berlin Heidelberg New York

springer.com

© Springer-Verlag Berlin Heidelberg 2009

Typesetting: Camera-ready by author, data conversion by Scientific Publishing Services, Chennai, India
Printed on acid-free paper   SPIN: 12748013   06/3180   5 4 3 2 1 0

Preface

TSD 2009 was the 12th event in the series of International Conferences on Text, Speech and Dialogue supported by the International Speech Communication Association (ISCA) and Czech Society for Cybernetics and Informatics (ČSKI). This year, TSD was held in Plzeň (Pilsen), in the Primavera Conference Center, during September 13–17, 2009 and it was organized by the University of West Bohemia in Plzeň in cooperation with Masaryk University of Brno, Czech Republic. Like its predecessors, TSD 2009 highlighted to both the academic and scientific world the importance of text and speech processing and its most recent breakthroughs in current applications. Both experienced researchers and professionals as well as newcomers to the text and speech processing field, interested in designing or evaluating interactive software, developing new interaction technologies, or investigating overarching theories of text and speech processing found in the TSD conference a forum to communicate with people sharing similar interests. The conference is an interdisciplinary forum, intertwining research in speech and language processing with its applications in everyday practice. We feel that the mixture of different approaches and applications offered a great opportunity to get acquainted with current activities in all aspects of language communication and to witness the amazing vitality of researchers from developing countries too.

This year's conference was partially oriented toward semantic processing, which was chosen as the main topic of the conference. All invited speakers (Frederick Jelinek, Louise Guthrie, Roberto Pieraccini, Tilman Becker, and Elmar Nöth) gave lectures on the newest results in the relatively broad and still unexplored area of semantic processing.

This volume contains a collection of submitted papers presented at the conference, which were thoroughly reviewed by three members of the conference reviewing team consisting of more than 40 top specialists in the conference topic areas. A total of 53 accepted papers out of 112 submitted, altogether contributed by 127 authors and co-authors, were selected by the Program Committee for presentation at the conference and for inclusion in this book. Theoretical and more general contributions were presented in common (plenary) sessions. Problem-oriented sessions as well as panel discussions then brought together specialists in limited problem areas with the aim of exchanging knowledge and skills resulting from research projects of all kinds.

We would like to gratefully thank the invited speakers and the authors of the papers for their valuable contributions, and the ISCA and ČSKI for their financial support. Last but not least, we would like to express our gratitude to the authors for providing their papers on time, to the members of the conference reviewing team and Program Committee for their careful reviews and paper selection, to the editors for their hard work preparing this volume, and to the members of the local Organizing Committee for their enthusiasm during the conference organization.

June 2009   Václav Matoušek

Organization

TSD 2009 was organized by the Faculty of Applied Sciences, University of West Bohemia in Plzeň (Pilsen), in cooperation with the Faculty of Informatics, Masaryk University in Brno, Czech Republic. The conference Website is located at:
http://www.kiv.zcu.cz/tsd2009/ or http://www.tsdconference.org.

Program Committee

Frederick Jelinek (USA), *General Chair*
Hynek Heřmanský (Switzerland), *Executive Chair*
Eneko Agirre (Spain)
Geneviève Baudoin (France)
Jan Černocký (Czech Republic)
Alexander Gelbukh (Mexico)
Louise Guthrie (UK)
Jan Hajič (Czech Republic)
Eva Hajičová (Czech Republic)
Patrick Hanks (UK)
Ludwig Hitzenberger (Germany)
Jaroslava Hlaváčová (Czech Republic)
Aleš Horák (Czech Republic)
Eduard Hovy (USA)
Ivan Kopeček (Czech Republic)
Steven Krauwer (The Netherlands)
Siegfried Kunzmann (Germany)
Natalija Loukachevitch (Russia)
Václav Matoušek (Czech Republic)
Hermann Ney (Germany)
Elmar Nöth (Germany)
Karel Oliva (Czech Republic)
Karel Pala (Czech Republic)
Nikola Pavešić, (Slovenia)
Vladimír Petkevič (Czech Republic)
Fabio Pianesi (Italy)
Roberto Pieraccini (USA)
Adam Przepiorkowski, (Poland)
Josef Psutka (Czech Republic)
James Pustejovsky (USA)
Léon J. M. Rothkrantz (The Netherlands)
Milan Rusko (Slovakia)
Ernst Günter Schukat-Talamazzini (Germany)

Pavel Skrelin (Russia)
Pavel Smrž (Czech Republic)
Petr Sojka (Czech Republic)
Marko Tadić (Croatia)
Tamás Varadi (Hungary)
Zygmunt Vetulani (Poland)
Taras Vintsiuk (Ukraine)
Yorick Wilks (UK)
Victor Zakharov (Russia)

Local Organizing Committee

Václav Matoušek *(Chair)*
Kamil Ekštein
Ivan Habernal
Jan Hejtmánek
Jana Hesová
Martin Hošna
Jana Klečková
Miloslav Konopík
Jana Krutišová
Pavel Mautner
Roman Mouček
Helena Ptáčková *(Secretary)*
Tomáš Pavelka

About Plzeň (Pilsen)

The New Town of Pilsen was founded at the confluence of four rivers – Radbuza, Mže, Úhlava and Úslava – following a decree issued by the Czech king, Wenceslas II. He did so in 1295. From the very beginning, the town was a busy trade center located at the crossroads of two important trade routes. These linked the Czech lands with the German cities of Nuremberg and Regensburg.

In the fourteenth century, Pilsen was the third largest town after Prague and Kutna Hora. It comprised 290 houses on an area of 20 ha. Its population was 3,000 inhabitants. In the sixteenth century, after several fires that damaged the inner center of the town, Italian architects and builders contributed significantly to the changing character of the city. The most renowned among them was Giovanni de Statia. The Holy Roman Emperor, the Czech king Rudolf II, resided in Pilsen twice between 1599 and 1600. It was at the time of the Estates revolt. He fell in love with the city and even bought two houses neighboring the town hall and had them reconstructed according to his taste.

Later, in 1618, Pilsen was besieged and captured by Count Mansfeld's army. Many Baroque style buildings dating to the end of the seventeenth century were designed by Jakub Auguston. Sculptures were made by Kristian Widman. The historical heart of the city – almost identical with the original Gothic layout – was declared a protected historic city reserve in 1989.

Pilsen experienced a tremendous growth in the first half of the nineteenth century. The City Brewery was founded in 1842 and the Skoda Works in 1859. With a population of 175,038 inhabitants, Pilsen prides itself on being the seat of the University of West Bohemia and Bishopric.

The historical core of the city of Pilsen is limited by the line of the former town fortification walls. These gave way, in the middle of the nineteenth century, to a green belt of town parks. Entering the grounds of the historical center, you walk through streets that still respect the original Gothic urban layout, i.e., the unique developed chess ground plan.

You will certainly admire the architectonic dominant features of the city. These are mainly the Church of St. Bartholomew, the loftiness of which is accentuated by its slim church spire. The spire was reconstructed into its modern shape after a fire in 1835, when it was hit by a lightening bolt during a night storm.

The placement of the church right within the grounds of the city square was also rather unique for its time. The church stands to the right of the city hall. The latter is a Renaissance building decorated with graffiti in 1908–12. You will certainly also notice the Baroque spire of the Franciscan monastery.

All architecture lovers can also find more hidden jewels, objects appreciated for their artistic and historic value. These are burgher houses built by our ancestors in the styles of the Gothic, Renaissance or Baroque periods. The architecture of these sights was successfully modeled by the construction whirl of the end of the nineteenth century and the beginning of the twentieth century.

Thanks to the generosity of the Gothic builders, the town of Pilsen was predestined for free architectonic development since its very coming to existence. The town has therefore become an example of a harmonious coexistence of architecture both historical and historicizing.

Sponsoring Institutions

International Speech Communication Association (ISCA)
Czech Society for Cybernetics and Informatics (CSKI)

Table of Contents

Speech

Dialog

Code Breaking for Automatic Speech Recognition

Frederick Jelinek

Center for Language and Speech Processing,
The Johns Hopkins University, Baltimore, USA

Abstract. Practical automatic speech recognition is of necessity a (near) real time activity performed by a system whose structure is fixed and whose parameters once trained may be adapted on the basis of the speech that the system observed during recognition.

However, in specially important situations (e.g., recovery of out-of-vocabulary words) the recognition task could be viewed as an activity akin to code–breaking to whose accomplishment can be devoted an essentially infinite amount effort. In such a case everything would be fair, including, for instance, the retraining of a language and/or acoustic model on the basis of newly acquired data (from the Internet!) or even a complete change of the recognizer paradigm.

An obvious way to proceed is to use the basic recognizer to produce a lattice or confusion network and then do the utmost to eliminate ambiguity. Another possibility is to create a list of frequent confusions (for instance the pair IN and AND) and prepare a appropriate individual decision processes to resolve each when it occurs in test data. We will report on our initial code breaking effort.

V. Matoušek and P. Mautner (Eds.): TSD 2009, LNAI 5729, p. 1, 2009.

The Semantics of Semantics in Language Processing

Louise Guthrie

Department of Computer Science, University of Sheffield
l.guthrie@dcs.shef.ac.uk

Abstract. In speech and language research, the semantics of an utterance always corresponds to the meaning of the utterance. Meaning however, is a concept that has been argued by philosophers for centuries, so in Language Processing, semantics has come to be used very differently in different applications. One can even make the case that although we believe we must program computers to represent the "semantics" of an utterance, we often have great difficulty as humans defining exactly what we want. The talk will give an overview of formal semantics, lexical semantics and conceptual semantics and then focus on how semantics is used in several application areas of Language Processing.

V. Matoušek and P. Mautner (Eds.): TSD 2009, LNAI 5729, p. 2, 2009.

Are We There Yet?
Research in Commercial Spoken Dialog Systems

Roberto Pieraccini, David Suendermann, Krishna Dayanidhi, and Jackson Liscombe

SpeechCycle, Inc., 26 Broadway, 11th Fl., New York, NY
{roberto,david,krishna,jackson}@speechcycle.com

Abstract. In this paper we discuss the recent evolution of spoken dialog systems in commercial deployments. Yet based on a simple finite state machine design paradigm, dialog systems reached today a higher level of complexity. The availability of massive amounts of data during deployment led to the development of continuous optimization strategy pushing the design and development of spoken dialog applications from an art to science. At the same time new methods for evaluating the subjective caller experience are available. Finally we describe the inevitable evolution for spoken dialog applications from speech only to multimodal interaction.

1 Introduction

Four years ago, a review of the state of the art of commercial and research spoken dialog systems (SDS) was presented at the 2005 SIGdial workshop on Discourse and Dialog in Lisbon, Portugal [2]. The main point of the review was that research and commercial endeavors within the SDS technology have different goals and therefore aim at different paradigms for building interactive systems based on spoken language. While usability, cost effectiveness, and overall automation rate—which eventually affect the ROI, or Return on Investment, provided by the application—have always been the primary goals of commercial SDS, the research community has always strived to achieve user's interaction naturalness and expression freedom. We also established that the latter—naturalness and expression freedom—do not necessarily lead to the former, i.e. usability, automation, and cost effectiveness. One of the reasons of that is that today's technology cannot imitate the human spoken communication process with the same levels of accuracy, robustness, and overall effectiveness. Speech recognition makes errors, language understanding makes errors, and so user interfaces are far from being as effective as humans. For all of this technology to work, one has to impose severe limitations on the scope of the applications, which require a great amount of manual work for the designers. These limitations are often not perceived by the users and that often creates a problem. The more anthropomorphic the system, the more the user tend to move into a level of comfort, freedom of expression, and naturalness which is eventually not supported by the application, creating a disconnect between the user and the machine that leads to an unavoidable communication breakdown.

V. Matoušek and P. Mautner (Eds.): TSD 2009, LNAI 5729, pp. 3–13, 2009.

This progression would follow a curve similar to that predicted by the uncanny valley [1] theory[1], proposed in 1970 by Masahiro Mori. Michael Phillips proposed a similar trajectory, shown in Figure 1, for the usability/flexibility curve in his keynote speech at Interspeech 2006.

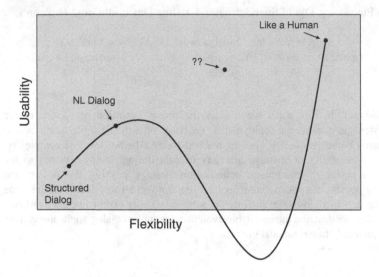

Fig. 1. The uncanny valley of usability[2]

Practically, as the flexibility (naturalness and freedom of expression) of a spoken dialog system increases, so does its usability, until a point when usability starts to decrease, to increase again when the system becomes almost indistinguishable from a human. Reaching the point identified by the question marks is very difficult, practically impossible, with today's technology, and the apparent naturalness provided by advanced research prototypes—often demonstrated in highly controlled environments—inevitably drives spoken dialog systems towards the uncanny valley of usability. According to the curve of Figure 1, today we are still in a place between structured—also known as directed—and natural language (NL) dialog.

One of the conclusions from the 2005 review is that, in order to fulfill the necessary usability requirements to drive automation and customer acceptance, commercial systems need to have completely defined Voice User Interfaces (VUI) following a principle called VUI completeness. An interface is VUI complete when its behavior is completely specified with respect to all possible situations that may arise during the interaction, including user inputs, and backend system responses. Today, in order to guarantee VUI completeness, it is best practice to come to a complete specification of

---

[1] The uncanny valley theory states that humans respond positively to machines with an increased human likeness up to a point where it would start causing an increasing feeling of repulsion. Repulsion will start decreasing again after a certain degree of humanness is achieved by the machine, after which a positive and empathic response is produced again. The interval of human likeness in which machines provoke a sense of repulsion is called the *uncanny valley*.

[2] This chart published with permission of Mike Phillips, Vlingo.

the interface during an initial design phase. The only dialog management paradigm today that allows for producing VUI complete interfaces in an economical and reproducible way is the functional one [2] embodied by the finite state controller representation known as *call flow*. Unfortunately, other more sophisticated mechanisms, such as those based on inference or statistical models, would require much higher a complexity and costs in order to guarantee VUI completeness.

In the rest of this paper we will describe the advancements we have witnessed in spoken dialog technology during the past four years. We will put particular emphasis on complex and sophisticated systems deployed for commercial use with a large volume of interactions—i.e. millions of calls per month. The main trend in this area is an increased use of data for improving the performance of the system. This is reflected by intense usage of statistical learning—even though with quite simplistic mechanisms—at the level of the dialog structure as well as at the level of the individual prompts and grammars. To that respect, the availability of large amounts of data and the ability to transcribe and annotate it at a reasonable cost, suggests deprecating the use of traditional *rule-based grammars*, widely used in the industry, and substituting them with statistical grammars. We will then discuss the use of data as the basis for the evaluation of subjective characterization of dialog systems, such as caller experience and caller cooperation. Finally, we will conclude with a mention of the current trends, including the evolution of SDS towards multimodal interaction.

2 RPA: Rich Phone Applications

There is not an established measure of the complexity of a dialog system. As software is often measured in *lines of code,* if a system is built using the traditional call flow paradigm one can use its size, in terms of nodes and arcs, as a measure of complexity. Most of the call flows built today are based on the notion of dialog modules. A dialog module is an object, typically associated with a node of the call flow, designed to collect a single piece of information, such as a date, a number, a name, a yes/no answer, or a free form natural language input. Dialog modules include retry, timeout, and the confirmation strategy necessary to prompt for and collect the information from the user even in presence of speech recognition errors, missed input, or low confidence. The number of dialog modules used in a system is a fair indicator of its complexity.

As discussed in [3], during the past decades, we have seen dialog systems evolve through three generations of increasingly more sophisticated applications: informational, transactional, and problem-solving, respectively. Correspondingly, their complexity evolved towards deployed systems that include several hundreds of dialog modules and span dozens of turns for several minutes of interactions. Yet, the call flow paradigm is still used to build these sophisticated and complex systems.

We associate the most sophisticated spoken dialog application today to a category called Rich Phone Applications, or RPAs. RPAs are characterized by a certain number of features, including:

a) **Composition and integration with other applications.** Non integrated, or blind, spoken dialog systems provide very limited automation at the expense of poor user experience, especially for customer care applications, in analogy with human agents who do not have any access to caller information, nor to account management tools, diagnostic tools, or external knowledge sources. The development cost of providing integration of a dialog system with external backends is often higher than that of building the voice user interface. For that purpose, an orchestration layer that allows the management of external object abstractions characterizes the most advanced dialog systems today, and it is a fundamental feature of RPAs.

b)   **Asynchronous interaction behavior.** Since RPAs can interact with multiple external backends, and at the same time interact with a user using spoken language, there is no reason not to carry on all these interactions, whenever appropriate, simultaneously. So, an RPA can, for instance, retrieve the caller's account from a database, check the status of his bill, and perform diagnostics on his home equipment while, at the same time, ask a few questions about the reason of the call. That will greatly speed up the interaction creating the basis for improved user experience and automation.

c)   **Continuous tuning.** Spoken dialog systems, once deployed, generally do not offer the best possible performance. The main reason is that their design is generally based on a number of assumptions which are not always verified until large amounts of data are available. First, the voice user interface is often based on what is considered *best practice,* but the reality of the particular context in which the application is deployed, the distribution of the user population, their peculiar attitude towards automated systems and the provider, their language, and many other variables may be different from what was envisioned at design time. The designed prompts may not solicit exactly the intended response, and the handcrafted rule-based grammars may not be able to catch the whole variety of expressions used by the callers. Besides that, things may be changing during the course of deployment. New situations may arise with new terms—e.g. think of marketing campaigns introducing products with new names and callers requesting assistance with that—and changes made to the system prompts or logic—it is not uncommon to have deployed spoken dialog systems updated monthly or more often by the IT departments who owns them. That may impose changes on the user interface that could invalidate the existing rule-based grammars. Thus, a feature of RPAs is the ability to undergo continuous data-driven tuning with respect to the logic and the speech resources, such as prompts and grammars.

d)   **Multimodal interaction.** Most of the issues related to the usability of an imperfect technology such as speech recognition would be greatly alleviated if the interaction would evolve with the aid of a second complementary modality. This has been known for a long time, and multimodal interaction has been the object of study for several decades. Most recently, W3C published a recommendation for a markup language, EMMA[3], targeted at representing the input to a multimodal system in a unified manner. However, the commercial adoption of multimodal systems with speech recognition as one of the modalities has been very limited. First of all the convenience of

---

[3] http://www.w3.org/TR/emma/

speech input on a PC with a regular keyboard is questionable. Speech can be useful indeed when small and impractical keyboards are the only choice, for instance on a smartphone or PDA. However, only recently, the adoption of powerful pocket devices became ubiquitous together with the availability of fast wireless data networks, such as 3G. Wireless multimodal interaction with speech and GUIs is a natural evolution of the current speech-only interaction.

3 Managing Complexity

One of the enablers of the evolution of complex and sophisticated RPAs is the availability of authoring tools that allow building, managing, and testing large call flows. We can say that the most advanced tools for authoring spoken dialog systems today are those that create a meta-language description that is eventually interpreted by a runtime system, which in turn generates dynamic markup language (i.e. VoiceXML) for the speech platform. Other types of authoring models, such as for instance direct VoiceXML authoring, does not allow managing the complexity necessary to create sophisticated systems of the RPA category. The question is how did authoring evolve to allow the creation and maintenance of call flows with hundreds of processes, dialog modules, prompts, and grammars. The answer to this question relies on the consideration that a call flow is not just a graph designed on a canvas, but it is software. At the beginning of the commercial deployment of spoken dialog systems, in the mid 1990s, the leading model was that of using very simplistic *state machine* representations of the call flow, associated with very poor programming models. Most often, the call flow design and development tools did not even offer hierarchical decomposition, relying on pasting together several pages of giant monolithic charts linked by *goto* arcs. Rather, if you consider a call flow as a particular embodiment of the general category of modern procedural software, you may take advantage of all the elements that allow reducing and managing the complexity of sophisticated applications. For instance, call flow development tools can borrow most, if not all, of the abstractions used today in modern languages, such as inheritance, polymorphism, hierarchical structure, event and exception handling, local and global variables, etc. The question then is why not develop dialog applications in regular Java or C# code, rather than by using the typical drag-n-drop graphical paradigm which is being adopted by most call flow development tools. There are several answers to that. First, call flows are large hierarchical state machines. Programming large finite state machines in regular procedural languages, like Java or C#, results in an unwieldy nest—each state corresponds to a nested statement—of case statements, which is quite unreadable, unintuitive, and difficult to manage. In addition to that we have also to consider that the modern development lifecycle calls for a reduction of the number of steps required creating, deploying, and maintaining spoken dialog applications. A few years ago, it was common practice to have a full dialog specification step, generally performed by a VUI designer on paper, and followed by an implementation step, generally carried on by a software developer. That is impractical today for complex RPAs. Rather, advanced VUI designers today fully develop the application on rapid development tools on their own, while software developers just provide connectivity with the backend systems. The availability of powerful visual development tools with embedded software abstractions perfectly fit this development paradigm.

4 Continuous Tuning

Even though relying on the best design practices and the most powerful authoring tools, designers need to make assumption that may not be verified in practice. Because of that, dialog systems typically underperform when deployed for the first time. Also, as we stated early, there are environmental changes that may happen during the lifecycle of a deployed system and that can affect its performance. Because of that, the ability to perform continuous tuning is an essential feature of RPAs. There are at least two elements of a dialog system that can be tuned, namely the VUI and the speech grammars.

4.1 Data Driven VUI Tuning

With the introduction of Partially Observable Markov Decision Processes (POMDPs) [4] there has been a fundamental evolution in dialog policy learning by taking into consideration the uncertainty derived by potential speech recognition errors. While dialog learning research is developing potential breakthrough technology, commercial deployment cannot yet take advantage of it [5] because of the inherent impracticalities of learning policies from scratch through real user interactions—in fact, research is using simulated users—and the difficulty of designing reward schemas, as opposed to explicitly specifying the interaction behavior as in standard call flow development.

However, there is a lot of interest to explore *partial* learning of VUI in the commercial world of spoken dialog systems. This interest spurs from the fact that there is a clear dependency between the performance of a system (e.g. overall automation rate, caller experience, abandon rate, etc.) and the specific VUI strategy and prompts. This dependency has been always considered an expression of the VUI designer *art*, rather than a *science*. Designers have a baggage of best practice experience on how to create prompts and strategies that would solicit certain behavior, but often they have to choose among different possibilities, without being certain which is the one that would perform best in a given situation. Using data is an effective approach to this issue. Whenever alternative strategies are possible from the design point of view, one could alternate among them in a random fashion, and then measure the individual contribution to an overall evaluation criterion, such as automation rate or caller experience. One can also make the random selection of alternative strategies depend on the current interaction parameters, such as time of the day, day of the week, or some type of caller's characterization (e.g. type of subscription account, area code, etc.). Once enough data is collected, one can decide which strategy offers the best performance for each set of parameters or chose an online optimization strategy that changes the frequency of usage of each alternative strategy based on an exploration/exploitation criterion similar to that used in reinforcement learning theory.

4.2 Speech Recognition Continuous Tuning

What makes a spoken dialog system a *spoken* dialog system is its capability to interact with a user by means of input and output speech. While output speech can be produced with highest intelligibility and quality using pre-recorded prompts combined

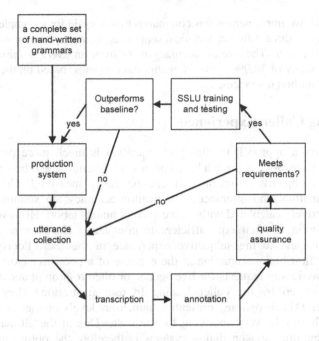

Fig. 2. The continuous speech recognition tuning cycle

using concatenative speech synthesis, the performance of speech recognition and understanding of these systems is often subject to criticism and user frustration.

The reason for non optimal performance of commercially deployed systems is based on the common practice to consider speech recognition and understanding as an art rather than science. Voice interaction designers and "speech scientists" design rules representing utterances they expect callers to speak at the various contexts of the interaction. This rule set is generally referred to as *rule-based grammar*. Sometimes, the designers and speech scientists may listen to some live calls, collect and transcribe a limited number of utterances and tweak the grammars, pronunciation dictionaries and noise models, set thresholds, sensitivities, and time-outs most often based on anecdotic experience or best practice.

In contrast to this common *handcrafting* approach, only recently did we present a framework [8] systematically replacing all rule-based grammars in spoken dialog systems by statistical language models and statistical classifiers trained on large amounts of data collected at each recognition context. This data collection involves partially automated transcriptions and semantic annotation of large numbers of utterances which are collected over the lifecycle of an application. The statistical grammars, that replace the hand written ones, are constantly retrained and tested against the baseline grammars currently used in the production system. Once the new grammars show significant improvement, they are released into production, providing improvement of its performance over time. Figure 2 shows an implementation of this continuous improvement cycle which also involves steps to assure highest reliability and consistency of the involved semantic annotations [6].

As an example, we implemented this continuous tuning cycle for a complex deployed system consisting of three different technical support applications interconnected via an open-prompt call router. The overall accuracy went from an average initial semantic classification accuracy of 78.0% to 90.5% within three months, based on the processing of more than 2.2 million utterances.

5   Evaluating Caller Experience

While a data-driven approach to dialog optimization is much more powerful and complete than an anecdotic approach based on human experience and best practices, it does require an objective criterion that can be easily measured. Call duration, successful automation, and utterance classification accuracy are suitable measures that can be effectively calculated without the need of human labor. However this type of objective criteria is often not sufficient to completely characterize a system's behavior with respect to the subjective experience of the user. Poorly designed systems may exhibit high completion at the expense of a poor user experience. For instance they may ignore requests for live agents, or often re-prompt and confirm the information produced by the caller leading to user frustration. They may also extensively offer DTMF options, presenting numerous levels of menus and yes/no questions, which may be very annoying for the user. One of the ultimate goals of designing and building spoken dialog systems is therefore the optimization of the *caller experience*. In contrast to impractical and often biased implementations of caller surveys we, introduced a method based on subjective evaluation of call recordings by a team of listeners. Each sample call is scored on a scale between 1 (very bad) and 5 (excellent). The average of these ratings across several calls and several listeners expresses the overall caller experience [7].

As an example, we tracked the caller experience of a complex call routing application over a three-month tuning cycle and showed that the average score improved from 3.4 to 4.6 as reported in Figure 3.

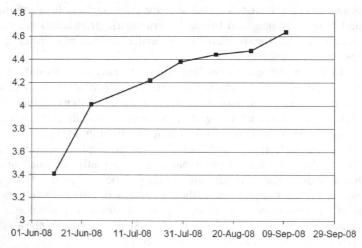

Fig. 3. Evolution of caller experience over multiple versions of a spoken dialog system

However, the nature of the caller experience score as introduced above being based on subjective ratings makes this measure impractical for automated system tuning or real-time reporting. To overcome this issue, research is directed towards the prediction of caller experience based on available objective and measureable parameters such as the number of no-matches, operator requests, call duration, etc. Experiments show that automatic prediction may achieve accuracy almost as high as human experts considering substantial inter-expert disagreement.

6 Wireless Multimodal Interactions

The trend is uncontestable. Soon everyone or nearly everyone will use a smartphone as their primary means of communication. Although this becomes immediately evident by walking around or riding the subway of a city like New York, the trend is supported by analyst research. For instance, a report by Harris Interactive on telephony usage for US adults shows that between Q4 2007 and Q1 2008 89% the adoption of mobile phones is 89%, compared with 79% of landline, and 15% of internet telephony. However, the most interesting piece of data is that, still according to Harris Interactive, 14% of the adult population is composed of *cordcutters*: people who do not have a fixed landline phone at home, and for which a mobile device is their only access to the telephone network. Another report shows this number to be 18% by the end of 2008.

As POTs, or plain old telephones, are quickly replaced by powerful pocket size devices with voice, data, and display capabilities, spoken dialog systems cannot dwell on the same old *"please press or say one"* voice + DTMF interaction paradigm. The ability to compress voice on the device, send it to a remote server through a fast 3G data link, and display information at the same time will shift the spoken dialog interaction paradigm towards wireless multimodal communication with machines. It is interesting to notice that just by adding a visual component to a speech recognition system one can reduce or remove most of the issues that are present in today spoken dialog interaction systems. As we know, speech recognition errors are one of the main causes of the slow adoption of speech systems, and copying with them is part of the VUI design of an application. Given the narrow bandwidth of the speech channel, information has to be presented to users in a sequential manner. Even though the accuracy of speech recognition may be close to 100% when considering alternative ranked hypotheses from the recognizer—what is typically called N-best—they have to be presented one by one, making the user experience quite unpleasant, and giving raise to the typical negative perception acquired by speech recognition interfaces. If you wanted to take advantage of the N-best list, the only possibility in a speech-only interface is to present the various alternatives in a sequential manner, as in the following example:

Caller: Boston
System: I think you said Austin, is that correct?
Caller: No, I said Boston
System: My mistake. Was it Houston?
Caller: No … Boston!
System: Sorry. I seem to have trouble understanding you. Maybe it was Boston?
Caller: YES!

Of course this strategy may lead to unacceptably long dialogs. In fact, most of the VUI designs today avoid using the N-best list and either re-prompt the caller, or move on to a backup dialog. The situation is even more exacerbated when dealing with open prompt input, since there are more semantic possibilities, and the difference between them may be more subtle than the acoustic similarity among words (as in Boston vs. Austin).

So, speech recognition today counts on the first recognition hypothesis to be correct, while we know very well that, with the current technology, the first hypothesis may not be the correct one in a small, but still significant, number of time (e.g. between 5% and 10%). Moving from speech-only to speech+GUI interfaces, we gain the advantage of eliminating the sequentiality constraint, and allow users to *browse* the results, and select the correct response, even though it is not the first best one.

7 Conclusions

Commercial spoken dialog application evolved during the past few years both in complexity and performance. The leading dialog management design paradigm is still based on the call-flow abstraction. However, the most sophisticated design tools have embraced a call-flow-as-software paradigm, where the interaction is still represented in a graphical way as a hierarchical finite state machine controller, but the programming model is enriched with most of the primitives of modern procedural programming languages. This design paradigm allows streamlining the development of complex applications such as those in the category of problem solving. A new category of spoken dialog applications, identified as RPAs, or Rich Phone Applications, is emerging as the voice interaction analog of Rich Internet Applications. The enhanced experience of RPA users is derived by a sophisticated interaction of the application with external services in an asynchronous manner, the capability of continually adapting the system, and the use of multimodal interaction whenever possible. In particular, while traditionally commercial systems were designed based on a best practice paradigm, we see an increased use of data driven optimization of spoken dialog system. The voice user interface can be optimized by selecting among competing strategies by using an exploitation/exploration paradigm. The speech grammars, which are commonly rule-based and built by hand, are being substituted by statistical grammars and statistical semantic classifiers which can be continuously trained from data collected while the system is deployed. Finally we discussed how speech only dialog systems are evolving towards multimodal interaction systems due to the growing adoption of smartphones as a unified communication device with voice, data, and visual display capabilities.

References

1. Mori, M.: Bukimi no tani The uncanny valley (K. F. MacDorman & T. Minato, Trans.). Energy 7(4), 33–35 (1970) (Originally in Japanese)
2. Pieraccini, R., Huerta, J.: Where do we go from here? Research and Commercial Spoken Dialog Systems. In: Proc. of 6th SIGdial Workshop on Discourse and Dialog, Lisbon, Portugal, September 2-3, pp. 1–10 (2005)

3. Acomb, K., Bloom, J., Dayanidhi, K., Hunter, P., Krogh, P., Levin, E., Pieraccini, R.: Technical Support Dialog Systems, Issues, Problems, and Solutions. In: HLT 2007 Workshop on Bridging the Gap, Academic and Industrial Research in Dialog Technology, Rochester, NY, April 26 (2007)
4. Thomson, B., Schatzmann, J., Young, S.: Bayesian Update of Dialogue State for Robust Dialogue Systems. In: Int. Conf. Acoustics Speech and Signal Processing ICASSP, Las Vegas (2008)
5. Paek, T., Pieraccini, R.: Automating spoken dialogue management design using machine learning: An industry perspective. Speech Communication 50, 716–729 (2008)
6. Suendermann, D., Liscombe, J., Evanini, K., Dayanidhi, K., Pieraccini, R.: C^5. In: Proc. of 2008 IEEE Workshop on Spoken Language Technology (SLT 2008), Goa, India, December 15-18 (2008)
7. Evanini, K., Hunter, P., Liscombe, J., Suendermann, D., Dayanidhi, K., Pieraccini, R.: Caller Experience: a Method for Evaluating Dialog Systems and its Automatic Prediction. In: Proc. of 2008 IEEE Workshop on Spoken Language Technology (SLT 2008), Goa, India, December 15-18 (2008)
8. Suendermann, D., Evanini, K., Liscombe, J., Hunter, P., Dayanidhi, K., Pieraccini, R.: From Rule-Based to Statistical Grammars: Continuous Improvement of Large-Scale Spoken Dialog Systems. In: Proceedings of the 2009 IEEE Conference on Acoustics, Speech and Signal Processing (ICASSP 2009), Taipei, Taiwan, April 19-24 (2009)

Semantic Information Processing
for Multi-party Interaction

Tilman Becker

DFKI GmbH, Stuhlsatzenhausweg 3, D-66123 Saarbrücken, Germany
Tilman.Becker@dfki.de

Abstract. We present ongoing research efforts using semantic representations and processing, combined with machine learning approaches to structure, understand, summarize etc. the multimodal information available from multi-party meeting recordings.

In the AMI and AMIDA projects[1], we are working with our partners on numerous aspects of analysing, structuring and understanding multimodal recordings of multi-party meetings. We are working on applications based on this understanding that allow browsing of meetings as well as supporting tools that can be used during meetings. The AMI and AMIDA corpora of well over 100 hours of meetings are freely available for research purposes.

Semantic Analysis: The main application of semantic analysis is through NLU components that work on the ASR output with semantic parsers. The target semantics are encoded in domain specific ontologies, used to represent the content of the discussion. Other semantic frameworks represent various aspects of the discourse, e.g., dialogue and negotiation acts. These are typically analysed by machine learning systems.

Semantic Processing: An important aspect of analysing the (meeting) documents are various levels of segmentation. On the highest level, these are topic segments that typically last several minutes. On the lowest level, these are dialogue acts that loosely correspond to utterances or sentences. The task of semantic processing is twofold: first, the underlying structure, e.g., statements and responses must be determined; second, a discourse model with inferences over these structures is used to determine the current state of the discussion and thus the final results of a meeting. As an example, we present a complete system that understands and summarizes the decisions made during a meeting.

Semantics–based Presentations: Based on the results of semantic processing, we can generate various summaries of meetings. As examples, we present short abstracts generated using NLG techniques as well as longer storyboards that combine extractive summarization technology with semantic discourse understanding. The latter can be presented as meeting browsers, allowing direct access to the entire meeting recording.

---

[1] This work was partly supported by the European Union 6th FWP IST Integrated Project AMIDA (Augmented Multi-party Interaction with Distance Access), IST-033812,
http://www.amidaproject.org

V. Matoušek and P. Mautner (Eds.): TSD 2009, LNAI 5729, p. 14, 2009.
© Springer-Verlag Berlin Heidelberg 2009

Communication Disorders and Speech Technology

Elmar Nöth[1], Stefan Steidl[1], and Maria Schuster[2]

[1] University of Erlangen-Nuremberg, Chair of Pattern Recognition, Erlangen, Germany
noeth@informatik-uni-erlangen.de
http://www5.informatik-uni-erlangen.de
[2] EuromedClinic, Fürth, Germany

Abstract. In this talk we will give an overview of the different kinds of communication disorders. We will concentrate on communication disorders related
to language and speech (i.e., not look at disorders like blindness or deafness).
Speech and language disorders can range from simple sound substitution to the
inability to understand or use language. Thus, a disorder may affect one or several linguistic levels: A patient with an articulation disorder cannot correctly produce speech sounds (phonemes) because of imprecise placement, timing, pressure, speed, or flow of movement of the lips, tongue, or throat. His speech may
be acoustically unintelligible, yet the syntactic, semantic, and pragmatic level are
not affected. With other pathologies, e.g. Wernicke-aphasia, the acoustics of the
speech signal might be intelligible, yet the patient is – due to mixup of words
(semantic paraphasia) or sounds (phonematic paraphasia) – unintelligible.

We will look at what linguistic knowledge has to be modeled in order to analyze different pathologies with speech technology, how difficult the task is, and
how speech technology is able to support the speech therapist for the tasks diagnosis, therapy control, comparison of therapies, and screening.

V. Matoušek and P. Mautner (Eds.): TSD 2009, LNAI 5729, p. 15, 2009.
© Springer-Verlag Berlin Heidelberg 2009

A Gradual Combination of Features for Building Automatic Summarisation Systems

Elena Lloret and Manuel Palomar

Department of Software and Computing Systems
University of Alicante
San Vicente del Raspeig, Alicante 03690, Spain
{elloret,mpalomar}@dlsi.ua.es

Abstract. This paper presents a Text Summarisation approach, which combines three different features (Word frequency, Textual Entailment, and The Code Quantity Principle) in order to produce extracts from newswire documents in English. Experiments shown that the proposed combination is appropriate for generating summaries, improving the system's performance by 10% over the best DUC 2002 participant. Moreover, a preliminary analysis of the suitability of these features for domain-independent documents has been addressed obtaining encouraging results, as well.

1 Motivation

Text Summarisation (TS) is a Natural Language Processing Task (NLP), which has experienced a great development in recent years, mostly due to the rapid growth of the Internet. More and more, users have to deal with lots of documents in different formats, especially if they want to find something on the Web. Consequently, we need methods and tools to present all information in a clear and concise way, allowing users to save time and resources. Text Summarisation is really useful to help achieve these goals, since its aim is to obtain a reductive transformation of source text to summary text through content condensation by selection and/or generalisation on what is important in the source [1]. Recently, many efforts to incorporate Automatic Summarisation within specific contexts and applications have been made. For instance, in [2] and [3] summaries of retrieved documents are provided by giving users more focused information and improving the effectiveness of Web search engines. Efforts have also concentrated on producing summaries for online documents [4], [5]. On the other hand, summaries are very appropriate within Opinion Mining tasks, in order to summarise, for instance, all the reviews about specific products and services [6], or movies [7].

The purpose of this paper is to analyse the impact of different features - Word Frequency, Textual Entailment and The Code Quantity Principle - on the TS task, in order to find the best approach that better helps identify the most relevant information within a document.

The remainder of this paper is organised as follows: the features suggested for the summarisation process and the preliminary system's architecture is explained in Section 2. Then, in Section 3 the experiments performed are shown, together with their results and analysis. Finally, Section 4 concludes this paper and discusses future work.

V. Matoušek and P. Mautner (Eds.): TSD 2009, LNAI 5729, pp. 16–23, 2009.

2 Combining Features within an Automatic Summarisation System

2.1 Word Frequency

This feature has been widely used in Text Summarisation. The first automatic system proposed in [8] was based on the word frequency count. Moreover, in [9] the impact of word frequency counting on automatic summarisation and its role in human summaries was analysed. It was shown that high frequency words from the original documents were very likely to appear in the human-made summaries, as well.

Regarding these observations, in [10] was hypothesised that sentences containing the most frequent words of a document, would be selected for the final summary, and it was also proved that this feature led to good results. Therefore, the relevance of a sentence with respect to its content words' frequency is defined as the first feature to our system.

2.2 Textual Entailment

The second feature of our summarisation approach is the use of a Textual Entailment (TE) module. Briefly, TE is a NLP task, which consists of determining if the meaning of one text snippet (the hypothesis) can be inferred by another one (the text) [11]. In the literature, we can find several approaches concerning the use of Summarisation and TE: (1) to evaluate summaries, i.e., to decide which summary (among several of them) can be best deduced from its original document [12]; (2) to segment the input by means of different algorithms which employ TE [13]; and (3) to avoid redundant information becoming part of the final summary [10].

In this paper, we integrate a TE module (similar to the one described in [14]) within a summarisation system, in order to deal with the redundancy problem. The identification of these entailment relations helps a summarisation system avoid incorporating repeated information in final summaries. We proceed as it is described in the approach suggested in [10], which has been proven to influence positively on single-document summarisation. On the one hand, this approach computes the entailment between a sentence and the one which follows, and forms groups of sentences in case there is no entailment relation. On the other hand, if there is an entailment relation the second sentence is discarded.

2.3 The Code Quantity Principle

The Code Quantity Principle [15] is a linguistic principle which proves the existence of a proportional relation between how important the information is, and the number of coding elements it has. This assumption is related to the way humans retain the information in their memory, depending on how salience and distinct it is (*the Code-quantity, Attention and Memory Principle* [15]). Moreover, a coding element can vary depending on the desired granularity. Therefore, syllables, as well as noun-phrases, for example, could both be considered coding elements of a text. This principle was first used in automatic summarisation tasks in [16] which considered noun-phrases as coding elements. In [17] it was found that the average length of complex noun-phrases in summary sentences was more than twice as long than those in non-summary sentences. Moreover,

Table 1. Analysis of noun-phrases found in input documents and human summaries

	News	News summaries	Fairy tales	Fairy tales summaries
# words NP	121105	47552	2334	335
# content words doc	155192	58847	3554	483
# total words doc	336008	119492	9598	1167
% words NP (content)	78%	81%	66%	69%
% words NP (total)	36%	40%	24%	29%

we decided to carry out an analysis of the composition of the documents used in this paper. This analysis consisted of calculating the percentage of noun-phrases both input documents and model summaries had, with respect to the total number of words (please see Table 1). In this way we could prove that the decision of selecting noun-phrases as coding units was appropriate, so there was no need to change the granularity, although it would be interesting to experiment with different granularities for future research. As far as the results of the table are concerned, words belonging to noun-phrases are predominant over other types of words in the texts, representing on average more than 70% of all content words, and 30% of the total words of documents.

Besides the approach taken in [16], we also took into consideration the frequency of a word to determine the importance of the words belonging to a noun-phrase. Consequently, the relevance of a sentence was computed adding up the frequency of the content words belonging to noun-phrases, and normalising by the total number of noun-phrases a sentence had.

2.4 System's Architecture

Figure 1 shows an overview of the system's architecture. Although there is a lot of room for improvement in the summarisation system, it serves as a framework which allows us to combine different features to analyse their suitability within the TS task. According to this scheme, the summarisation process can be divided into three stages, the second one being the core of our system. These stages involve: (1) a pre-process stage of the source document, where all unnecessary tags which could introduce noisy

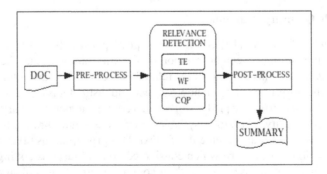

Fig. 1. Overview of the system's architecture

or irrelevant information to the final summary are removed; (2) a summarization process, aiming is to detect the relevant information within a document. At this point, the most important sentences of a document are selected using the features previously described in Section 2. One important remark about the TE module is that it could also be considered as a pre-processing task because when combined with other features, it is always performed first. However, we decided to treat it as a feature since it helps to select important sentences and discard redundant ones. Finally, (3) a post-process stage is performed, which deals with the order of sentences, presenting the selected sentences in the same order as they appeared in the original document. At the moment, the system produces generic single-document extracts in English.

3   Evaluation Environment

We carried out experiments within two different domains: newswire documents, and short stories, particularly well-known fairy tales for children. On the one hand, we followed the guidelines for task 1 within the DUC 2002 conference[1], and we used the documents provided by the organisation (a total of 533 documents) as the input for the system. The reason we used such data was because that edition had a specific task for single-document summarisation, and since then, single-document summarisation was no longer included as a task in any of the following DUC editions[2]. On the other hand, an additional preliminary experiment with fairy tales was performed in order to see whether the studied features could be applied to other kinds of documents (please see Section 3.1). The features suggested were explained in Section 2. Firstly, we selected each feature individually and we ran the system over the 533 newswire documents from DUC. It is important to mention that the Textual Entailment feature is not analysed as a single feature because the TE module was used to remove redundant sentences (by discarding those ones which held an entailment relation). The way sentences were processed in our approach would have led to obtaining a summary very similar to the baseline, due to the fact that the TE module started with the first sentence of a document, and continued with the second, third, fourth, and so on until all sentences were processed. Further on, we combined two features, having three derived approaches: TE+WF, TE+CQP, WF+CQP. Finally, all of the features were combined into a single approach: TE+WF+CQP. In the end, 533 automatic summaries were generated for each of the approaches. An example of a summary automatically generated combining all the features (TE+WF+CQP) can be seen in Figure 2.

Concerning the evaluation process, we used the ROUGE tool [18], which is a state-of-the-art tool for evaluating summaries automatically. Basically, this tool computes the number of different kinds of overlap n-grams between an automatic summary and a human-made summary. For our evaluation, we computed ROUGE-1 (unigrams), ROUGE-2 (bigrams), ROUGE-SU4 (it measures the overlap of skip-bigrams between a candidate summary and a set of reference summaries with a maximum skip distance

---

[1] Document Understanding Conference:
http://www-nlpir.nist.gov/projects/duc/guidelines/2002.html
[2] Currently, the DUC conferences have changed into TAC (Text Analysis Conference) which includes more tasks apart from summarisation.

What do Charlie Chaplin, Greta Garbo, Cary Grant, Alfred Hitchcock and Steven Spielberg have in common?
They have never won Academy Awards for their individual achievements.
Oscar's 60–year history is filled with examples of the film world's highest achievers being overlooked by the Academy of Motion Picture Arts and Sciences.
The honorary award has also proved useful to salve the Academy's conscience.
Douglas Fairbanks, Judy Garland, Noel Coward, Ernst Lubitsch, Fred Astaire, Gene Kelly, Harold Lloyd, Greta Garbo, Maurice Chevalier, Stan Laurel, Cary Grant, Lillian Gish, Edward G. Robinson, Groucho Marx, Howard Hawks and Jean Renoir are others who have received honorary awards.

Fig. 2. Automatic summary for document AP880325-0239

Table 2. Results for individual, groups of two, and all features

Approach	ROUGE-1	ROUGE-2	ROUGE-SU4	ROUGE-L
WF	0.42951	0.16970	0.19618	0.38951
CQP	0.41751	0.17127	0.19261	0.38039
TE+WF	0.44491	0.18726	0.21056	0.40360
TE+CQP	0.42853	0.17885	0.19892	0.38722
WF+CQP	0.44153	0.18565	0.20795	0.39920
TE+WF+CQP	**0.45611**	**0.20252**	**0.22200**	**0.41382**

Table 3. Comparison with the best DUC 2002 participant systems and the baseline

Approach	ROUGE-1	ROUGE-2	ROUGE-SU4	ROUGE-L
TE+WF+CQP	**0.46008**	**0.20431**	**0.22399**	**0.41744**
S28	0.42776	0.21769	0.17315	0.38645
S21	0.41488	0.21038	0.16546	0.37543
DUC baseline	0.41132	0.21075	0.16604	0.37535

of 4), and ROUGE-L (Longest Common Subsequence between two texts). Table 2 shows the results obtained when using the suggested features on its own, or through a combination between them. Although the ROUGE tool allows us to get recall, precision, and F-measure values, only the F-measure is shown in the table, because it is a measure that combines both precision and recall.

Moreover, Table 3 shows a comparison among the top participant systems of the DUC 2002 conference's edition, according to the recall value. The participants' results were reported in [19] where a post-competition evaluation was carried out with the ROUGE tool. In addition, a baseline was also provided within the DUC 2002 conference. This baseline simply generated a summary taking the first 100 words of a document.

3.1 Expansion to Different Domains

Summarization research work has not only focused on newswire domain, but also in biomedical literature [20], legal documents [21], or even in literary fiction [22]. In order to prove that the suggested features could be used in other domains different from newswire, we wanted to carry out an additional experiment with a group of well-known fairy tales for children. For this experiment, we did not take into account the TE module,

Table 4. Results for the fairy tales domain

Approach	ROUGE-1	ROUGE-2	ROUGE-SU4	ROUGE-L
WF	0.39709	0.07939	0.14093	0.31498
CQP+WF	**0.41797**	**0.10267**	**0.15898**	**0.33742**

due to the fact that fairy tales narrate a sequence of events, and normally these events are not repeated. Therefore, as it can be seen in Table 4, we just experimented with WF and CQP+WF features. We set up the same evaluation environment as it was for evaluating the newswire documents, using also the ROUGE tool for evaluating the generated summaries obtaining the F-measure value. It was not possible to gather a high number of fairy tales[3] with its corresponding human-made summaries[4]. So, in the end, we experimented with five different well-known fairy tales, which were: *Cinderella*, *Little Red Riding Hood*, *Jack and the Beanstalk*, *Rumpelstilskin*, and *Hansel and Gretel*.

3.2 Discussion

As can be seen from the experiments previously performed, the approach taking into account all proposed features leads to the best results for summarising newswire documents. When these three features are combined together, an increase of 7% and 12% on average is obtained with respect to two and one features only, respectively. This means that all features together have a positive influence on the summarisation process because when we integrate them in the same approach, the results improve. Furthermore, if we compare this approach with the participant systems in the DUC 2002 conference, we realise that the results outperform the best system by approximately 10%, and an increase of 14% is obtained over the baseline. Giving a general overview of the performance of each feature, it is worth mentioning that the Code Quantity Principle is the feature that obtains the lowest results when experimenting with newswire documents. This may be due to the fact that in this type of document, the information given tends to be more concise without having very long noun-phrases.

Regarding the experiments performed with fairy tales, results show that the combination of features improves the results as well. In this case, when the word frequency count and the Code Quantity Principle are combined, results increase by 14% approximately, with respect to the word frequency used on its own. Again, it is shown that, even changing the domain of the documents, the features perform well and when they are integrated together, the system's performance improves considerably, obtaining very encouraging results for further research.

4   Conclusions and Future Work

In this paper, we presented three approaches (Word Frequency, Textual Entailment and The Code Quantity Principle) that could be used for building an automatic summarisation system. Firstly, we explained each of the features on their own, and then, we

---

[3] Fairy tales text: http://www.surlalunefairytales.com/
[4] Fairy tales summaries: http://www.comedyimprov.com/music/schmoll/tales.html

combined them with the assumption that when used together in the same process, they could lead to better results than used separately. Different analyses were performed, combining all possible features, and finally it was shown that the best approach was the one grouping all the features together, improving the baseline's results by 14%. In a preliminary attempt to determine whether the features considered were appropriate to different types of domains, we also experimented with a small number of fairy tales for children, obtaining very promising results.

For future research, we plan to analyse the suitability of the features presented in other types of domains, such as medical articles. Furthermore, we want to include other types of features, such as semantic knowledge, taking profit of the Wordnet lexical database for English, or other NLP tasks. Our final goal is to integrate the automatic summarisation system within a real application, for instance to produce summaries from products' opinions.

Acknowledgments

This paper has been supported by the FPI grant (BES-2007-16268) from the Spanish Ministry of Science and Innovation, under the project TEXT-MESS (TIN2006-15265-C06-01).

References

1. Spärck Jones, K.: Automatic summarising: The state of the art. Information Processing & Management 43, 1449–1481 (2007)
2. Dunlavy, D.M., O'Leary, D.P., Conroy, J.M., Schlesinger, J.D.: QCS: A system for querying, clustering and summarizing documents. Information Processing & Management 43, 1588–1605 (2007)
3. Yang, X.P., Liu, X.R.: Personalized multi-document summarization in information retrieval. In: International Conference on Machine Learning and Cybernetics, vol. 7, pp. 4108–4112 (2008)
4. Radev, D.R., Otterbacher, J., Winkel, A., Blair-Goldensohn, S.: Newsinessence: summarizing online news topics. Communications of the ACM 48, 95–98 (2005)
5. Steinberger, J., Jezek, K., Sloup, M.: Web topic summarization. In: Proceedings of the 12th International Conference on Electronic Publishing (ELPUB), pp. 322–334 (2008)
6. Titov, I., McDonald, R.: A joint model of text and aspect ratings for sentiment summarization. In: Proceedings of ACL 2008: HLT, Columbus, Ohio, pp. 308–316 (2008)
7. Stede, M., Bieler, H., Dipper, S., Suriyawongkul, A.: Summar: Combining linguistics and statistics for text summarization. In: 17th European Conference on Artificial Intelligence, vol. 141, pp. 827–828 (2006)
8. Luhn, H.P.: The automatic creation of literature abstracts. In: Mani, I., Maybury, M. (eds.) Advances in Automatic Text Summarization, pp. 15–22. MIT Press, Cambridge (1958)
9. Nenkova, A., Vanderwende, L., McKeown, K.: A compositional context sensitive multi-document summarizer: exploring the factors that influence summarization. In: Proceedings of the 29th annual international ACM SIGIR conference on Research and development in information retrieval, pp. 573–580 (2006)
10. Lloret, E., Ferrández, O., Muñoz, R., Palomar, M.: A Text Summarization Approach Under the Influence of Textual Entailment. In: Proceedings of the 5th International Workshop on Natural Language Processing and Cognitive Science (NLPCS 2008), pp. 22–31 (2008)

11. Glickman, O.: Applied Textual Entailment. PhD thesis. Bar Ilan University (2006)
12. Harabagiu, S., Hickl, A., Lacatusu, F.: Satisfying information needs with multi-document summaries. Information Processing & Management 43, 1619–1642 (2007)
13. Tatar, D., Tamaianu-Morita, E., Mihis, A., Lupsa, D.: Summarization by logic segmentation and text entailment. In: Gelbukh, A. (ed.) CICLing 2008. LNCS, vol. 4919, pp. 15–26. Springer, Heidelberg (2008)
14. Ferrández, O., Micol, D., Muñoz, R., Palomar, M.: A perspective-based approach for solving textual entailment recognition. In: Proceedings of the ACL-PASCAL Workshop on Textual Entailment and Paraphrasing, pp. 66–71 (2007)
15. Givón, T.: A functional-typological introduction, II. John Benjamins, Amsterdam (1990)
16. Lloret, E., Palomar, M.: Challenging issues of automatic summarization: Relevance detection and quality-based evaluation. Informatica (Forthcoming 2009)
17. Mittal, V., Kantrowitz, M., Goldstein, J., Carbonell, J.: Selecting text spans for document summaries: heuristics and metrics. In: Proceedings of the 16th national conference on Artificial intelligence and the eleventh Innovative applications of artificial intelligence conference innovative applications of artificial intelligence, pp. 467–473 (1999)
18. Lin, C.Y.: Rouge: a package for automatic evaluation of summaries. In: Proceedings of ACL Text Summarization Workshop, pp. 74–81 (2004)
19. Steinberger, J., Poesio, M., Kabadjov, M.A., Ježek, K.: Two uses of anaphora resolution in summarization. Information Processing & Management 43, 1663–1680 (2007)
20. Plaza, L., Díaz, A., Gervás, P.: Concept-graph based biomedical automatic summarization using ontologies. In: Proceedings of the 3rd Textgraphs workshop on Graph-based Algorithms for Natural Language Processing, Manchester, UK, pp. 53–56 (2008)
21. Cesarano, C., Mazzeo, A., Picariello, A.: A system for summary-document similarity in notary domain. In: International Workshop on Database and Expert Systems Applications, pp. 254–258 (2007)
22. Kazantseva, A.: An approach to summarizing short stories. In: Proceedings of the Student Research Workshop at the 11th Conference of the European Chapter of the Association for Computational Linguistics (2006)

Combining Text Vector Representations
for Information Retrieval

Maya Carrillo[1,2], Chris Eliasmith[3], and A. López-López[1]

[1] Coordinación de Ciencias Computacionales, INAOE,
Luis Enrique Erro 1, Sta.Ma. Tonantzintla, 72840, Puebla, Mexico
[2] Facultad de Ciencias de la Computación, BUAP,
Av. San Claudio y 14 Sur Ciudad Universitaria, 72570 Puebla, Mexico
{cmaya,allopez}@inaoep.mx
[3] Department of Philosophy, Department of Systems Design Engineering,
Centre for Theoretical Neuroscience, University of Waterloo,
200 University Avenue West Waterloo, Canada
celiasmith@uwaterloo.ca

Abstract. This paper suggests a novel representation for documents that is intended to improve precision. This representation is generated by combining two central techniques: Random Indexing; and Holographic Reduced Representations (HRRs). Random indexing uses co-occurrence information among words to generate semantic context vectors that are the sum of randomly generated term identity vectors. HRRs are used to encode textual structure which can directly capture relations between words (e.g., compound terms, subject-verb, and verb-object). By using the random vectors to capture semantic information, and then employing HRRs to capture structural relations extracted from the text, document vectors are generated by summing all such representations in a document. In this paper, we show that these representations can be successfully used in information retrieval, can effectively incorporate relations, and can reduce the dimensionality of the traditional vector space model (VSM). The results of our experiments show that, when a representation that uses random index vectors is combined with different contexts, such as document occurrence representation (DOR), term co-occurrence representation (TCOR) and HRRs, the VSM representation is outperformed when employed in information retrieval tasks.

1 Introduction

The vector space model (VSM) [1] for document representation supporting search is probably the most well-known IR model. The VSM assumes that term vectors are pair-wise orthogonal. This assumption is very restrictive because words are not independent. There have been various attempts to build representations for documents and queries that are semantically richer than only vectors based on the frequency of terms occurrence. One example is Latent Semantic Indexing (LSI), a word space model, which assumes that there is some underlying latent semantic structure (concepts) that can be estimated by statistical techniques. The traditional word space models produce a high dimensional vector space storing co-occurrence data in a matrix M known as co-occurrence matrix, where each row M_w represents a word and each column M_c

V. Matoušek and P. Mautner (Eds.): TSD 2009, LNAI 5729, pp. 24–31, 2009.

a context (a document or other word). The entry M_{wc} records the co-occurrence of word w in the context c. The M_w rows are vectors, whose size depends on the number of contexts, and are known as 'context vectors' of the words because they represent the contexts in which each word appears. Thus, an algorithm that implements a word space model has to handle the potentially high dimensionality of the context vectors, to avoid affecting its scalability and efficiency. Notably, the majority of the entries in the co-occurrence matrix will be zero given that most words occur in limited contexts.

The problems of very high dimensionality and data sparseness have been approached using dimension reduction techniques such as singular value decomposition (SVD). However, these techniques are computationally expensive in terms of memory and processing time. As an alternative, there is a word space model called Random Indexing [4], which presents an efficient, scalable, and incremental method for building context vectors. Here we explore the use of Random Indexing to produce context vector using document occurrence representation (DOR), and term co-occurrence representation (TCOR). Both DOR and TCOR can be used to represent the content of a document as a bag of concepts (BoC), which is a recent representation scheme based on the perception that the meaning of a document can be considered as the union of the meanings of its terms. This is accomplished by generating term context vectors from each term within the document, and generating a document vector as the weighted sum of the term context vectors contained within that document [4].

In DOR, the meaning of a term is considered as the sum of contexts in which it occurs. In this case, contexts are defined as entire documents. In TCOR the meaning of a term t is viewed as the sum of terms with which it co-occurs, given a window centered in t.

In addition to random indexing, we explore the use of linguistic structures (e.g., compound terms as: *operating system, information retrieval*; and binary relations as: subject-verb and verb-object) to index and retrieve documents. The traditional methods that include compound terms first extract them and then subsequently include these compound terms as new VSM terms. We explore a different representation of such structures, which uses a special kind of vector binding (called holographic reduced representations (HRRs) [3]) to reflect text structure and distribute syntactic information across the document representation. This representation has the benefit, over adding new terms, of preserving semantic relations between compounds and their constituents (but only between compounds to the extent that both constituents are similar). In other words, HRRs do not treat compounds as semantically independent of their constituents. A processing text task where HRRs have been used together with Random Indexing is text classification, where they have shown improvement under certain circumstances, using BoC as baseline [2].

The remainder of this paper is organized as follows. In Section 2 we briefly review Random Indexing. Section 3 introduces the concept of Holographic Reduced Representations (HRRs). Section 4 presents how to use HRRs to add information displaying text structure to document representations. Section 5 explains how different document representations were combined, aiming to improve precision. Section 6 describes the experiments performed. Section 7 shows the results that were obtained in experimental collections. Finally, Section 8 concludes the paper and gives some directions for further work.

2   Random Indexing

Random Indexing (RI) [4] is a vector space methodology that accumulates context vectors for words based on co-occurrence data. First, a unique random representation known as index vector is assigned to each context (either document or word), consisting of a vector with a small number of non-zero elements, which are either +1 or -1, with equal amounts of both. For example, if the index vectors have twenty non-zero elements in a 1024-dimensional vector space, they have ten +1s and ten -1s. Index vectors serve as indices or labels for words or documents. Second, index vectors are used to produce context vectors by scanning through the text and every time a target word occurs in a context, the index vector of the context is added to the context vector of the target word. Thus, with each appearance of the target word t with a context c the context vector of t is updated as follows:

$$ct+ = ic \tag{1}$$

where ct is the context vector of t and ic is the index vector of c. In this way, the context vector of a word keeps track of the contexts in which it occurred.

3   Holographic Reduced Representation

Two types of representation exist in connectionist models: localist, which uses particular units to represent each concept (objects, words, relationships, features); and distributed, in which each unit is part of the representation of several concepts. HRRs are a distributed representation and have the additional advantage that they allow the expression of structure using a circular convolution operator to bind terms (without increasing vector dimensionality). The circular convolution operator ($\otimes$) binds two vectors $\vec{x} = (x_0, x_1, ..., x_{n-1})$ and $\vec{y} = (y_0, y_1, ..., y_{n-1})$ to produce $\vec{z} = (z_0, z_1, ..., z_{n-1})$ where $\vec{z} = \vec{x} \otimes \vec{y}$ is defined as:

$$z_i = \sum_{k=0}^{n-1} x_k y_{i-k} \qquad i = 0 \text{ to } n-1 \text{ (subscripts are module-}n\text{)} \tag{2}$$

A finite-dimensional vector space over the real numbers with circular convolution and the usual definition of scalar multiplication and vector addition form a commutative linear algebra system, so all the rules that apply to scalar algebra also apply to this algebra [3]. We use this operator to combine words and represent compound terms and binary relations.

4   HRR Document Representation

We adopt HRRs to build a text representation scheme in which the document syntax can be captured and can help improve retrieval effectiveness. To define an HRR document representation, the following steps are done: a) we determine the index vectors for the vocabulary by adopting the random indexing method, described earlier; b) all documents are indexed adding the index vectors of the single terms they contain (IVR);

c) for each textual relation in a document, the index vectors of the involved words are bound to their role identifier vectors (using HRRs); d) The tf.idf-weighted sum of the resulting vectors is taken to obtain a single HRR vector representing the textual relation; e) HRRs of the textual relations, multiplied by an attenuating factor α, are added to the document vector (formed with the addition of the single term vectors), to obtain a single HRR vector representing the document, which is then normalized.

For example, given a compound term: $R = $ *information retrieval*. This will be represented using the index vectors of its terms *information* ($\overrightarrow{r}_1$) and *retrieval* ($\overrightarrow{r}_2$), as each of them plays a different role in this structure (right noun/left noun). To encode these roles, two special vectors (HRRs) are needed: $\overrightarrow{role_1}, \overrightarrow{role_2}$. Then, the *information retrieval* vector is:

$$\overrightarrow{R} = (\overrightarrow{role_1} \otimes \overrightarrow{r}_1 + \overrightarrow{role_2} \otimes \overrightarrow{r}_2) \tag{3}$$

Thus, given a document D, with terms $t_1, t_2, \ldots, t_{x1}, t_{y1}, \ldots, t_{x2}, t_{y2}, \ldots, t_n$, and relations R_1, R_2 among terms $t_{x1}, t_{y1}; t_{x2}, t_{y2}$, respectively, its vector is built as:

$$\overrightarrow{D} = \langle \overrightarrow{t}_1 + \overrightarrow{t}_2 + \ldots + \overrightarrow{t}_n + \alpha((\overrightarrow{role_1} \otimes \overrightarrow{t}_{x1} + \overrightarrow{role_2} \otimes \overrightarrow{t}_{y1}) +$$
$$(\overrightarrow{role_1} \otimes \overrightarrow{t}_{x2} + \overrightarrow{role_2} \otimes \overrightarrow{t}_{y2}))\rangle \tag{4}$$

where $\langle \rangle$ denotes a normalized vector and α is a factor less than one intended to lower the impact of the coded relations. Queries are represented in the same way.

5 Combining Representations

We explored several representations: index vector representation (IVR), which uses index vectors as context vectors, DOR, TCOR with a one-word window (TCOR1), and TCOR with a ten-word window (TCOR10). These four document representations were created using BoC. We then combined the similarities obtained from the different representations to check if they took into account different aspects that can improve precision. This combination involves adding the similarity values of each representation and re-ranking the list. Thus, IVR-DOR is created by adding the IVR similarity values to their corresponding values from DOR and re-ranking the list, where documents are now ranked according to the relevance aspects conveyed by both IVR and DOR. We create IVR-TCOR1 using the same process as described above, but now with the similarity lists generated by IVR and TCOR1. Finally, the two similarity lists IVR and TCOR10 are added to form IVR-TCOR10.

In addition, the similarity list obtained with HRR document representations, denoted as IVR+PHR, is also combined with DOR, TCOR1 and TCOR10 similarity lists to produce the IVR+PHR-DOR, IVR+PHR-TCOR1 and IVR+PHR-TCOR10 similarity lists, respectively. These combinations are performed to include varied context information. The following section outlines the experiments performed.

6 Experiments

The proposed document representation was applied to two collections: CACM, with 3,204 documents and 64 queries and NPL, with 11,429 documents and 93 queries. The traditional vector space model (VSM) was used as a baseline, implemented using tf.idf weighting scheme and cosine function to determine vector similarity. We compared this against our representations, which used random indexing, the cosine as a similarity measure, and the same weighting scheme. We carried out preliminary experiments intended to assess the effects of dimensionality, limited vocabulary, and context definition; the following experiments were done using vectors of 4,096 dimensionality, removing stop words, and doing stemming, in the same way as for VSM. The experimental setup is described in the following sections.

6.1　First Set of Experiments: Only Single Terms

CACM and NPL collections were indexed using RI. The number of unique index vectors generated for the former was 6,846 (i.e. terms) and 7,744 for the latter. These index vectors were used to generate context vectors using DOR, TCOR1 and TCOR10. We consider four experiments: a) IVR b) IVR-DOR c) IVR-TCOR1 d) IVR-TCOR10 as described in section 5. It is worth mentioning that the results independently obtained with DOR and TCOR alone were below VSM precision by more than 20%.

6.2　Second Set of Experiment: Noun Phrases

Compound terms were extracted after parsing the documents with Link Grammar [5], doing stemming, and selecting only those consisting of pairs of collocated words. The compound terms obtained for CACM were 9,373 and 18,643 for NPL. These compound terms were added as new terms to the VSM (VSM+PHR). The experiments performed for comparison to this baseline were: a) IVR+PHR, which represents documents as explained in section 4, using the term index vectors, and HRRs to encode compound terms, taking α equal to 1/6 in (4). b) IVR+PHR-DOR, c) IVR+PHR-TCOR1, and d) IVR+PHR-TCOR10, as described in section 5.

6.3　Third Set of Experiments: Binary Relations

The relations to be extracted and included in this vector representation were: compound terms (PHR), verb-object (VO) and subject-verb (SV). These relationships were extracted from the queries of the two collections using Link Grammar and MontyLingua 2.1 [6]. The implementation of the Porter Stemmer used in the experiments came from the Natural Language Toolkit 0.9.2. In this experiment, all stop words were eliminated and stemming was applied to all the relations. If one of the elements of composed terms or SV relations had more than one word, only the last word was taken. The same criterion was applied for the verb in the VO relation; the object was built only with the first set of words extracted and the last word taken, but only if the first word of the set was neither a preposition nor a connective.

Afterwards, a similarity file using only simple terms was generated (IVR). Following this, the HRRs for PHR relations were built for documents and queries and another similarity file was defined. This process was repeated to generate two additional similarity files, but now using SV and VO relations. Then, three similarity files for the extracted relations were built. The IVR similarity file was then added to the PHR similarity, multiplied by a constant of less than one, and the documents were sorted again according to their new value. Afterwards, the SV and VO similarity files were added and the documents once again sorted. Therefore, the similarity between a document d and a query q is calculated with (5), where β, δ, γ are factors less than 1.

$$similarity(q,d) = IVRsimilarity(q,d) + \beta \, PHRsimilarity(q,d) +$$
$$\delta \, SVsimilarity(q,d) + \gamma \, VOsimilarity(q,d) \qquad (5)$$

7   Results

In Tables 1 and 2, we present the calculated mean average precision (MAP - a measure to assess the changes in the ranking of relevant documents), for all our experiments. IVR when considering single terms or compound terms with TCOR reaches higher MAP values than VSM in all cases. For NPL collection, IVR combined with DOR also surpasses the VSM MAP; even the MAP for IVR+PHR is higher than the MAP obtained for VSM+PHR. For CACM, the results obtained with IVR-TCOR10 were found to be statistically significant in a 93.12% confidence interval. For NPL, however, the results for IVR-TCOR10 were significant in a 99.8% confidence interval. IVR+ PHR-TCOR1 was significant in a 97.8% confidence interval, and finally IVR+PHR-DOR and IVR+PHR-TCOR10 were significant in a 99.99% confidence interval.

Finally, the experimentation using binary relations was done after extracting the relations for the queries of each collection. Table 3 shows the number of queries for

Table 1. MAP comparing VSM against IVR and IVR-DOR

	\multicolumn{5}{c}{Single terms}				
	VSM	IVR	%of change	IVR-DOR	% of change
CACM	0.2655	0.2541	-4.28	0.2634	-0.81
NPL	0.2009	0.1994	-0.76	0.2291	**14.02**
	\multicolumn{5}{c}{Terms including compound terms}				
	VSM+PHR	IVR+PHR	% of change	IVR+ PHR-DOR	% of change
CACM	0.2715	0.2538	-6.54	0.2631	-3.10
NPL	0.1857	0.1988	**7.05**	0.2291	**23.40**

Table 2. MAP comparing VSM against IVR-TCOR1 and IVR-TCOR10

	\multicolumn{5}{c}{Single terms}				
	VSM	IVR-TCOR1	% of change	IVR-TCOR10	% of change
CACM	0.2655	0.2754	**3.72**	0.3006	**13.22**
NPL	0.2009	0.2090	**4.01**	0.2240	**11.48**
	\multicolumn{5}{c}{Terms including compound terms}				
	VSM+ PHR	IVR+ PHR TCOR1	% of change	IVR+ PHR TCOR10	% of change
CACM	0.2715	0.2743	**1.04**	0.3001	**10.54**
NPL	0.1857	0.2088	**12.43**	0.2232	**20.22**

Table 3. Number of queries with selected relations per collection

Collection	Compound terms	Subject-Verb	Object-Verb
CACM	48	28	33
NPL	66	1	3

Table 4. MAP comparing the VSM with IVR after adding all the relations

VSM	IVR	% of change	IVR+ PHR	% of change
	0.2570	**0.28**	0.2582	**0.74**
0.2563	IVR+PHR+SV	% of change	IVR+PHR+SV+VO	% of change
	0.2610	**1.82**	0.2693	**5.07**

each collection that had at least one relation of the type specified in the column. NPL queries had very few relations other than compound terms. Consequently, we only experimented using CACM. For this collection, we worked with 21 queries, which had all the specified relations. The value given to β in (5) was 1/16 and to δ and γ 1/32, determined by experiments.

The MAP reached by VSM and the proposed representation with the relations joined is shown in table 4, where the average percentage of change goes from 0.27% for IVR to 5.07% after adding all the relations.

8   Conclusion and Future Research

In this paper, we have presented a proposal for representing documents and queries using random indexing. The results show that this approach is feasible and is able to support the retrieval of information, while reducing the vector dimensionality when compared to the classical vector model. The document representation, using index vector generated by random indexing and the HRRs to encode textual relations, captures some syntactical details that improve precision, according to the experiments. The semantics expressed by contexts either using DOR or TCOR added to our representation also improves the retrieval effectiveness, seemingly by complementing the terms coded alone, something that, as far as we know, has not been experimented on before.

The representation can also support the expression of other relations between terms (e.g. terms forming a named entity). We are in the process of further validating the methods in bigger collections, but we require collections with sufficient features (i.e. queries with binary relations) to fully assess the advantages of our model.

Acknowledgements

The first author was supported by scholarship 217251/208265 granted by CONACYT, while the third author was partially supported by SNI, Mexico.

References

1. Salton, G., Wong, A., Yang, C.S.: A vector space model for automatic indexing. Communications of the ACM 18(11), 613–620 (1975)
2. Fishbein, J.M., Eliasmith, C.: Integrating structure and meaning: A new method for encoding structure for text classification. In: Macdonald, C., Ounis, I., Plachouras, V., Ruthven, I., White, R.W. (eds.) ECIR 2008. LNCS, vol. 4956, pp. 514–521. Springer, Heidelberg (2008)
3. Plate, T.A.: Holographic Reduced Representation: Distributed representation for cognitive structures. CSLI Publications (2003)
4. Sahlgren, M., Cöste, R.: Using Bag-of-Concepts to Improve the Performance of Support Vector Machines in Text Categorization. In: Procs. of the 20th International Conference on Computational Linguistics, pp. 487–493 (2004)
5. Grinberg, D., Lafferty, J., Sleator, D.: A Robust Parsing Algorithm for Link Grammars, Carnegie Mellon University, Computer Science Technical Report CMU-CS-95-125 (1995)
6. Liu, H.: MontyLingua: An end-to-end natural language processor with common sense. web.media.mit.edu/ hugo/montylingua (2004)

Detecting and Correcting Errors in an English Tectogrammatical Annotation

Václav Klimeš

Charles University in Prague, Faculty of Mathematics and Physics,
Institute of Formal and Applied Linguistics
klimes@ufal.mff.cuni.cz

Abstract. We present our first experiments with detecting and correcting errors in a manual annotation of English texts, taken from the Penn Treebank, at the dependency-based tectogrammatical layer, as it is defined in the Prague Dependency Treebank. The main idea is that errors in the annotation usually result in an inconsistency, i. e. the state when a phenomenon is annotated in different ways at several places in a corpus. We describe our algorithm for detecting inconsistencies (it got positive feedback from annotators) and we present some statistics on the manually corrected data and results of a tectogrammatical analyzer which uses these data for its operation. The corrections have improved the data just slightly so far, but we outline some ways to more significant improvement.[1]

1 Introduction

Wall Street Journal collection (WSJ) is the largest subpart of the Penn Treebank ([1]). It consists of texts from the Wall Street Journal, its volume is one million words and it is syntactically annotated using constituent syntax.

The Prague Dependency Treebank (PDT), now in version 2.0 ([2]), is a long-term research project aimed at complex, linguistically motivated manual annotation of Czech texts. It was annotated at three layers: morphological, analytical, and tectogrammatical. The Functional Generative Description theory ([3]) is the main guidance for principles and rules of annotation of PDT.

In Sect. 1, we outline the tectogrammatical layer of PDT and the way of annotating the WSJ. In Sect. 2, the algorithm for detecting and correcting errors in annotation is described. An indirect method of evaluation of error corrections is given in Sect. 3 and we propose our closing remarks and possible improvements in Sect. 4.

1.1 Tectogrammatical Layer of the Prague Dependency Treebank

At the *tectogrammatical* layer of annotation ([4]), shortly *t-layer*, the sentence is represented as a rooted tree. Edges usually represent the relation of dependency between two nodes: the governor and the dependent. As some edges are of technical nature

---

[1] This research was supported by the Information Society projects of the Grant Agency of the Academy of Sciences of the Czech Republic No. 1ET101470416 and No. 1ET201120505.

V. Matoušek and P. Mautner (Eds.): TSD 2009, LNAI 5729, pp. 32–39, 2009.

(e. g. those capturing coordination and apposition constructions), we denote the adjacent nodes with general terms "parent" and "child".

The tectogrammatical layer captures the deep (underlying) structure of the sentence. Nodes represent only autosemantic words (i. e. words with their own meanings); synsemantic (auxiliary) words and punctuation marks can only affect values of attributes of the autosemantic words they belong to. Nodes may be created or copied, e. g. when rules of valency "dictate" to fill ellipses. Relative position of nodes at the t-layer can differ from the position of their counterparts on the surface of a sentence (if they exist). At the t-layer, as many as 39 attributes can be assigned to nodes. Some of the most important attributes follow.

- *(Deep) functor* captures the tectogrammatical function of a node relative to its parent, i. e. the type of the modification. Special functors are used for child-parent relations that are of technical character.
- The `t_lemma` attribute means "tectogrammatical lemma" and it arises from the morphological lemma of the corresponding token; when a node does not have its counterpart on the surface, the attribute has a special value indicating that it represents e. g. a general participant, or an "empty" governing verb predicate.
- For connecting with the lower layers, a link to the corresponding autosemantic word (if any) is stored in the `a/lex.rf` attribute, and links to corresponding synsemantic words (if any) are stored in the `a/aux.rf` attribute.
- In the `val_frame.rf` attribute, the identifier of the corresponding valency frame, which is kept in a separate valency lexicon, is stored.
- The `is_member` attribute states whether the node is a member of a coordination or apposition.
- The `is_generated` attribute expresses whether the node is new at the t-layer. When set, it does not necessarily mean that the node has no counterpart at lower layers – when it is a "copy" of a t-layer node, it refers to the same lower-layer node through its `a/lex.rf` attribute as the original node does.
- The `deepord` attribute stores left-to-right order of nodes at the t-layer.

1.2 Dependency Annotation of the Penn Treebank

Presently, texts from WSJ are being annotated at the t-layer ([5]), which is very similar to the Czech t-layer of PDT – the annotation guidelines are presented in [6]. In order to be able to process English constituency-annotated data, we converted and annotated the data by the respective sequence of components of *TectoMT* ([7]) – the version used is from March 2009 – an framework associating tools for language processing, in the following way.

Hand-written rules appoint the head of each phrase and the phrase structure is then converted into the dependency structure: roughly speaking, the head of a phrase becomes the parent of the other nodes in the phrase. The t-layer is created and partial tectogrammatical annotation is performed using hand-written rules: they delete synsemantic nodes and copy some information from them into the governing autosemantic nodes; they assign functors on the basis of lemmas, morphological tags and function tags; and they add new nodes in a few cases. Human annotators correct and complete only the tectogrammatical annotation, although the annotation at the lower layers exist in the data as well.

2   Detecting and Correcting Errors

2.1   The General Idea

Some errors remain in annotation even in human-checked data, though. Guided by the effort to obtain the most correctly annotated data (from which tools using them for their training could benefit as well), we developed a procedure for searching for errors in an annotation. Inspired by [8], places where inconsistencies in annotation occur are suspected of errors, i. e. an error probably occurs if the same phenomenon is annotated in a different way at several places. The big advantage of this approach is that such places can be found easily.

While the idea presented in [8] remains, our implementation differs for the following reasons. The author of [8] developed his system primarily for checking the consistency of annotations where a label is assigned to each token in the surface representation of a text, e. g. during part-of-speech tagging. Our field of interest is a tree representation, indeed, and in this case the method has to be made more complicated. Moreover, the author stated that inconsistencies in labels of boundary tokens of his surface sequence of tokens need not denote an error, because the boundary tokens may belong to a different phrase than the rest of the sequence. This is why he had to do extra work with examining which boundary tokens are unreliable for the detection of inconsistencies.

2.2   The Way of Detecting Errors

Due to the reasons indicated above, we have chosen another approach: we do not search for tree representations belonging to a sequence of tokens; we determine the sequence of tokens for every tree structure in the data – thus no boundary tokens have to be considered. (A slight disadvantage is that the inconsistency is not revealed when the set of nodes corresponding to a sequence of tokens making one phrase does not form a tree by mistake.) When there are more different trees corresponding to a sequence of tokens, all the trees are regarded as potentially erroneous and they are written out for manual checking. It should be noted that a similar idea was used during checking the data of PDT 2.0 before their release.

A real example of the listing for the sequence *amount of $ 1 million or more* follows and Fig. 1 depicts the appropriate trees (in the same order as in the listing).

- *amount(or*,CONJ⟨ID⟩(*$*,⟨up⟩|*of(million*,RSTR(*1*,RSTR)) *more*,⟨up⟩))
- *amount(or*,DISJ⟨RSTR⟩(*$*,⟨up⟩|*of(million*,RSTR(*1*,RSTR)) *more*,⟨up⟩|*of*))
- *amount(or*,CONJ⟨RSTR⟩(*$*,⟨up⟩(*million*,RSTR|*of(1*,RSTR)) *more*,⟨up⟩))
- *amount(or*,CONJ⟨EXT⟩(*$*,⟨up⟩|*of(million*,RSTR(*1*,RSTR)) *$*,⟨up⟩(*more*,EXT)))

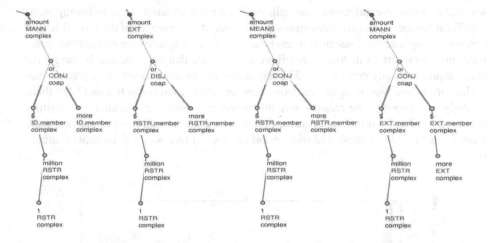

Fig. 1. Trees in the data corresponding to the sequence *amount of $ 1 million or more*

Children of a node are listed behind it in parentheses in their deep order and they are separated by spaces. Description of a node consists of its morphological lemma, a comma, its functor and optionally the '|' sign followed by morphological lemmas of its synsemantic nodes. The functor is thus the only attribute that is explicitly checked for consistency presently (although preliminary experiments with checking the consistency of other attributes were performed as well); however, we can see that the checking is affected by values of attributes is_member, a/lex.rf, a/aux.rf, and deepord as well.

The functor of the root of each tree is omitted from the listing. The reason is that it depends on the context which the tree occurs in, because the functor, expressing the relation between the root and its parent, is in fact outside the tree and not omitting it would result in generating false inconsistencies. For the same reason there are labels in angle brackets in the listing. Because of the manner of capturing the coordination and apposition constructions, functors of members of such construction have to be shifted up (denoted by ⟨up⟩ in the place of the functor) to the root of the construction (denoted by an extra functor in angle brackets), so that it could be omitted when the node is the root of the tree.

2.3   Error Corrections and Statistics

The algorithm described above was applied on the manually annotated data from WSJ containing 13,389 sentences with 228,198 nodes. 1,300 inconsistently annotated sequences were found there. Scatter of number of trees corresponding to these sequences is depicted in Fig. 2. Now we had the listing of suspected trees and we wished to correct errors in the data. (It should be noted that an idea of error corrections in syntactically annotated data was already used e. g. in [9].) Our idea was to manually choose the correct tree structure corresponding to each sequence from those found in the data. Since it was not possible to assign correct trees to such a high number of sequences, only those sequences where the judgement can help the most were selected in the following manner. Total number of occurrences minus the number of occurrences of the most frequent corresponding tree was assigned to each sequence. This quantity expresses the minimum number of errors in consistency if we hypothesize that only one tree is correct for each sequence in any context. For 331 sequences where this quantity was greater than 1, the correct tree was being chosen from those occurring; and for as few as 152 of them it really was chosen. The reasons why no correct tree could be chosen are that either it was not present among the occurring ones, or the annotation rules could not make a decision.[2] For 37 of these 152 sequences, the correct tree was not the most frequent one.

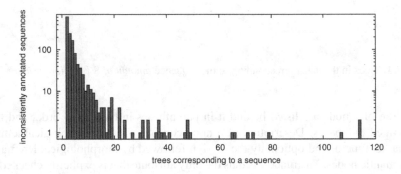

Fig. 2. Numbers of inconsistently annotated sequences depending on numbers of trees corresponding to a sequence. One sequence with 297 corresponding trees is omitted for the sake of clarity of the graph.

Because the trees corresponding to mere 26 of 152 sequences differ in structure (trees corresponding to the rest of sequences differ in functors only), we decided to implement the correction of functors only in the first stage. In the data, 2,371 occurrences of inconsistently annotated sequences were found and 843 corrections of a functor were made (0.0037 corrections per node).

In order to estimate the potential of the corrections, the most frequent tree corresponding to each sequence having no correct tree assigned was denoted as correct. The corrections were made once again: there were 6,351 occurrences of inconsistently annotated sequences in the data and 2,396 corrections of a functor were made (0.0105 corrections per node). We remark that the potential of the method is significantly higher because it was the most frequent tree which was considered correct, and thus the number of corrections was the lowest possible.

The most frequent inconsistently annotated sequences with numbers of corresponding trees of particular types are *New York*: 281+10, *stock market*: 53+52+9, *last year*: 87+26, *Wall Street*: 92+13+1, *big board*: 77+2, and *Dow Jones*: 38+35.

---

[2] This holds true for valency of nouns mainly – the annotation rules are not complete yet; moreover, they are still in progress.

3 Indirect Evaluation through TBLa2t

In order to be able to evaluate how the data improved by the corrections of inconsistencies, we used them for training and testing of a tectogrammatical analyzer called *TBLa2t*.

3.1 What Is TBLa2t

TBLa2t is developed for the tectogrammatical annotation of PDT-like annotated texts in virtually any language, given it (e. g.) the partial tectogrammatical annotation described in Subsect. 1.2. It is based on machine-learning method (to be specific, on transformation-based learning), which means that it uses training data for creation of its language model. It operates in several phases and it is able to create, copy and delete nodes, to assign values of attributes and to perform complex transformations of tree structures. Its English clone is described in [10].

3.2 Evaluation

Since two tectogrammatical trees constructed over the same sentence do not necessarily contain the same number of nodes, a node from a tree from the test annotation is paired with a node from the corresponding tree from the correct annotation in the first step of evaluation and each node is a part of at most one such pair. Only after performing this step, attributes of the paired nodes can be compared.

We define *precision* in an attribute to be the number of pairs where both nodes have a matching value of the attribute divided by the total number of nodes in the test annotation; and we define *recall* to be the same number divided by the total number of nodes in the correct annotation. For our results, we report *F-measure*: it is the equally weighted harmonic mean of precision and recall.

When we want to compare the structure, we have to modify this approach. We define a node to be correctly placed if the node and its parent are in a pair and the counterpart of the parent of the node in question is the parent of the counterpart of the node in question.[3] Then the numerator of the fractional counts from the paragraph above is the number of correctly placed nodes.

The a/aux.rf attribute is a set attribute, and thus it requires special treatment. We evaluate it en bloc and consider it to match if the whole set of links matches its counterpart.

3.3 Results

The table below summarizes the F-measure of results of our experiments with TBLa2t. For training, we used 10,994 sentences with 187,985 nodes; for testing, we used 1,116 sentences with 18,689 nodes.

In the first and second set of columns, the results obtained on the data before and after processing by TBLa2t are captured, respectively. The first column of each set describes

---

[3] Informally, the "same" node has to depend on the "same" parent in both trees.

the data without corrections of inconsistencies; the second column is related to the data with manually made corrections; the third one is related to data where the most frequent tree structure was considered correct when human judgement was missing (see Subsect. 2.3).

attribute	Before TBLa2t			After TBLa2t		
	without	manual	all	without	manual	all
structure		82.6%			87.5%	
functor	56.6%	56.4%	56.5%	79.3%	79.5%	79.6%
val_frame.rf		87.3%			93.2%	
t_lemma		86.0%			90.0%	
a/lex.rf		93.6%		96.4%	96.4%	96.5%
a/aux.rf		78.7%			81.2%	
is_member		90.7%			93.8%	
is_generated		93.0%			95.2%	

Fig. 3. The F-measure of structure and attributes in the data before and after processing by TBLa2t – without corrections of annotation inconsistencies, with manual corrections, and with all corrections

We can see that the results differ only slightly: for manual corrections we got 1.0% error reduction in functors. We might expect better results for data with all corrections; however, these were definitely made worse by the way the judgements were made: for similar cases the choice of the "correct" tree structure might have been done inconsistently, and thus it could not improve the data.

As a side effect, we got the best results for tectogrammatical analysis of English: in [10], there was reported F-measure of 85.0% for structure and 71.9% for functors. The reasons for improvement are (in estimated descending order of importance) higher amount of data for training, different guidelines of annotation, and improvements of the code of the analyzer.

4   Closing Remarks and Possible Improvements

We introduced a working method for detecting and correcting inconsistencies in the human-revised tectogrammatical annotation. The listing of inconsistencies got very positive feedback from annotators and its importance will grow together with the amount of annotated data, because new sequences will be less frequent.

The benefit of corrections is modest; however, it will rise as the annotation guidelines will be made clearer and more complete, because for more sequences, one of the trees will be able to be denoted correct – and listing of inconsistencies can help with this significantly, since it points out the annotation rules which are not clear. By the time the number of inconsistencies will be relatively small, other ideas from [8] will be possible to employ, e. g. automatic selection of the correct tree from those existing (it often does not exist presently).

Another idea is to implement regular expressions for the description of particular tokens of sequences: when manually choosing the correct trees, very little amount of extra work could result in much higher coverage.

Even the presentation of inconsistencies can be improved, e. g. for the sake of clarity, those subtrees of trees corresponding to a sequence can be omitted from the listing, which no difference exists in. However, such decisions should always emerge from the needs of annotators.

References

1. Marcus, M.P., Santorini, B., Marcinkiewicz, M.A.: Treebank-2, Linguistic Data Consortium, Philadelphia, LDC Catalog No. LDC95T7 (1995),
   http://www.cis.upenn.edu/~treebank/
2. Hajič, J., et al.: Prague Dependency Treebank 2.0, Linguistic Data Consortium, Philadelphia, LDC Catalog No. LDC2006T01 (2006), http://ufal.mff.cuni.cz/pdt2.0/
3. Sgall, P., Hajičová, E., Panevová, J.: The Meaning of a Sentence in Its Semantic and Pragmatic Aspects Academia. Kluwer, Praha (1986)
4. Mikulová, M., et al.: Annotation on the tectogrammatical level in the Prague Dependency Treebank, ÚFAL Technical Report no. 2007/3.1, Charles University in Prague, Prague (2007)
5. Hajič, J., et al.: Prague English Dependency Treebank 1.0, Charles University in Prague, Prague (2009) (CD-ROM) ISBN 978-80-904175-0-2
6. Cinková, S., et al.: Annotation of English on the Tectogrammatical Level. ÚFAL Technical Report No. TR-2006-35, Charles University in Prague, Prague (2006)
7. Žabokrtský, Z., Ptáček, J., Pajas, P.: TectoMT: Highly Modular MT System with Tectogrammatics Used as Transfer Layer. In: ACL 2008 WMT: Proceedings of the Third Workshop on Statistical Machine Translation, Association for Computational Linguistics, Columbus, Ohio, pp. 167–170 (2008)
8. Dickinson, M.: Error detection and correction in annotated corpora. PhD thesis. The Ohio State University (2006), http://ling.osu.edu/~dickinso/papers/diss/
9. Štěpánek, J.: Závislostní zachycení větné struktury v anotovaném syntaktickém korpusu (nástroje pro zajištění konzistence dat) [Capturing a Sentence Structure by a Dependency Relation in an Annotated Syntactical Corpus (Tools Guaranteeing Data Consistence)], PhD thesis, Charles University in Prague (2006)
10. Klimeš, V.: Transformation-Based Tectogrammatical Dependency Analysis of English. In: Matoušek, V., Mautner, P. (eds.) Proceedings of Text, Speech and Dialogue 2007, pp. 15–22. Springer Science+Business Media Deutschland GmbH, Heidelberg (2007)

Improving the Clustering of Blogosphere with a Self-term Enriching Technique

Fernando Perez-Tellez[1], David Pinto[2], John Cardiff[1], and Paolo Rosso[3]

[1] Social Media Research Group, Institute of Technology Tallaght, Dublin, Ireland
fernandoperez@itnet.ie, John.Cardiff@ittdublin.ie
[2] Benemerita Universidad Autónoma de Puebla, Mexico
dpinto@cs.buap.mx
[3] Natural Language Engineering Lab. – EliRF, Dept. Sistemas Informáticos y Computación,
Universidad Politécnica Valencia, Spain
prosso@dsic.upv.es

Abstract. The analysis of blogs is emerging as an exciting new area in the text processing field which attempts to harness and exploit the vast quantity of information being published by individuals. However, their particular characteristics (shortness, vocabulary size and nature, etc.) make it difficult to achieve good results using automated clustering techniques. Moreover, the fact that many blogs may be considered to be narrow domain means that exploiting external linguistic resources can have limited value. In this paper, we present a methodology to improve the performance of clustering techniques on blogs, which does not rely on external resources. Our results show that this technique can produce significant improvements in the quality of clusters produced.

1 Introduction

In recent years the use of World Wide Web has changed considerably. One of its most prominent new features is that it has become a tool in the process of socialization with services like blogs, wikis, and file sharing tools. Blogs have become a particularly important decentralized publishing medium which allows a large number of people to share their ideas and spread opinions on the internet. In order to harness the huge volume of information being published, it is essential to provide techniques which can automatically analyze and classify semantically related content, in order to help information retrieval systems to fulfill specific user needs.

The task of automatic classification of blogs is complex. Firstly, we consider it is necessary to employ clustering techniques rather than categorization, since the latter would require us to provide the number and tags of categories in advance. While we could expect to achieve better results using categorization since tags are required a priori, the dynamic categories found in blogs make clustering the best choice. Secondly, clustering techniques typically can produce better results when dealing with wide domain full-text documents where more discriminative information is available. In most cases however, blogs can be considered to be "short texts", i.e., they are not extensive documents and exhibit undesirable characteristics from a clustering

V. Matoušek and P. Mautner (Eds.): TSD 2009, LNAI 5729, pp. 40–47, 2009.

perspective such as low frequency terms, short vocabulary size and vocabulary overlapping of some domains.

In our approach, we treat blog content purely as raw text. While there exist initiatives such as the SIOC project [2] that provide the means to interconnect this information using implicit relations, they require information to be tagged in advance. By dealing purely with raw text, our approach can be applied to any blog.

The main contribution in this paper is the presentation and application of a novel approach to improve the clustering of blogs entitled "S-TEM" (Self-Term Expansion Methodology). This methodology consists of two steps. Firstly, it improves the representation of short documents by using a term enriching (or term expansion) procedure. In this particular case, external resources are not employed because we consider that it is quite difficult to identify appropriate linguistic resources for information of this kind (for instance, blogs often cover very specific topics, and the characteristics of the content may change considerably over time). Moreover, we intend to exploit intrinsic properties of the same corpus to be clustered in an unsupervised way. In other words, we take the same information that will be clustered to perform the term expansion; we called this technique self-term expansion.

The second step consists of the Term Selection Technique (TST). It selects the most important and discriminative information of each category thereby reducing the time needed for the clustering algorithm, in addition to improving the accuracy/precision of the clustering results.

Our contention is that S-TEM can help improve the automatic classification of blogs by addressing some of the factors which make it difficult to achieve strong results with clustering techniques. We show the feature analysis and clustering, using corpora extracted from the popular blogs Boing Boing ("B-B") and Slashdot[1]. In order to demonstrate that the benefits are not confined to a single clustering algorithm, we perform tests using two different clustering methods: K-star [13] and K-means [6]. The obtained results are compared and analyzed using the F-Measure, which is widely used in the clustering field [15].

The remaining of this paper is organized as follows. In the next section the corpora and preprocessing techniques are presented. In Section 3 the techniques and algorithms used in the experiments are introduced. Section 4 describes the experiments using the proposed methodology. Finally in Section 5 conclusions and future work are discussed.

2 Corpora

Table 1 presents some properties of the two datasets used in the experiments. These properties include running words (number of words in the corpus), vocabulary length, number of post categories, number of discussion lines, and the total number of posts.

The gold standard (i.e., the manually constructed expert classification) was created by Perez[2]. A discussion of the corpora is presented in [9], along with a detailed analysis of the features such as shortness degree, domain broadness, class imbalance, stylometry and structure of the corpora.

---

[1] Boing Boing http://boingboing.net; Slashdot http://slashdot.org. A preprocessed version of each dataset is available at http://www.dsic.upv.es/grupos/nle/downloads.html (June 2009)

[2] Available at http://www.dsic.upv.es/grupos/nle/downloads.html (June 2009)

Table 1. Properties of the blogs datasets

Corpus property	Boing Boing	Slashdot
Running words	75935	25779
Vocabulary length	15282	6510
Post categories	4	3
Discussion lines	12	8
Posts	1005	8

3 Improving the Performance of the Blogosphere Clustering Task

In order to test the effectiveness of S-TEM, we have carried out a number of experiments with both corpora. Firstly we applied the methodology on the corpora, and then performed clustering on the enriched corpora. We compared the results with the clusters obtained on the original corpora (ie. without application of the enrichment methodology). For comparative purposes, we use two separate clustering algorithms, K-means [6] and K-star [13].

3.1 Description of the Clustering Algorithms

K-means [6] is one of the most popular iterative clustering algorithms, in which the number of clusters k has to be fixed a-priori. In general terms, the idea is to choose k different centroids. As different locations will produce different results, the best approach is to place these centroids as far away from each other as possible. Next, choose given data and associate each point to the nearest centroid, then k new centroids will be calculated and the process is repeated.

K-star [13] is an iterative clustering method that begins by building a similarity matrix of the documents to be clustered (corpus), In contrast with K-means, K-star does not need to know the k value a priori, and instead it automatically proposes a number of clusters in a totally unsupervised way. K-star is a considerably faster algorithm than K-means and it also obtains reasonably good results when it is applied to short text corpora. For practical purposes a minimum of document similarity threshold was established in order to permit the K-star clustering method to determine whether or not two documents belong to the same cluster. This threshold was defined as the similarity averaged among all the documents.

3.2 The Self-term Expansion Methodology

The Self-term expansion methodology ("S-TEM") [12] comprises a twofold process: the self-term expansion technique, which is a process of replacing terms with a set of co-related terms, and a Term Selection Technique with the role of identifying the relevant features.

The idea behind Term Expansion is not new; it has been studied in previous works such as [12] and [5] in which external resources have been employed. Term expansion has been used in many areas of natural language processing as in word disambiguation in [1], in which WordNet [4] is used in order to expand all the senses of a word. However, as we previously mentioned, we use only the information being clustered to perform the term expansion, i.e., no external resource is employed.

The technique consists of replacing terms of a post with a set of co-related terms. A co-occurrence list will be calculated from the target dataset by applying the Pointwise Mutual Information (*PMI*) [7] as described in Eq. (1):

$$PMI(x, \ y) = \log_2(N \ \frac{fr(x, \ y)}{fr(x) \cdot fr(y)}), \tag{1}$$

where *fr(x,y)* is the frequency in which both words *x* and *y* appear together; *fr(y)* and *fr(x)* are the word frequency of *x* and *y*, respectively, and *N* is a normalization factor equal to the total number of words in the vocabulary.

PMI provides a value of relationship between two words; however, the level of this relationship must be empirically adjusted for each task. In this work, we found *PMI* equal or greater than 2 to be the best threshold. This threshold was established by experience, analyzing the performance of clustering algorithms with different samples of the datasets. In other experiments [11], [12], a threshold equal to 6 was used; however, in the case of blog corpora documents, correlated terms are rarely found.

The Self-Term Expansion Technique is defined formally in [10] as follows:

Let $D = \{d_1, d_2, \ldots, d_n\}$ be a document collection with vocabulary $V(D)$. Let us consider a subset of $V(D) \times V(D)$ of co-related terms as $RT = \{(t_i, t_j) | t_i, t_j \in V(D)\}$ *The RT* expansion of *D* is $D' = \{d'_1, d'_2, \ldots, d'_n\}$, such that for all $d_i \in D$, it satisfies two properties: 1) if $t_j \in d_i$ then $t_j \in d'_i$, and 2) if $t_j \in d_i$ then $t'_j \in d'_i$, with $(t_j, t'_j) \in RT$. If *RT* is calculated by using the same target dataset, then we say that *D'* is the self-term expansion version of *D*.

The Term Selection Technique helps us to identify the best features for the clustering process. However, it is also useful to reduce the computing time of the clustering algorithms. In particular, we used Document Frequency (DF) [14], which assigns the value DF(*t*) to each term *t*, where DF(*t*) means the number of posts in a collection, where *t* occurs.

The Document Frequency technique assumes that low frequency terms will rarely appear in other documents; therefore, they will not have significance on the prediction of the class of a document. This assumption is completely valid for all textual datasets, including blog corpora.

4 Experimental Results

In this section the performance of the two clustering algorithms in conjunction with S-TEM is presented. The goal is to show one strategy that could be used in order to deal with some problems related with blogs such as, low frequency terms, short vocabulary size and vocabulary overlapping of some domains.

We apply S-TEM in order to replace terms in blogs with a list of co-related terms; this list may be obtained by using general purpose and external knowledge resources; however, due to the topic specificity of blogs, there is a lack of linguistic resources of this kind. Intrinsic information in the target dataset should be exploited together with a selection of terms in order to use the most important and relevant information needed for the clustering task.

The results we obtained with and without the Self-Term Expansion Methodology can give us an overview of the level of improvement that may be obtained by applying this methodology. In order to be objective with the results, we have used a well-known measure to evaluate the performance of the clustering algorithms, which is named F-Measure [15] and it is described as follows:

$$F - Measure = \frac{2 * precision * recall}{precision + recall} \tag{2}$$

The clustering results of applying S-TEM to the B-B and Slashdot corpora are shown in Figures 1 and 2 respectively. Different vocabulary percentages of the enriched corpus were selected by the Term Selection Technique (from 30% to 90%).

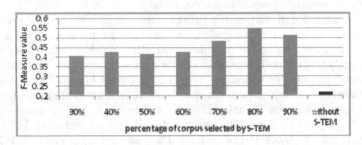

Fig. 1. Applying S-TEM to B-B and clustering with K-means

In general the best clustering results were obtained by selecting (using the DF technique) 80% of the total number of terms belonging to the B-B enriched corpus vocabulary, and using the value of *k=13* in K-means algorithm; this parameter *k* (number of clusters) was estimated by experience checking the best performance of the clustering algorithm, whereas with Slashdot in Figure 2, the best result was obtained selecting 40% of its vocabulary.

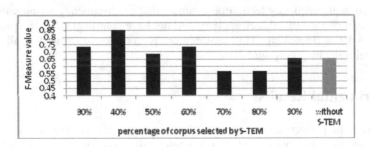

Fig. 2. Applying S-TEM to Slashdot and clustering with K-means

In the case of B-B (Figure 1), the quality of clusters shows a significant improvement when S-TEM is applied. The enriched corpus obtained a F-Measure of 0.55, whereas the baseline (non-enriched) version obtained F-Measure equal to 0.25. These results confirm the usefulness of the methodology with short-text and narrow domain corpora.

The clustering results applied to the Slashdot corpus are shown in Figure 2. Again, the impact of the S-TEM in the clustering of Slashdot produces a considerable improvement in the quality of results, even though the number of documents of the corpus is considerably smaller than B-B. The best F-Measure value (0.85) is obtained with 40% of the vocabulary of the enriched corpus, and k of K-means equal to five; this parameter was estimated in the same form as in B-B. In contrast with the best value obtained by K-means without using S-TEM that is 0.65.

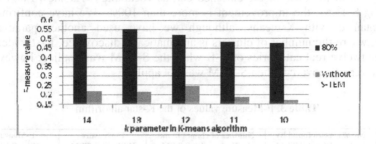

Fig. 3. Varying k parameter in K-means for B-B corpus

In Figures 3 and 4 the results using different values for the k parameter in the K-means algorithm are presented (from 10 to 14 for B-B and from 2 to 7 for Slashdot). The importance of varying the value of k is to confirm the number of clusters that can produce the best results after applying S-TEM to the dataset which will be clustered.

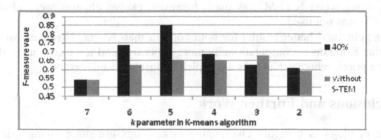

Fig. 4. Varying k parameter in K-means for Slashdot corpus

Additionally, Figure 3 shows a considerable improvement obtained with $k=13$ by the clustering algorithm using S-TEM, whereas in Figure 4 the best value was with $k =5$. The improvement in the results is smaller but still considerable, due to the number of documents and discriminative information used by the K-means algorithm.

In order to more easily understand the obtained results, we have calculated the percentage growth of the enriched corpora with respect to the original data. In Table 2 we see that B-B is the best benefited with a very high percentage of enriching terms. We consider this fact was fundamental for the improvement obtained by using S-TEM, when the results are compared to the baselines.

Table 2. Increase of vocabulary after applying the Self-term Expansion Technique

Boing Boing	Slashdot
+ 1665.8 %	+ 70.29 %

In order to analyze the performance of S-TEM in blogs with different kind of clustering algorithms, we compare the performance improvement of both corpora with the K-star clustering algorithm before and after applying S-TEM. We focused our analysis with the 80% of the enriched corpus vocabulary. As shown in Table 3, when using S-TEM, the improvement is more than 100%. We have confirmed the results obtained with K-means, although we have used another paradigm of clustering, where the number of clusters is automatically discovered. K-star identified nine of the twelve real categories of B-B when the S-TEM was applied; however, 30 categories were discovered when S-TEM was not applied.

Table 3. F-Measure values of the K-star algorithm

Corpus	No. of classes found by K-star without S-TEM	Number of classes found by K-star with S-TEM	F-Measure Without S-TEM	F-Measure With S-TEM
B-B	30	9	0.2	0.41
Slashdot	12	2	0.44	0.45

The values shown in Table 3 by the K-star when clustering the Slashdot corpus with and without S-TEM are quite similar. We consider this is the result of the poor vocabulary of this particular corpus, in addition to noisy words and high overlapping vocabulary between classes. These issues have a notable effect on the clustering task. The K-star algorithm suggested that only two clusters should be considered in the Slashdot corpus when S-TEM was used, however, twelve clusters were identified when S-TEM was not used.

K-means obtained better results for both corpora than K-star. However, we must consider that K-means is computationally more expensive and it requires us to fix the value of k a priori, whereas K-star is a completely unsupervised method.

5   Conclusions and Further Work

Clustering of blogs is a highly challenging task. Blogs are often characterized as narrow domain short-texts and, are therefore unsuitable for augmentation by using external resources. In this paper, we presented a text enrichment technique, the aim of which is to improve the quality of the corpora with respect to the clustering task. The novel feature of this approach, the Self-Term Expansion Methodology (S-TEM) is that it does not rely on external linguistic resources, but it uses the corpus to be clustered itself. We presented a comparison of using the S-TEM with two blog datasets. The results obtained in the experiments show a significant improvement of the enriched corpora with respect to its baseline (non-enriched version).

In future work we plan to use other co-relation measures like those described in [3] and [8], which will be applied in order to find better relationships between terms, with the goal of providing information which is beyond the lexical level to the clustering algorithms.

Acknowledgements. The work of the first author is supported by the HEA under grant PP06TA12. The work of the fourth author is supported by the CICYT TIN2006-15265-C06 research project. The authors wish to express their thanks to Prof. Mikhail Alexandrov for his invaluable feedback on this research.

References

1. Banerjee, S., Pedersen, T.: An adapted lesk algorithm for word sense disambiguation using wordNet. In: Gelbukh, A. (ed.) CICLing 2002. LNCS, vol. 2276, pp. 136–145. Springer, Heidelberg (2002)
2. Bojars, U., Breslin, J.G., Passant, A.: SIOC Browser Towards a richer blog browsing experience. In: The 4th BlogTalk Conference (2006)
3. Daille, B.: Qualitative terminology extraction. In: Bourigault, D., Jacquemin, C., et l'homme, M.-C. (eds.) Recent Advances in Computational Terminology. Natural Language Processing, vol. 2, pp. 149–166. John Benjamins (2001)
4. Fellbaum, C.: WordNet: An Electronic Lexical Database. MIT Press, Cambridge (1998)
5. Grefenstette, G.: Explorations in Automatic Thesaurus Discovery. Kluwer Academic Publishers, Dordrecht (1994)
6. MacQueen, J.B.: Some methods for classification and analysis of multivariate observations. In: Proc. of the 5th Berkeley Symposium on Mathematical Statistics and Probability, pp. 281–297. University of California Press, Berkeley (1967)
7. Manning, D.C., Schütze, H.: Foundations of statistical natural language processing. MIT Press, Cambridge (1999)
8. Nakagawa, H., Mori, T.: A Simple but Powerful Automatic Term Extraction Method, International Conference on Computational Linguistics. In: COLING 2002 on COMPUTERM 2002: second international workshop on computational terminology, vol. 14 (2002)
9. Perez-Tellez, F., Pinto, D., Rosso, P., Cardiff, J.: Characterizing Weblog Corpora. In: 14th International Conference on Applications of Natural Language to Information Systems (2009)
10. Pinto, D., Rosso, P., Jiménez-Salazar, H.: UPV-SI: Word Sense Induction using Self-Term Expansion. In: 4th Workshop on Semantic Evaluations - SemEval 2007, Association for Computational Linguistics (2007)
11. Pinto, D.: On Clustering and Evaluation of Narrow Domain Short-Text Corpora, PhD dissertation, Universidad Politécnica de Valencia, Spain (2008)
12. Qiu, Y., Frei, H.P.: Concept based query expansion. In: Proc. of the 16th annual international ACM SIGIR conference on Research and development in information retrieval, pp. 160–169. ACM Press, New York (1993)
13. Shin, K., Han, S.Y.: Fast clustering algorithm for information organization. In: Gelbukh, A. (ed.) CICLing 2003. LNCS, vol. 2588, pp. 619–622. Springer, Heidelberg (2003)
14. Spärck, J.K.: A statistical interpretation of term specificity and its application in retrieval. Journal of Documentation 28(1), 11–21 (1972)
15. Van Rijsbergen, C.J.: Information Retireval. Butterworths, London (1979)

Advances in Czech – Signed Speech Translation*

Jakub Kanis and Luděk Müller

Univ. of West Bohemia, Faculty of Applied Sciences, Dept. of Cybernetics
Univerzitní 8, 306 14 Pilsen, Czech Republic
{jkanis,muller}@kky.zcu.cz

Abstract. This article describes advances in Czech – Signed Speech translation. A method using a new criterion based on minimal loss principle for log-linear model phrase extraction was introduced and it was evaluated against two another criteria. The performance of phrase table extracted with introduced method was compared with performance of two another phrase tables (manually and automatically extracted). A new criterion for semantic agreement evaluation of translations was introduced too.

1 Introduction

In the scope of this paper, we are using the term Signed Speech (SS) for both the Czech Sign Language (CSE) and Signed Czech (SC). The CSE is a natural and adequate communication form and a primary communication tool of the hearing-impaired people in the Czech Republic. It is composed of the specific visual-spatial resources, i.e. hand shapes (manual signals), movements, facial expressions, head and upper part of the body positions (non-manual signals). It is not derived from or based on any spoken language. On the other hand the SC was introduced as an artificial language system derived from the spoken Czech language to facilitate communication between deaf and hearing people. SC uses grammatical and lexical resources of the Czech language. During the SC production, the Czech sentence is audibly or inaudibly articulated and simultaneously the CSE signs of all individual words of the sentence are signed.

2 Phrase-Based Machine Translation

The goal of the machine translation is to find the best translation $\hat{\mathbf{t}} = w_1, ..., w_I$ of the given source sentence $\mathbf{s} = w_1, ..., w_J$. The state of the art solution of this problem is using log-linear model [1]:

$$Pr(\mathbf{t}|\mathbf{s}) = p_{\lambda_1^M}(\mathbf{t}|\mathbf{s}) = \frac{\exp(\sum_{m=1}^{M} \lambda_m h_m(\mathbf{t}, \mathbf{s}))}{\sum_{\mathbf{t}'} \exp(\sum_{m=1}^{M} \lambda_m h_m(\mathbf{t}', \mathbf{s}))} \tag{1}$$

* This research was supported by the Grant Agency of Academy of Sciences of the Czech Republic, project No. 1ET101470416 and by the Ministry of Education of the Czech Republic, project No. MŠMT LC536.

V. Matoušek and P. Mautner (Eds.): TSD 2009, LNAI 5729, pp. 48–55, 2009.

There are feature models $h_m(\mathbf{t}, \mathbf{s})$, which model a relationship between the source and the target language and its weights λ_m. If we want to have the best translation we should choose the one with the highest probability, thus:

$$\hat{\mathbf{t}} = \underset{\mathbf{t}}{\operatorname{argmax}} \left\{ \frac{\exp(\sum_{m=1}^M \lambda_m h_m(\mathbf{t}, \mathbf{s}))}{\sum_{\mathbf{t}'} \exp(\sum_{m=1}^M \lambda_m h_m(\mathbf{t}', \mathbf{s}))} \right\} = \underset{\mathbf{t}}{\operatorname{argmax}} \left\{ \sum_{m=1}^M \lambda_m h_m(\mathbf{t}, \mathbf{s}) \right\},$$
(2)

where we have disregarded the denominator of the Equation 2. In the log-linear model we can use a portion of different feature models. The source sentence $\mathbf{s}$ is segmented into a sequence of K phrases $\bar{s}_1, ..., \bar{s}_K$ which we call phrase alignment (all possible segmentations have the same probability) in the case of phrase-based translation. We define the phrase of a given **length** l as a continual word sequence: $\bar{s}_i = w_j, ..., w_{j+l}, j = 1, ..., J - l$. Each source phrase $\bar{s}_i, i = 1, ..., K$ is translated into a target phrase $\bar{t}_i$ in the decoding process. This particular ith translation is modeled by a probability distribution $\phi(\bar{s}_i|\bar{t}_i)$. The target phrases can be reordered to get more precise translation. The reordering of the target phrases can be modeled by a relative distortion probability distribution $d(a_i - b_{i-1})$ as in [3], where a_i denotes the start position of the source phrase which was translated into the ith target phrase, and b_{i-1} denotes the end position of the source phrase translated into the $(i - 1)th$ target phrase. The basic feature models are: the both direction translation models ϕ, distortion model d, n-gram based language model p_{LM} and phrase p_{PhP} and word p_{WP} penalty models. The mostly used method for the weight adjustment is minimum error rate training (MERT) [2], where the weights are adjusted to minimize the error rate of the resulting translation:

$$\hat{\lambda}_1^M = \underset{\lambda_1^M}{\operatorname{argmin}} \left\{ \sum_{n=1}^N \sum_{k=1}^K E(r_n, \mathbf{t}_{n,k}) \delta(\hat{\mathbf{t}}(s_n, \lambda_1^M), \mathbf{t}_{n,k}) \right\}$$
(3)

$$\hat{\mathbf{t}}(\mathbf{s}_n, \lambda_1^M) = \underset{\mathbf{t} \in C_n}{\operatorname{argmax}} \left\{ \sum_{m=1}^M \lambda_m h_m(\mathbf{t}, \mathbf{s}_n) \right\}$$
(4)

$$\delta(\hat{\mathbf{t}}(\mathbf{s}_n, \lambda_1^M), \mathbf{t}_{n,k}) = \begin{cases} 1 \ if & \hat{\mathbf{t}}(\mathbf{s}_n, \lambda_1^M) = \mathbf{t}_{n,k} \\ 0 \ else \end{cases},$$

where N is number of sentence pairs in a training corpus, E error criterion which is minimized, r_n is reference translation of the source sentence $\mathbf{s}_n$ and $C_n = \{\mathbf{t}_{n,1}, ..., \mathbf{t}_{n,K}\}$ is a set of K different translations $\mathbf{t}_n$ of each source sentence $\mathbf{s}_n$.

3   Phrase Extraction Based on Minimal Loss Principle

The main source of the SMT system is a phrase table with bilingual pairs of phrases. State of the art methods for the phrase extraction are based on alignment modeling (especially on the word alignment modeling). The word alignment can be modeled by probabilistic models of different complexity (Models 1 – 6 [7]). The model complexity directly influences the alignment error rate and thus the translation accuracy: the more complexity model, the better translations. However, more complicated models are computationally challenging. For example, the task of finding the Viterbi alignment for the Models 3 – 6 is an NP-complete problem [7]. Only a suboptimal solution can be found with usage of approximations. In addition, it was founded that the next reduction of word alignment errors does not have to lead to better translations [8]. Because of problems with word alignment models we have proposed using of the log-linear model for the phrase extraction, which can be optimized directly to the translation precision. Our solution is similar to the one in work [9] with some differences. Firstly, we are using different set of features without using of any alignment modeling. Secondly, we introduce a new criterion for phrase extraction based on a minimal loss principle.

Method Description: Our task is to find for each source phrase $\bar{s}$ its translation, i.e. the corresponding target phrase $\bar{t}$. We suppose that we have a sentence aligned bilingual corpus (pairs of the source and target sentences). We start with the source sentence $\mathbf{s} = w_1, ..., w_J$ and the target sentence $\mathbf{t} = w_1, ..., w_I$ and generate a bag β of all possible phrases up to the given length l: $\beta\{\mathbf{s}\} = \{\bar{s}_m\}_{m=1}^l, \{\bar{s}_m\} = \{w_n, ..., w_{n+m-1}\}_{n=1}^{J-m+1}, \beta\{\mathbf{t}\} = \{\bar{t}_m\}_{m=1}^l, \{\bar{t}_m\} = \{w_n, ..., w_{n+m-1}\}_{n=1}^{I-m+1}$. The source phrases longer than one word are keeping for next processing only if they have been seen in the corpus at least as much as given threshold τ (reasonable threshold is five). All target phrases are keeping regardless of the number of their occurrence in the corpus. Each target phrase is considered to be a possible translation of each kept source phrase $\forall \bar{s} \in \beta\{\mathbf{s}\} : N(\bar{s}) \geq \tau : \bar{s} \rightarrow \beta\{\mathbf{t}\}$, where $N(\bar{s})$ is number of occurrences of phrase $\bar{s}$ in the corpus. Now for each possible translation pair $(\bar{s}, \bar{t}) : \bar{t} \in T(\bar{s}), T(\bar{s}) = \{\bar{t}\} : \bar{s} \rightarrow \tilde{t}$ we compute its corresponding score:

$$c(\bar{s}, \bar{t}) = \sum_{k=1}^{K} \lambda_k h_k(\bar{s}, \bar{t}), \tag{5}$$

where $h_k(\bar{s}, \bar{t}), k = 1, 2, ..., K$ is set of K features, which describe the relationship between the pair of phrases $(\bar{s}, \bar{t})$. The MERT training can be used for weights λ_k optimization. The resulting scores $\mathbf{c} = \{c\}$ are stored in a hash table, where the source phrase $\bar{s}$ is the key and all possible translations $\bar{t} \in T(\bar{s})$ with its score $c(\bar{s}, \bar{t})$ are the data. We process the whole training corpus and store the scores for all possible translation pairs.

The next step is choosing only "good" translations $\bar{t}_G$ from all possible translations $T(\bar{s})$ for each source phrase $\bar{s}$, i.e. we get a set of translations $T_G(\bar{s}) = \{\bar{t}_G\} : \bar{s} \rightarrow \bar{t}_G$. For each sentence pair we generate the bag of all phrases up to the given length l for both sentences. Then for each $\bar{s} \in \beta(\mathbf{s})$ we compute a **translation loss $\mathbf{L_T}$** for each $\bar{t} \in T(\bar{s}) = \beta(\mathbf{t})$. The translation loss L_T for the source phrase $\bar{s}$ and its possible translation $\bar{t}$ is defined as:

$$L_T(\bar{s}, \bar{t}) = \frac{\sum_{\bar{s}_i \in \beta(\mathbf{s}), \bar{s}_i \neq \bar{s}} c(\bar{s}_i, \bar{t})}{c(\bar{s}, \bar{t})} \tag{6}$$

We compute how much probability mass we lost for the rest of source phrases from the bag $\beta(\mathbf{s})$ if we translate $\bar{s}$ as $\bar{t}$. For each $\bar{s}$ we store all translation losses $L_T(\bar{s}, \bar{t})$ for all $\bar{t} \in \beta(\mathbf{t})$. The "good" translation $\bar{t}_G$ for $\bar{s}$ is the one (or more) with the lowest translation loss $L_T(\bar{s}, \bar{t})$:

$$\bar{t}_G = \operatorname*{argmin}_{\bar{t}} L_T(\bar{s}, \bar{t}) \tag{7}$$

and all the other translations are discarded. We process all sentence pairs and get a new phrase table. This table comprises source phrases $\bar{s}$, corresponding "good" translations $\bar{t}_G \in T_G(\bar{s})$ only, and the numbers of how many times a particular translation $\bar{t}$ was determined as a "good" translation $\bar{t}_G$. These information can be then used for example for calculation of translation probabilities ϕ.

Used Features: We used only features based on number of occurrences of translation pairs and particular phrases in the training corpus. We collect these numbers: number of occurrences of each considered source phrase $N(\bar{s})$, number of occurrences of each target phrase $N(\bar{t})$, number of occurrences of each possible translation pair $N(\bar{s}, \bar{t})$ and number of how many times was given source or target phrase considered as translation $N_T(\bar{s})$ and $N_T(\bar{t})$ (it corresponds to the number of all phrases for which was given phrase considered as their possible translation in all sentence pairs). These numbers are used to compute the following features: translation probability ϕ, probability p_T that given phrase is a translation - all for both translation directions and translation probability p_{MI} based on mutual information. The translation probability ϕ is defined on base of relative frequencies as [3]:

$$\phi(\bar{s}|\bar{t}) = \frac{N(\bar{s}, \bar{t})}{N(\bar{t})} \quad \phi(\bar{t}|\bar{s}) = \frac{N(\bar{s}, \bar{t})}{N(\bar{s})} \tag{8}$$

Probability p_T, that given phrase is a translation, i.e. it appears together with considered phrase as its translation, is defined as:

$$p_T(\bar{s}|\bar{t}) = \frac{N(\bar{s}, \bar{t})}{N_T(\bar{t})} \quad p_T(\bar{t}|\bar{s}) = \frac{N(\bar{s}, \bar{t})}{N_T(\bar{s})} \tag{9}$$

Translation probability p_{MI} based on mutual information is defined as [10] (we can use both numbers N and N_T for computing):

$$p_{MI}(\bar{s}, \bar{t}) = \frac{MI(\bar{s}, \bar{t})}{\sum_{\bar{t} \in T(\bar{s})} MI(\bar{s}, \bar{t})} \quad p_{MI_T}(\bar{s}, \bar{t}) = \frac{MI_T(\bar{s}, \bar{t})}{\sum_{\bar{t} \in T(\bar{s})} MI_T(\bar{s}, \bar{t})} \tag{10}$$

$$MI(\bar{s}, \bar{t}) = p(\bar{s}, \bar{t}) \log \frac{p(\bar{s}, \bar{t})}{p(\bar{s}) \cdot p(\bar{t})} \quad MI_T(\bar{s}, \bar{t}) = p_T(\bar{s}, \bar{t}) \log \frac{p_T(\bar{s}, \bar{t})}{p_T(\bar{s}) \cdot p_T(\bar{t})} \tag{11}$$

$$p(\bar{s}, \bar{t}) = \frac{N(\bar{s}, \bar{t})}{N_S} \quad p_T(\bar{s}, \bar{t}) = \frac{N(\bar{s}, \bar{t})}{N_T} \tag{12}$$

$$p(\bar{s}) = \frac{N(\bar{s})}{N_S} \quad p(\bar{t}) = \frac{N(\bar{t})}{N_S} \quad p_T(\bar{s}) = \frac{N_T(\bar{s})}{N_T} \quad p_T(\bar{t}) = \frac{N_T(\bar{t})}{N_T}, \tag{13}$$

where N_S is the number of all sentence pairs in the corpus and N_T is the number of all possible considered translations, i.e. if source sentence length is five and target sentence length nine then we add 45 to N_T. Finally we have six features: $\phi(\bar{s}|\bar{t})$, $\phi(\bar{t}|\bar{s})$, $p_T(\bar{s}|\bar{t})$, $p_T(\bar{t}|\bar{s})$, $p_{MI}(\bar{s}, \bar{t})$ and $p_{MI_T}(\bar{s}, \bar{t})$ for the phrase extraction.

4   Tools and Evaluation Methodology

Data: The main resource for the statistical machine translation is a parallel corpus which contains parallel texts of both the source and the target language. Acquisition of such corpus in case of SS is complicated by the absence of the official written form of both the CSE and the SC. Therefore we have used the Czech to Signed Czech (CSC) parallel corpus [4] for all experiments. For the purpose of experiments we have split the CSC corpus into training, development and testing part, which are described in Table 1 in more details.

Evaluation Criteria: We have used the following well known criteria for evaluation of our experiments. The first criterion is the **BLEU** score: it counts modified n-gram precision for output translation with respect to the reference translation. The second criterion is the **NIST** score: it counts similarly as BLEU modified n-gram precision, but uses arithmetic mean and weighing by information gain of each n-gram. Next criterion is **Sentence Error Rate** (**SER**): it is a ratio of the number of incorrect sentence translations to the number of all translated sentences. The **Word Error Rate** (**WER**) criterion is adopted from ASR area: is defined as the Levensthein edit distance between the produced translation and the reference translation in percentage (a ratio of the number of all deleted, substituted and inserted produced words to the total number of reference words). The third error criterion is **Position-independent Word Error Rate** (**PER**): it compares two sentences without regard to their word order. These criteria however evaluate only lexical agreement between the reference and the resulting translation. But in the automatic translation we need to find out if two different word constructions have the same meaning, i.e. are semantically identical, because there are always equally correct different translations of each source sentence (for example there are mostly more reference translations of each source sentence in the corpus). We have proposed a new **Semantic Dimension Overlap** (**SDO**) criterion to evaluate semantic similarity of the translations between Czech and SC. The SDO criterion is based on the overlap between semantic annotation of the reference translation and semantic annotation of the resulting translation. The semantic annotation is created by HVS (Hidden Vector State) parser [5], which is trained on the CSC corpus data (the CSC corpus contains semantic annotation layer needed for the HVS parser training). A lower values of the three error criteria: SER, WER, PER and a higher values of the three precision criteria: BLEU, NIST, SDO indicates better, i.e. more precise translation.

Decoders: Two different phrase-based decoders were used in our experiments. The first decoder is freely available state-of-the-art factored phrase-based beam-search decoder - **MOSES**[1] [6], which uses log-linear model (MERT training). The training tools for extraction of phrases from the parallel corpus are also available, i.e. the whole translation system can be constructed given a parallel corpus only. For the language modeling was used the SRILM[2] toolkit.

The second decoder is our implementation of monotone phrase-based decoder - **SiM-PaD**, which already uses log-linear model (MERT training). The monotonicity means

---

[1] http://www.statmt.org/moses/

[2] http://www.speech.sri.com/projects/srilm/download.html

Table 1. Dividing of the CSC corpus into training, development and testing part

	Training data		Development data		Testing data	
	CZ	SC	CZ	SC	CZ	SC
Sent. pairs	12 616		1 578		1 578	
# words	86 690	86 389	10 700	10 722	10 563	10 552
Vocab. size	3 670	2 151	1 258	800	1 177	748
# singletons	1 790	1 036	679	373	615	339
OOV(%)	–	–	240 (2.24)	122 (1.14)	208 (1.97)	105 (1.00)

using the monotone reordering model only, i.e. no phrase reordering is permitted during the search. SiMPaD uses SRILM^2 language models and the Viterbi algorithm for the decoding, which defines generally n-gram dependency between translated phrases.

5 Experiments and Conclusion

Phrase Extraction Based on Minimal Loss Principle: In the first experiment we compared the new criterion based on minimal loss principle (ML) proposed in Section 3 with two another criteria for the phrase extraction. All six features defined in

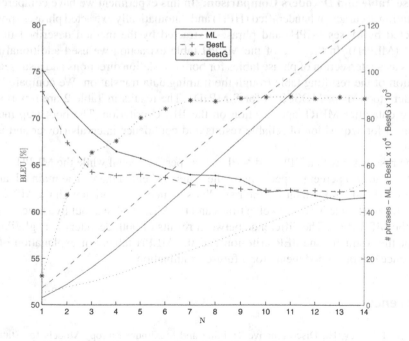

Fig. 1. Comparison of different criteria for the phrase extraction

Table 2. Comparison of different phrase tables and decoders

	HPH		MPH		MLPH	
Size	5 325		65 494		11 585	
	M	S	M	S	M	S
Bleu[%]	$\mathbf{81.29^{1.27}_{1.29}}$	$81.22^{1.31}_{1.31}$	$80.87^{1.31}_{1.31}$	$81.08^{1.27}_{1.32}$	$80.20^{1.28}_{1.33}$	$80.21^{1.32}_{1.36}$
NIST	$\mathbf{11.65^{0.13}_{0.14}}$	$\mathbf{11.65^{0.13}_{0.13}}$	$11.57^{0.13}_{0.14}$	$11.58^{0.14}_{0.14}$	$11.47^{0.14}_{0.14}$	$11.44^{0.14}_{0.14}$
SER[%]	$\mathbf{38.15^{3.49}_{3.30}}$	$38.53^{3.42}_{3.30}$	$38.21^{3.42}_{3.42}$	$38.59^{3.49}_{3.36}$	$40.56^{3.49}_{3.36}$	$42.90^{3.55}_{3.36}$
WER[%]	$13.14^{1.33}_{1.29}$	$\mathbf{13.06^{1.32}_{1.25}}$	$13.43^{1.36}_{1.31}$	$13.43^{1.31}_{1.25}$	$14.48^{1.41}_{1.35}$	$14.88^{1.42}_{1.33}$
PER[%]	$\mathbf{11.64^{1.22}_{1.17}}$	$11.72^{1.20}_{1.13}$	$11.85^{1.21}_{1.16}$	$11.93^{1.20}_{1.13}$	$12.95^{1.21}_{1.18}$	$13.24^{1.26}_{1.16}$
SDO[%]	$92.08^{1.95}_{2.37}$	$\mathbf{92.25^{1.96}_{2.30}}$	$92.12^{2.03}_{2.30}$	$92.11^{2.01}_{2.39}$	$90.84^{2.07}_{2.49}$	$90.82^{2.13}_{2.51}$

Section 3 was used in log-linear model. The first one (BestG) is criterion used in the work [9] which selects all translation pairs for each sentence pair with score c higher than maximal score $c_m - threshold$ τ. The second one (BestL) criterion is criterion which selects only the translation pair with the highest score c_m for each source phrase in the sentence pair. The results are in Figure 1, where N means a number of first N best scores c selected for each source phrase. The ML criterion performs best (75.09), the second is BestL criterion (72.83) and the last is BestG criterion (72.31).

Phrase Table and Decoders Comparison: In this experiment we have compared the translation accuracy of handcrafted (HPH) and automatically extracted phrases (phrases extracted by Moses (MPH) and phrases extracted by the method described in Section 3 (MLPH)). In the case of the MLPH table extraction we used additional techniques as a intersection of phrase tables for both translation directions and a subsequent filtration of the resulting table trough the training data translation. We compared both decoders too (M for MOSES, S for SiMPaD). The results in Table 2 are reported for testing data after MERT optimization on the BLEU criterion. The bootstrap method was used for acquisition of reliable results and confidence intervals (lower and upper indexes).

The results show that HPH and MPH tables perform equal while the MLPH table is about one to two percent depending on the criterion behind them. The main advantage of the HPH and MLPH tables is their smaller size in confrontation with the MPH table size. The HPH table is about twelve times and the MLPH table about five times smaller than the MPH table. The difference between results of both decoders is negligible too except the result for the SER criterion and the MLPH table. An explanation of this difference can be a good theme for a future examination.

References

1. Och, F.J., Ney, H.: Discriminative Training and Maximum Entropy Models for Statistical Machine Translation. In: Proc. 40th Annual Meeting of the ACL, Philadelphia, PA, July 2002, pp. 295–302 (2002)

2. Och, F.J.: Minimum Error Rate Training in Statistical Machine Translation. In: Proc. 41st Annual Meeting of the ACL, Sapporo, Japan (July 2003)
3. Koehn, P., et al.: Statistical Phrase-Based Translation. In: HLT/NAACL (2003)
4. Kanis, J., et al.: Czech-sign speech corpus for semantic based machine translation. In: Sojka, P., Kopeček, I., Pala, K. (eds.) TSD 2006. LNCS (LNAI), vol. 4188, pp. 613–620. Springer, Heidelberg (2006)
5. Jurčíček, F., et al.: Extension of HVS semantic parser by allowing left-right branching. In: International Conference on Acoustics, Speech, and Signal Processing, Las Vegas, USA (2008)
6. Koehn, P., et al.: Moses: Open Source Toolkit for Statistical Machine Translation. In: Annual Meeting of the ACL, Prague, Czech Republic (June 2007)
7. Och, F.J., Ney, H.: A Systematic Comparison of Various Statistical Alignment Models. Computational Linguistics 29(1), 19–51 (2003)
8. Zens, R.: Phrase-based Statistical Machine Translation: Models, Search, Training. PhD thesis, RWTH Aachen University, Aachen, Germany (February 2008)
9. Deng, Y., et al.: Phrase Table Training for Precision and Recall: What Makes a Good Phrase and a Good Phrase Pair? In: Proceedings of ACL 2008: HLT, Columbus, Ohio, June 2008, pp. 81–88 (2008)
10. Lavecchia, C.: el al.: Phrase-Based Machine Translation based on Simulated Annealing. In: Proceedings of the Sixth International Conference on Language Resources and Evaluation, Marrakech, Morocco, ELRA (2008)

Improving Word Alignment
Using Alignment of Deep Structures*

David Mareček

Institute of Formal and Applied Linguistics,
Charles University in Prague
marecek@ufal.mff.cuni.cz

Abstract. In this paper, we describe differences between a classical word alignment on the surface (word-layer alignment) and an alignment of deep syntactic
sentence representations (tectogrammatical alignment). The deep structures we
use are dependency trees containing content (autosemantic) words as their nodes.
Most of other functional words, such as prepositions, articles, and auxiliary verbs
are hidden. We introduce an algorithm which aligns such trees using perceptron-
based scoring function. For evaluation purposes, a set of parallel sentences was
manually aligned. We show that using statistical word alignment (GIZA++) can
improve the tectogrammatical alignment. Surprisingly, we also show that the tectogrammatical alignment can be then used to significantly improve the original
word alignment.

1 Introduction

Alignment of parallel texts is one of the well-established tasks in NLP (see [1] and
[2]). It can be used for various purposes, such as for creating training data for machine
translation (MT) algorithms, for extracting bilingual dictionaries, and for projections of
linguistic features from one language to another.

In tectogrammatics (as introduced in Functional Generative Description by Sgall [3],
and implemented in the Prague Dependency Treebank [4]), each sentence is represented by a tectogrammatical tree (t-tree for short). T-tree is a rooted dependency deep-
syntactic tree. Example of an English t-tree is shown in Figure 2. Unlike in the surface
syntax, only content (autosemantic) words have their own nodes in t-trees. Function
words such as auxiliary verbs, subordinating conjunctions, articles, and prepositions
are represented differently: for instance, there is no node representing auxiliary verbs
has and *been* in the t-tree example, but one of the functions they convey is reflected by
attribute *tense* attached to the autosemantic verb's node (*set*). Other attributes describe
several cognitive, syntactic and morphological categories.

Our motivation for developing tectogrammatical alignment system is following.
First, we need aligned dependency trees for experimenting with statistical dependency-
based MT. Second, we want to collect evidence for the hypothesis that typologically

---

* The work on this project was supported by the grants GAUK 9994/2009, GAČR 201/09/H057,
and GAAV ČR 1ET101120503.

V. Matoušek and P. Mautner (Eds.): TSD 2009, LNAI 5729, pp. 56–63, 2009.

different languages look more similar on the tectogrammatical layer, and thus the alignment of tectogrammatical trees should be "simpler" (in the sense of higher inter-annotator agreement and higher achievable performance of automatic aligners evaluated on manually aligned data). Third, we show that the alignment gained using the tectogrammatical analysis can be helpful for improving quality of the automatic alignment back on the word layer.

Several works already came with the idea of aligning content words only. Haruno and Yamazaki [5] argued that in structurally different languages (Japanese and English in their case) it is not feasible to align functional words, because their functions are often very different. Menezes and Richardson [2] use so called "logical forms" (LFs) – an unordered graphs representing the relations among the meaningful elements of sentences. These structures are very similar to tectogrammatical trees. Nodes are identified by lemmas of the content words and labeled arcs indicate the underlying semantic dependency relations between nodes. Watanabe, Kurohashi, and Aramaki [6] have also similar sentence structures and use an algorithm which finds word correspondences by consulting a bilingual dictionary.

The remainder of the text is structured as follows. In Section 2, we describe the manually annotated data. The evaluation metric we use is in Section 3. The GIZA++ alignment is described in Section 4. Our tectogrammatical aligner is presented in Section 5. The way how to improve GIZA++ word alignment using our tectogrammatical aligner is the subject of Section 6. Section 7 concludes and discusses future work.

2   Manual Word Alignment

For the purpose of training and evaluating alignment systems, we compiled a manually annotated data set. It consists of 2,500 pairs of sentences[1] from several different domains (newspaper articles, E-books, short stories and also from EU law). Quantitative properties of the annotated data sets are summarized in Table 1. PCEDT parallel corpus [7] contains a subset of English sentences from Penn Treebank corpus which were translated into Czech (sentence by sentence) for the MT purposes.

Each sentence pair was manually aligned on the word level independently by two annotators. The task was to make connections (links) between Czech and English corresponding tokens. Following Bojar and Prokopová [8] we used three types of connections: *sure* (individual words match), *phrasal* (whole phrases correspond, but not literally), and *possible* (used especially to connect words that do not have a real equivalent in the other language but syntactically clearly belong to a word nearby). An example is presented in Figure 1.

The "golden" alignment annotation was created from the two parallel annotations according to the following rules: a connection is marked as *sure* if at least one of the annotators marked it as *sure* and the other also supported the link by any connection

---

[1] Boundaries of Czech and English sentences did not always match. (i.e., there were not only 1:1 sentence relations). In such cases we either split the sentence in one language or join several sentences in the other language in order to have only 1:1 sentence relations in the annotated data.

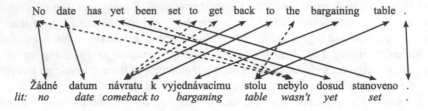

Fig. 1. Word-layer alignment (possible links are dashed)

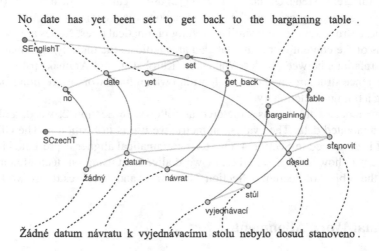

Fig. 2. Tectogrammatical-layer alignment (only lemmas are depicted with the nodes)

type. In all other cases (at least one annotator makes any type of link), the connection is marked as *possible*.

We made no extra annotations for aligning tectogrammatical trees. Each node represents one content word on the surface. Thus we can transfer the manual word alignment to the alignment of tectogrammatical trees. Links from/to functional words (the tokens that do not have their own tectogrammatical nodes) are disregarded.

The transfer of alignment is illustrated in Figure 2. Each node has one reference to the original content word on the surface, which it got its lexical meaning from. Such references are depicted in Figure 2 by dashed lines.

Inter-annotator agreement (IAA) reflects the reliability of manual annotations. For evaluating IAA on the alignment task, we use the following formula:

$$IAA(A, B) = \frac{2 \cdot |L_A \cap L_B|}{|L_A| + |L_B|},$$

where L_A and L_B are sets of alignment connections made by annotator A and annotator B respectively. Since we have different types of connections, we can either distinguish the types in the evaluation (i.e., $L_A \cap L_B$ contains only connections that both annotators labeled equally), or disregard them. The third alternative is to regard only the sure

Table 1. Size of the manually aligned data

domain	sent. pairs	words/sent. CS.	EN
newspapers articles	984	21	24
E-books, short stories	500	18	20
EU-laws	501	21	27
PCEDT corpus	515	24	25

Table 2. Inter-annotator agreement

IAA [%]	all	content	funct.
types distinguished	83.3	90.9	76.8
types not dist.	89.6	94.6	84.2
sure links only	92.9	94.8	88.8

connections. The results of the three evaluation variants are summarized in Table 2. Connections between content and functional words are included within functional connections (the third column).

We can see that the alignment of the content words only is obviously less problematic for annotators compared to functional words. This evidence is in the same direction as that of Haruno and Yamazaki [5].

3 Evaluation Metrics

We randomly split the golden aligned data into two equal-sized parts, one for evaluating the performance of all aligners, and the other for training our tectogrammatical aligner.

For evaluating the quality of aligners we use alignment error rate (AER) described in [1], which combines precision and recall. Obviously, asserting connections that were neither sure nor possible causes lower precision, whereas omitting sure connections causes lower recall.

$$Prec = \frac{|(P \cup S) \cap A|}{|A|}, \quad Rec = \frac{|S \cap A|}{|S|}, \quad AER = 1 - \frac{|(P \cup S) \cap A| + |S \cap A|}{|S| + |A|}$$

where S is the set of *sure* links, P is the set of the *possible* links and A is the set of links suggested by the evaluated automatic aligner.

4 GIZA++ Word Alignment

We applied GIZA++[2] alignment tool [1] on data composed of two parts: 3,500,000 sentence pairs from the Czech-English parallel corpus CzEng [9], and 1,250 sentence pairs from the above mentioned evaluation data set. All texts were lemmatized[3] in both languages by lemmatizers available in TectoMT [10].

We ran GIZA++ in both directions (English to Czech and Czech to English) and symmetrized the outputs using intersection, grow-diag-final and union symmetrization, as described by Och and Ney [1]. The results are given in Table 3.

---

[2] Default settings, IBM models and iterations: $1^5 3^3 4^3$.

[3] Bojar and Prokopová [8] showed that lemmatization of the input text reduces the Czech vocabulary size to a half. Thus the vocabulary sizes of Czech and English become comparable. The data are thus not so sparse, which helps alignment error rate by about 10% absolute.

Table 3. GIZA++ evaluation table for all words (%)

symmetrization	precision	recall	AER
intersection	95.8	79.0	**13.2**
grow-diag-final	71.5	92.0	20.3
union	68.5	93.2	22.1

Table 4. GIZA++ evaluation table for content words only (%)

symmetrization	precision	recall	AER
intersection	97.8	82.2	**10.6**
grow-diag-final	78.5	93.6	14.7
union	74.3	94.7	17.1

We can see that while the *intersection* symmetrization is too sparse (precision is much higher than recall), the *grow-diag-final* and *union* symmetrizations have the opposite problem. We did not present *grow* and *grow-diag* symmetrizations – they are very similar to the *grow-diag-final* and have also much higher recall than precision. There is no symmetrization with the density of connections similar to our manually aligned data, the nearest one is the *intersection*.

We also measured how successful is GIZA++ on the content words only. We transferred the word alignment generated by GIZA++ to the tectogrammatical trees in the same way as we did it for the manual alignment (Figure 2). Table 4 shows that the best AER is achieved using the intersection symmetrization again.

5   Tectogrammatical Alignment

This section describes our approach to aligning content words using tectogrammatical tree structures.

At first we have to build the trees. All sentences in both languages are automatically parsed up to the t-layer using TectoMT[4] [10]. Czech sentences are first tokenized, morphologically analyzed and disambiguated by the morphological tagger shipped with Prague Dependency Treebank [4]. After that, the syntactic analysis realized by McDonald's MST parser [11] comes. The resulting analytical trees are then automatically converted (mostly by rule-based scripts) into tectogrammatical trees. English sentences are tokenized in the Penn Treebank style, tagged by the TnT tagger [12]. The analysis continues in the same way as for Czech – McDonald's MST parser and conversion to the tectogrammatical trees. Finally, after applying GIZA++ on the lemmatized surface sentences, the data are prepared for the tectogrammatical alignment.

```
foreach (CT, ET) ∈ TreePairs do
    foreach cnode ∈ CT do
        counterpart(cnode) = argmax_enode ∑ w_i^{c2e} · f_i(cnode, enode);
    foreach enode ∈ ET do
        counterpart(enode) = argmax_cnode ∑ w_i^{e2c} · f_i(cnode, enode);
        if counterpart(counterpart(enode)) = enode then Align(cnode,enode);
```

Fig. 3. Pseudo-code for the first phase of t-alignment

---

[4] TectoMT is a software framework for developing machine translation systems.

Table 5. T-aligner evaluation (%)

alignment tool	prec.	recall	AER
GIZA++ (intersection)	97.8	82.2	10.6
T-aligner without GIZA	92.7	86.8	10.3
T-aligner using GIZA	96.0	89.7	**7.3**

Table 6. Improved GIZA++ word alignment evaluation (%)

alignment tool	prec.	recall	AER
GIZA++ (intersection)	95.8	79.0	13.2
improved GIZA++	94.3	84.6	**10.7**

Table 7. Features, their types, and weights in both directions

feature name	type	cs2en	en2cs
identical t-lemmas	binary	1.41	1.04
equal to 1 if Czech t-lemma is the same string as the English one			
verb, *adjective* **position similarity**	real	2.66	3.12
difference between relative linear positions of t-nodes			
t-lemma pair in dictionary	binary	1.88	2.06
equal to 1 if the pair of t-lemmas occurs in the translation dictionary			
3 letter match	binary	2.86	2.53
equal to 1 if the three-letter prefixes of Czech and English t-lemmas are identical			
equal number prefix	binary	9.58	7.00
Czech and English t-lemmas start with the same sequence of digits.			
aligned by GIZA++, direction en2cs	binary	1.20	2.31
equal to 1 if the corresponding surface words were aligned by GIZA++ (left)			
aligned by GIZA++, direction cs2en	binary	2.29	0.07
equal to 1 if the corresponding surface words were aligned by GIZA++ (right)			
aligned by GIZA++, intersection symmetrization	binary	1.57	0.92
equal to 1 if the corresponding surface words were aligned by GIZA++ (intersection)			
translation probability from dictionary	real	1.02	1.34
probability of Czech t-lemma, if the English t-lemma is given			
equal semantic part of speech	binary	1.61	1.06
equals to 1 if both semantic parts of speech (e.g. *noun*, *verb*, and *adjective*) are equal			
position similarity of parents nodes	real	0.76	0.92
difference between relative linear positions of parents			
parents aligned by GIZA++, intersection sym.	binary	0.21	0.28
equal to 1 if the surface words corresponding to parents were aligned by GIZA++			

The presented algorithm is based on a linear model and consists of three phases. First, we connect each English node with its most probable Czech counterpart. Second, we do the same for the opposite direction – we find the most probable English node for each Czech node. Finally, we make an intersection of these two alignments and declare it as our result. There is a pseudocode is in Figure 3. The most probable counterpart is the node with the highest score – the scalar product of vector of feature values and vector of feature weights. The weights for the Czech-to-English direction (w_i^{c2e}) are different from those for the opposite direction (w_i^{e2c}). We train the weights on the training part of manual aligned data and use an implementation of the discriminative reranker described by Collins in [13] (basically a modification of averaged perceptron).

Features are individual measurable properties of a pair of Czech and English node. We use features concerning similarities of lemmas and other t-node attributes, similarity in relative linear position of t-nodes within the sentences, and similarities of their child and parent nodes. There are several features taking into account whether GIZA++ aligned the examined pair on the surface or not; some features carry information from the probabilistic translation dictionary. This dictionary was compiled from parallel corpora PCEDT [7] and subsequently extended by word pairs acquired from the parallel corpus CzEng [9] aligned on the surface. All features are listed in Table 7. Their type and weights obtained from the reranker are also included.

The T-aligner evaluation and comparison with the results of GIZA++ (*intersection* symmetrization) in Table 5. Our T-aligner helps GIZA++ to decrease AER by 3.3% absolute. We also ran T-aligner without using GIZA++, AER reached 10.3 %, the error rate is therefore comparable to GIZA++. The advantage is that the combination of these two different approaches reaches lower AER.

6   Improving Surface Word Alignment

Our T-aligner has better results on the content words, when evaluated using our manual alignment. We can also measure how it can increase GIZA++ alignment on the surface. Since the GIZA++ word alignment has higher precision and lower recall (Table 3), we add all connections that were made by our T-aligner (it does not matter whether there was a connection made by GIZA++ or not) and no connections to delete. The results are in Table 6. The alignment error rate decreases by 2.5% absolute.

7   Conclusions

We described the algorithm for alignment of deep-syntactic dependency trees (tectogrammatical trees), where only content (autosemantic) words are aligned. We showed that alignment of content words is "simpler" both for the people (inter-annotator agreement is 5 % absolute higher) and for the automatic tools (GIZA++ AER decreases by 2.6 %). Our T-aligner outperformed GIZA++ in AER, however, the results are not straightforwardly comparable since GIZA++ could not be trained using the manually aligned data. We use aligned dependency trees for experimenting with statistical dependency-based MT. If we merge the best acquired alignment of content words with the GIZA++ surface word alignment, the resulting error-rate is lower by 2.5 %. It remains an open question how this "better" word alignment can improve phrase-based machine translation systems.

References

1. Och, F.J., Ney, H.: A Systematic Comparison of Various Statistical Alignment Models. Computational Linguistics 29(1), 19–51 (2003)
2. Menezes, A., Richardson, S.D.: A best-first alignment algorithm for automatic extraction of transfer mappings from bilingual corpora. In: Proceedings of the workshop on Data-driven methods in machine translation, vol. 14, pp. 1–8 (2001)
3. Sgall, P.: Generativní popis jazyka a česká deklinace. Academia, Prague (1967)

4. Hajič, J., Hajičová, E., Panevová, J., Sgall, P., Pajas, P., Štěpánek, J., Havelka, J., Mikulová, M.: Prague Dependency Treebank 2.0. Linguistic Data Consortium, LDC Catalog No.: LDC2006T01, Philadelphia (2006)
5. Haruno, M., Yamazaki, T.: High-performance Bilingual Text Alignment Using Statistical and Dictionary Information. In: Proceedings of the 34th conference of the Association for Computational Linguistics, pp. 131–138 (1996)
6. Watanabe, H., Kurohashi, S., Aramaki, E.: In: Finding Translation Patterns from Paired Source and Target Dependency Structures, pp. 397–420. Kluwer Academic, Dordrecht (2003)
7. Cuřín, J., Čmejrek, M., Havelka, J., Hajič, J., Kuboň, V., Žabokrtský, Z.: Prague Czech-English Dependency Treebank, Version 1.0. Linguistics Data Consortium, Catalog No.: LDC2004T25 (2004)
8. Bojar, O., Prokopová, M.: Czech-English Word Alignment. In: Proceedings of the Fifth International Conference on Language Resources and Evaluation (LREC 2006), ELRA, May 2006, pp. 1236–1239 (2006)
9. Bojar, O., Janíček, M., Žabokrtský, Z., Češka, P., Beňa, P.: CzEng 0.7: Parallel Corpus with Community-Supplied Translations. In: Proceedings of the Sixth International Language Resources and Evaluation (LREC'08), Marrakech, Morocco, ELRA (May 2008)
10. Žabokrtský, Z., Ptáček, J., Pajas, P.: TectoMT: Highly Modular MT System with Tectogrammatics Used as Transfer Layer. In: Proceedings of the 3rd Workshop on Statistical Machine Translation, ACL (2008)
11. McDonald, R., Pereira, F., Ribarov, K., Hajič, J.: Non-Projective Dependency Parsing using Spanning Tree Algorithms. In: Proceedings of Human Langauge Technology Conference and Conference on Empirical Methods in Natural Language Processing (HTL/EMNLP), Vancouver, BC, Canada, pp. 523–530 (2005)
12. Brants, T.: TnT - A Statistical Part-of-Speech Tagger. In: Proceedings of the 6th Applied Natural Language Processing Conference, Seattle, pp. 224–231 (2000)
13. Collins, M.: Discriminative Training Methods for Hidden Markov Models: Theory and Experiments with Perceptron Algorithms. In: Proceedings of EMNLP, vol. 10, pp. 1–8 (2002)

Trdlo, an Open Source Tool for Building Transducing Dictionary

Marek Grác

Natural Language Processing Centre at the Faculty of Informatics,
Masaryk University, Brno

Abstract. This paper describes the development of an open-source tool named Trdlo. Trdlo was developed as part of our effort to build a machine translation system between very close languages. These languages usually do not have available pre-processed linguistic resources or dictionaries suitable for computer processing. Bilingual dictionaries have a big impact on quality of translation. Proposed methods described in this paper attempt to extend existing dictionaries with inferable translation pairs. Our approach requires only 'cheap' resources: a list of lemmata for each language and rules for inferring words from one language to another. It is also possible to use other resources like annotated corpora or Wikipedia. Results show that this approach greatly improves effectivity of building Czech-Slovak dictionary.

1 Introduction and Motivation

Machine translation is among those problems expected to be already solved using computers. Development in this area passes through periods of both prosperity and diminished research investments. In recent years, unrealistic expectations seem to have faded. Investments are logically focused on translations among languages with commercial potential: English–Japanese, Arabic or Spanish. Minor languages including Czech and Slovak lack both commercial potential and resources. Therefore, translation among those is generally neglected.

One of the basic building blocks needed for creation of a machine translation tool is a translation dictionary. In case of close languages, we can usually work with a simple dictionary that only contains translation pairs and with a smaller, more robust dictionary, for more complex cases (e.g. context-dependent word translations). In this paper, we will focus on a method for efficient creation of a dictionary with translation pairs only.

2 Slovak and Czech Languages

Czech and Slovak are two closely related languages which together form the Czech-Slovak sub-group of West Slavonic languages [1]. Despite their similarity and nearly universal mutual intelligibility, the literary languages are clearly differentiated. The effect is that among other things they are each based on a different dialect due to their separate standardization. Although Czechs and Slovak lived together in one state for a relatively short period of 68 years (1918–1938 and 1945–1992), compared to their literary tradition, their linguistic relations have been very close since the Middle Age.

V. Matoušek and P. Mautner (Eds.): TSD 2009, LNAI 5729, pp. 64–69, 2009.

Similarities among these languages allow us to simplify the groundwork required for a translation dictionary of words and phrases. In most cases, it is not needed to include more than a lemma for translated word as word sense is not changed at all. Dictionaries for these languages published as books tends to simplify the task further. These do not cover words which have same spelling and meaning in both languages – such dictionaries are called differential. Other useful language resources for Czech and Slovak are single-language corpora (e.g. ČNK [2] and SNK [3]), morphological databases and lists of lemmata from other sources (e.g. spell-checkers). There are more resources that could be useful in the future but either they are not available in both languages (e.g. lexicon of verb valencies) or their size is not sufficient (e.g. Czech-Slovak parallel corpora [4] containing 7 documents with only about 350 000 tokens).

3 Rule Based Method

Basic resources for all of our methods are lists of lemmata. We were looking for such that are freely available and thus we took the lists used for spell-checking ([5] and [6]). Unfortunately it is not easy to determine lexical categories for lemmata in these lists but they fulfill our need for a simple list of available words for each language. As a reference dictionary, we use PC Translator[1] dictionary which we have available in digital form. All the translations we have created were tested against this dictionary. Our wordlist for Slovak (source language) was stripped so that we had a translation in the reference dictionary available for every word. The wordlist for the target language was not modified at all and also contained words that are not used in reference dictionary.

Due to similarities among the two languages, we have decided to devise rules for rewriting letters contained in a word according to context (in the direction Slovak → Czech) [7]. We have created three sets of rules based on different assumptions. All of these rules were written using the formal language of regular expressions.

First set B provides no rules at all. It covers only words that have exactly same spelling in both languages and will be used as a baseline. The second set K contains 50 rules (examples are in table 1) while the third set G contains just 15 rules. The differences between sets K and G are not only in the number of rules but also in the approach. On average, set K generates four times as many candidates for Czech words as the set G.

Table 1. Example transformation rules

ô → ů
ia → a, e
[dtn][eo] → [dtn]ě

Translation candidates created from Slovak words are looked up in the Czech list of words and, if found, we assume the translation pair to be valid. It goes without saying

---

[1] The dictionary has been heavily reduced to roughly 80 thousand pairs, but quality of data still remains a problem.

Table 2. Results of method based on set of rules

Set of Rules	Recall	Precision
B — 0 rules	18.22%	99.31%
G — 15 rules	36.32%	98.98%
K — 50 rules	49.73%	97.07%

that such transduction introduces a risk of erroneous translation. Introduced errors are specifically words that have different meaning but after applying the rules are spelt the same (e.g. *(sk) kel → (cz) kapusta, (sk) kapusta → (cz) zelí*). For very close languages, the set of such words is usually very well covered by existing (mostly paper-based) resources.

Results of this method are presented in table 2. We have achieved very good precision for this automatic method but we still have problems with recall that is below our expectation.

4 Method Based on Edit Distance

Another way how to create translation pairs is to take advantage of their similarity. As we expect only basic language resources, we don't have access to prefixes or stems. We can only rely on characters itself. There are several metrics used for measuring the amount of difference between two sequences (so called edit distance). We have chosen Levenshtein distance [8], which computes the number of changes (addition, deletion and replacement of a single character) that are necessary to perform to transform a word into another (e.g. *l(kitten, sitting) = 3, since kitten → sitten → sittin → sitting*). Another important property of this metric is that it can be used with words of different length.

After computation on the reference dictionary, we have found that 75 % of the translation pairs have Levenshtein distance in the range from zero to three. Edit distance between the words in the translation pair is generally not related to the frequency of the word and differences among the distribution of words with different distance among 1000, 5000, 10000 and 20000 most frequent words are, according to our research, just above statistical error.

Finding translation pairs is done in two independent steps. First, we need to find all the words that are most similar to our source word. In the second step, we have to choose the correct translation between words obtained in the previous step.

We have implemented the search itself by picking the words with minimal distance to each Slovak word. This operation is fairly computationally intensive since for each pair the computation has to start over and we have to count the distance between the source and all its possible translations.

Selection of the translation pair was facilitated using three basic methods: *First-Match, AnyMatch, JustOne*. First of them, FirstMatch, takes as the valid pair the pair that is first in alphabetic order, and it has been used as a baseline for comparison. The AnyMatch method accepted the translation pair to be valid if at least one of the pairs was present in the dictionary. Since this method never removes valid results, we have used it as an upper bound. The JustOne method picks the translation pair only if there

Table 3. Results of the method based on the edit distance

Method	Edit Distance 1		Edit Distance 2		Edit Distance 3	
	Recall	Precision	Recall	Precision	Recall	Precision
AnyMatch	40.35 %	93.04 %	68.64 %	83.75 %	90.35 %	79.34 %
FirstMatch	40.35 %	84.27 %	68.64 %	71.63 %	90.35 %	62.23 %
JustOne	33.62 %	94.46 %	50.04 %	88.89 %	57.63 %	86.47 %

is exactly one candidate pair. The results for different distance thresholds are presented in table 3.

5   Corpora Based Methods

Instead of previous methods that are usable also with limited resources, we present another technique that is only applicable to languages with available corpora. Employing parallel corpora will surely help us a lot, although as we have already mentioned, they are currently insufficient because of their small size [4]. We have not tested this option.

We focus our effort on using large monolingual corpora ([2], [3]) which are balanced in a similar manner and thus we can talk about them as comparable corpora. Both of these corpora are morphologically annotated ([9], [10]) and hence we can obtain the lemma for each word.

The non-existence of large parallel corpora makes it more difficult for us as we are unable to use the corpora directly. In our case, comparable corpora can be used to confirm or reject candidates for translation pairs. We will use translation candidates which were created using previous methods. At the beginning we have the original word and its translation candidates (e.g. for *kotol (sk)* (boiler, melting pot) we have translation candidates *kotoul (cz)* (roll, somersault) and *kotel (cz)* (boiler, melting pot). For each of the candidates we will obtain lemmata which occur in $-3/+3$ context in the appropriate corpora. In our experiment only 20 most frequent are taken into account. Lemmata for source and target word in a translation pair are considered the same if the translation for them already exists or they are at least translation candidates. If we have enough matched lemmata we consider the translation pair to be correct. Example of most frequent words for source lemma *kotol* and its translation candidates are available in table 4.

Our attempt to use similar method in order to reject translation pairs has not led to success. The main problem has been that even for relatively frequent lemmas ($>$ 1000

Table 4. Example of most frequent lemma in context

kotol (sk)	kotoul (cz)	kotel (cz)
plynový (gas)	udělat (to do)	fluidní (fluid)
vriaci (boiling)	svatý (saint)	plynový (gas)
fluidný (fluid)	vpřed (fore)	parní (steam)
taviaci (melting)	žiněnka (mattress)	výkon (power)
parný (steam)	vydechovat (breathe out)	dva (two)

occurrences), there were too many words that did not have any of the frequent words in common.

This method, even if it cannot be used alone, proves that it can confirm translation pair with a very high reliability (more than 98 %). Coverage is relatively small comparing to edit distance methods but can be improved with larger corpora.

6   Combined Methods

The described approaches have different combinations of coverage and precision. We can focus on best recall, suitable for semi-automatic approach, or on best precision, suitable for automatic approach.

We started to use the method with best precision and used others approaches afterwards. After usage of each method we will remove words contained in translation pairs from our lists so that every word will be used just once. We believe that if is are more than one correct translation then only one of them will be similar enough.

Rule based methods have the best precision and therefore they were used first [1]. After this step we counted edit distance and used corpora based methods [2] on nontranslated words. After this step we can choose whether we need better precision (and use JustOne for disambiguation) or we want better recall at expense of ambiguity in translation pairs (but we will know that at least one translation candidate is correct). The second option can be used in situations where skilled user will choose the correct translation as this is more error-proof and faster than just typing it down. Selected results are presented in table 5.

It is interesting that the difference in recall between rule sets G and K rapidly decreases when they are used in combined methods. Differences in coverage between rule sets G and K have diminished after using edit distance methods from 18 to 2 percentage point. It means that we don't have to spend too much time writing transduction rules because less frequently used rules can be replaced by edit distance methods.

It is possible to use several edit distance methods e.g. JustOne and FirstMatch and get better recall. Unfortunately such effort did not prove to be useful as it can improve quality in less than one percent of cases.

We have also experimented with using interlingual links in Wikipedia [11]. This resource should be very close to quality of human translation. Unfortunately most of the terms in Slovak and Czech language are not lemmas hence they cannot be used

Table 5. Results of combined methods

Sequence of Actions	Precision	Recall
rules G	99.07 %	36.43 %
rules G, corpora	98.85 %	46.32 %
rules G, corpora, distance ≤ 1 – JustOne	97.04 %	68.58 %
rules G, corpora, distance ≤ 20 – JustOne	92.40 %	82.40 %
rules G, corpora, distance ≤ 1 – AnyMatch	96.71 %	69.94 %
rules G, corpora, distance ≤ 20 – AnyMatch	84.54 %	97.66 %

directly. We decided to accept only those words that are in our word lists. We obtained 544 translation pairs and thus the coverage is really poor 0.73 %.

7 Conclusions

This project has shown that it is possible to create a Czech-Slovak dictionary even without extensive resources. Since high precision is expected in dictionaries, the presented method is not suitable for fully automatic dictionary creation. The achieved precision, however, gives hope that the method will be useful, accompanied with existing differential dictionary, which should contain all kinds of problematic words. This approach is unlikely to be useful for languages with deeper differences, but there is possibility to apply it e.g. for creation of dialectological dictionaries or modelling derivation of words.

Acknowledgments

This work has been partly supported by the Academy of Sciences of Czech Republic under the project T100300419, by the Ministry of Education of CR within the Center of basic research LC536 and in the National Research Programme II project 2C06009 and by the Czech Science Foundation under the project 201/05/2781.

References

1. Nábělková, M.: Closely-related languages in contact: Czech, slovak, czechoslovak. International Journal of the Sociology of Language, 53–73 (2007)
2. Český národní korpus SYN2006PUB: Ústav Českého národního korpusu FF UK (2006), http://www.korpus.cz
3. Slovenský národný korpus – prim-3.0-public-all: Jazykovedný ústav L'. Štúra SAV (2007), http://korpus.juls.savba.sk
4. Intercorp Parallel Corpora: Ústav Českého národního korpusu FF UK (2009), http://ucnk.ff.cuni.cz/intercorp/
5. Kolář, P.: Czech dictionary for ispell (2006), http://www.kai.vslib.cz/~kolar/rpms.html
6. Podobný, Z.: Slovak dictionary for ispell (2006), http://sk-spell.sk.cx
7. Bémová, A., Kuboň, V.: Czech-to-russian transducing dictionary. In: Proceedings of the 13th conference on Computational linguistics, Morristown, NJ, USA, pp. 314–316. Association for Computational Linguistics (1990)
8. Levenshtein, V.I.: Binary codes capable of correcting deletions, insertions, and reversals. Soviet Physics Doklady 10, 707–710 (1966)
9. Hajič, J.: Disambiguation of rich inflection: computational morphology of Czech. Karolinum (2004)
10. Spoustova, D., Hajic, J., Votrubec, J., Krbec, P., Kveton, P.: The best of two worlds: Cooperation of statistical and rule-based taggers for Czech. In: Proc. of the ACL Workshop on Balto-Slavonic Natural Language Processing (2007)
11. Wikipedia contributors: Wikipedia (2009), http://www.wikipedia.org

Improving Patient Opinion Mining
through Multi-step Classification

Lei Xia[1], Anna Lisa Gentile[2], James Munro[3], and José Iria[1]

[1] Department of Computer Science, The University of Sheffield, UK
{l.xia,j.iria}@dcs.shef.ac.uk
[2] Department of Computer Science, University of Bari, Italy
al.gentile@di.uniba.it
[3] Patient Opinion
james.munro@patientopinion.org.uk
http://www.patientopinion.org.uk/

Abstract. Automatically tracking attitudes, feelings and reactions in on-line forums, blogs and news is a desirable instrument to support statistical analyses by companies, the government, and even individuals. In this paper, we present a novel approach to polarity classification of short text snippets, which takes into account the way data are naturally distributed into several topics in order to obtain better classification models for polarity. Our approach is multi-step, where in the initial step a standard topic classifier is learned from the data and the topic labels, and in the ensuing step several polarity classifiers, one per topic, are learned from the data and the polarity labels. We empirically show that our approach improves classification accuracy over a real-world dataset by over 10%, when compared against a standard single-step approach using the same feature sets. The approach is applicable whenever training material is available for building both topic and polarity learning models.

1 Introduction

Opinion mining or extraction is a research topic at the crossroads of information retrieval and computational linguistics concerned with enabling automatic systems to determine human opinion from text written in natural language. Recently, opinion mining has gained much attention due to the explosion of user-generated content on the Web. Automatically tracking attitudes and feelings in on-line blogs and forums, and obtaining first hand reactions to the news on-line, is a desirable instrument to support statistical analyses by companies, the government, and even individuals.

Patient Opinion[1] is a social enterprise pioneering an on-line review service for users of the British National Health Service (NHS). The aim of the site is to enable people to share their recent experience of local health services online, and, ultimately, help citizens change their NHS. Online feedback is sent to health service managers and clinicians, but, due to "big numbers", it is difficult for a single person to read them all. Dedicated staff are needed to classify the comments received, both according to topic

---

[1] http://www.patientopinion.org.uk

V. Matoušek and P. Mautner (Eds.): TSD 2009, LNAI 5729, pp. 70–76, 2009.

and polarity (i.e., as either positive, negative or neutral), and to provide managers a summary report. However, the cost of running such staff quickly becomes prohibitive as the amount of comments grows, and this greatly limits the usefulness of all the collected data.

Careful analysis of the on-line comments from Patient Opinion indicates that the polarity distribution of the comments is not independent from their topic, and that topic-specific textual cues provide valuable evidence for deciding about polarity. For example, patient comments about "parking" almost always express a negative opinion, presumably because when parking does not pose a problem people tend to forget to mention it; and in comments about "staff", the presence words like "care" and "good" is highly indicative of positive experiences involving human contact with health care staff, whereas this is not necessarily the case with comments about something else. Unfortunately, to the best of our knowledge, no previous work has exploited the topic distribution of the data in the design of automatic polarity classification systems.

In this paper, we present a novel machine learning-based approach to polarity classification of short text snippets, which takes into account the way data are naturally distributed into several topics in order to learn better classification models for polarity. Our approach is multi-step, where in the initial step a standard topic classifier is learned from the data and the topic labels, and in the ensuing step several polarity classifiers, one per topic, are learned from the data and the polarity labels. We empirically show that our approach improves classification accuracy over a real-world dataset from Patient Opinion by over 10%, when compared against a standard single-step approach using the same feature sets. We also compare against a baseline multi-step approach in which the first step is unsupervised, which turned out to perform the worst of all approaches, further confirming our claim that distribution according to topic is relevant for polarity classification. Our approach is applicable whenever training material is available for building both topic and polarity learning models.

The rest of the paper is structured as follows. A review of related work is given in the next section, including an examination of application domains and commonly used techniques for the polarity classification task. In Section 3 we describe our approach in detail. A complete description of the experimental setting is given in Section 4, and a discussion of the results obtained is presented in Section 5. Our conclusions and plans for future work close the paper.

2   Related Work

The research field of *Sentiment Analysis* deals with the computational treatment of sentiments, subjectivity and opinions within a text. *Polarity Classification* is a subtask of sentiment analysis which reduces the problem to identifying whether the text expresses positive or negative sentiment.

Approaches to the polarity classification task mainly fall into two categories: *symbolic* techniques and *machine learning* techniques. Symbolic techniques are commonly based on rules and manually tuned thresholds, e.g., [1], and tend to rely heavily on external lexicons and other structured resources. Machine learning techniques have more recently been widely studied for this task, mostly with the application of supervised methods such as Support Vector Machines, Naïve Bayes and Maximum Entropy [2,3,4,5,6], but also unsupervised methods, such as clustering, e.g., [7]. In this paper, we use an off-the-shelf supervised learning algorithm within the context of a meta-learning strategy which we propose in order to take advantage of the availability of labeled data along two dimensions (topic and polarity) to improve classifier accuracy.

Our meta-learning strategy resembles work on hierarchical document classification [8,9]. There, the idea is to exploit the existence of a document topic hierarchy to improve classification accuracy, by learning classification models at several levels of the hierarchy instead of one single "flat" classifier. Inference of a document's category is performed in several steps, using the sequence of classifiers from the top to the bottom of the hierarchy. The key insight is that each of the lower-level classifiers is solving a simpler problem than the original flat classifier, because it only needs to distinguish between a small number of classes. Our approach is similar in that we perform classification in two steps, building separate polarity classifiers per topic which are also solving an easier problem than a single topic-unaware polarity classifier.

Machine learning-based work on polarity classification typically formulates a binary classification task to discriminate between positive and negative sentiment text, e.g., [4]. However, some works consider neutral text, hence addressing the additional problem of discriminating between subjective and objective text before classifying polarity, e.g., [10]. Niu et al. [11] even consider a fourth class, namely "no-outcome". In this work, we follow the traditional formulation and consider the problem of classifying subjectivity vs. objectivity to be outside the scope of the paper.

A significant amount of work has also been done on studying which features work well for the polarity classification task. Abbassi et al. [2] provides a categorization of the features typically employed, dividing them into four top level categories: *syntactic*, *semantic*, *link-based* and *stylistic*. Syntactic features are the most commonly used and range from simply the word token to features obtained by natural language processing, e.g., lemma. Semantic features result from additional processing of the text, typically relying on external resources or lexicons designed for this task, such as SentiWordnet [12]. Link-based features [13,14] are derived from the link structure of the (hyperlinked) documents in the corpus. Examples of stylistic features include vocabulary richness, greetings, sentence length, word usage frequency [2]. In this paper, it suffices to make use of standard syntactic features in the literature, since the focus is rather in drawing conclusions about the proposed multi-step approach.

Sentiment analysis has been applied to numerous domains, including news articles [15], web forums [14] and product reviews [4,10]. Our problem domain is more similar to the latter, since we also analyse reviews, but rather deal with the feedback provided by patients of the British NHS, spanning topics such as staff, food and parking quality.

3   Multi-step Polarity Classification

We introduce a novel multi-step approach to polarity classification, which exploits the availability of labels along the dimensions of topic and polarity for the same data. Following the intuition that polarity is not statistically independent from topic, our hypothesis is that learning topic-specific polarity classification models can help improve the accuracy of polarity classification systems.

The proposed multi-step method is the following. Given a document collection $\mathcal{D} = \mathcal{D}_T \cup \mathcal{D}_P$, a set of topics $\mathcal{T}$, and training labels given by $\mathcal{L}_T : \mathcal{D} \to \mathcal{T}$ and $\mathcal{L}_P : \mathcal{D} \to \{-1, 1\}$, for topic and polarity respectively, do:

1. Learn a topic classifier from $\mathcal{D}_T$ using labels $\mathcal{L}_T$, by approximating a classification function of the form $f_T : \mathcal{D} \to \mathcal{T}$;
2. Apply f_T to $\mathcal{D}_P$, thereby splitting $\mathcal{D}_P$ by topic, that is, creating sub-datasets $\mathcal{D}_{Pt} = \{d \in \mathcal{D}_P : f_T(d) = t\}, \forall t \in \mathcal{T}$;
3. For each $t \in \mathcal{T}$, learn a polarity classifier from $\mathcal{D}_{Pt}$ using labels $\mathcal{L}_P$, by approximating a classification function of the form $f_{Pt} : \mathcal{D} \to \{-1, 1\}$;
4. Classify any previously unseen input document d by $\arg\max_P f_{Pf_T(d)}(d)$, that is, topic classification determines which polarity classifier to run.

We use a standard multinomial Naïve Bayes approach for learning the classifiers. In the Naïve Bayes probabilistic framework, a document is modeled as an ordered sequence of word events drawn from a vocabulary $\mathcal{V}$, and the assumption is that the probability of each word event is independent of the word's context and position in the document. Each document $d_i \in \mathcal{D}$ is labeled with a class $c_j \in \mathcal{C}$ and drawn from a multinomial distribution of words with as many independent trials as its length. This yields the familiar "bag of words" document representation. The probability of a document given its class is the multinomial distribution [16]:

$$P(d_i|c_j; \theta) = P(|d_i|)|d_i|! \prod_{t=1}^{|\mathcal{V}|} \frac{P(w_t|c_j; \theta)^{N_{it}}}{N_{it}!}, \tag{1}$$

where N_{it} is the count of the number of times word w_t occurs in document d_i, and θ are the parameters of the generative model. Estimating the parameters for this model from a set of labeled training data consists in estimating the probability of word w_t in class c_j, as follows (using Laplacian priors):

$$P(w_t|c_j; \theta) = \frac{1 + \sum_{i=1}^{\mathcal{D}} N_{it} P(c_j|d_i)}{|\mathcal{V}| + \sum_{s=1}^{|\mathcal{V}|} \sum_{i=1}^{|\mathcal{D}|} N_{is} P(c_j|d_i)}. \tag{2}$$

The class prior parameters are set by the maximum likelihood estimate:

$$P(c_j|\theta) = \frac{\sum_{i=1}^{|\mathcal{D}|} P(c_j|d_i)}{|\mathcal{D}|}. \tag{3}$$

Given these estimates, a new document can be classified by using Bayes rule to turn the generative model around and calculate the posterior probability that a class would have generated that document:

$$P(c_j|d_i; \theta) = \frac{P(c_j|\theta)P(d_i|c_j; \theta)}{\sum_{j=1}^{|\mathcal{C}|} P(c_j|\theta)P(d_i|c_j; \theta)}. \tag{4}$$

Classification becomes a simple matter of selecting the most probable class.

4　Experiments

A set of experiments was designed and conducted to validate our hypothesis, using 1200 patient comments provided by Patient Opinion. The comments consist of short texts manually annotated by Patient Opinion experts with an indication of polarity (either positive or negative) and one of the following eight topics: Service, Food, Clinical, Staff, Timeliness, Communication, Environment, and Parking. Each comment is represented as a vector of frequencies of word stem unigrams. The word stems are obtained by running the *OpenNLP*[2] tokeniser, removing stop words and applying the *Java Porter Stemmer*[3].

The accuracy of the proposed method is compared against three baseline methods:

- *Single-step.* This learns $f_P : \mathcal{D} \to \{-1, 1\}$ directly from $\mathcal{D}$, without any knowledge about topic.
- *Topic as feature.* Same as single-step, but introduces knowledge about the topic from the labels $\mathcal{L}_T$ as a feature for learning.
- *Clustering as first step.* Here an approximation of the true $f_T : \mathcal{D} \to \mathcal{T}$ is created in an unsupervised way by clustering $\mathcal{D}$ according to the traditional cosine similarity between the vector representations of the comments, without any knowledge about topic.

In order to understand the influence of the topic classifier, used in the first step of our proposed method, on the overall system accuracy, we simulate a noise process over the topic labels $\mathcal{L}_T$. Concretely, given a noise ratio $r \in [0, 1]$, let $\mathcal{S}$ be a random subset of $\mathcal{D}$ such that $|\mathcal{S}| = r \times |\mathcal{D}|$. We define *noisy topic labels* $\mathcal{L}'_T : \mathcal{D} \to \mathcal{T}$ as:

$$\mathcal{L}'_T(d) = \begin{cases} \mathcal{L}_T(d) & \text{, if } d \in \mathcal{D} \setminus \mathcal{S}; \\ t \in \mathcal{T} \setminus \{\mathcal{L}_T(d)\} & \text{, if } d \in \mathcal{S}. \end{cases}$$

By varying r, it is possible to study the effect of the quality of the underlying topic classifier, since $\mathcal{L}'_T$ mimics the (imperfect) output of a learned f_T.

Finally, for the Naïve Bayes algorithm we used the off-the-shelf implementation in *LingPipe*[4], while for clustering we used our implementation of the well-known *k-means* algorithm, setting $k = 8$ to match the number of topics.

5　Results and Discussion

We measure systems' accuracy using the standard information retrieval measures of precision, recall and F-measure. All the results presented in Figure 1 consist of the micro-averaged F-measure over 10 trials with 60/40 random splits (60% of the comments used for training and 40% used for testing).

The single-step approach achieved a F1 value of 65.6%. The attempt to improve it by introducing the topic from $\mathcal{L}_T$ as an additional feature for the learning task

---

[2] http://opennlp.sourceforge.net/
[3] http://tartarus.org/~martin/PorterStemmer/java.txt
[4] http://alias-i.com/lingpipe/

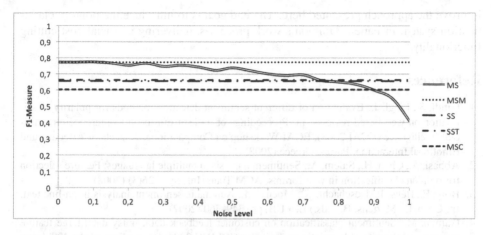

Fig. 1. F-measure, micro-averaged over all polarity classes and trials, obtained by the proposed multi-step approach (**MS**) by varying the ratio of noise introduced into the topic labels. Indicated for the purposes of comparison are also the results obtained by the "single-step" approach (**SS**), the "topic as feature" approach (**SST**) and the "clustering as first step" approach (**MSC**).

yielded a small improvement of roughly 1%. The proposed multi-step approach provides a much more significant improvement over the previous ones: the upper bound, given by the use of a perfect topic classifier, is at 77% F1, but even considering a noise ratio of up to 60% the multi-step approach obtains better results than the single-step approaches. Thus, we can conclude that it is sufficient to use a reasonable accuracy topic classifier, easily obtainable with the current state-of-the-art, in order to get a significant improvement in polarity classification.

The "clustering as a first step" approach, which clusters documents according to cosine similarity, performed worse than the original single-step approach, obtaining roughly 60% F1, while the lower bound, at around 40%, is given by a fully random allocation of documents to topics. This further confirms our claim that distribution according to topic is relevant for polarity classification.

6 Conclusions and Future Work

Exploiting the availability of training material for building both topic and polarity learning models, in this paper we proposed a meta-learning strategy which performs classification in two steps: classifying documents according to topic as a first step and then building separate polarity classifiers per topic. Our experimental results show that our approach, evaluated against three baseline methods, improves conspicuously polarity classification accuracy, thus proving our intuition that the topic distribution has significant influence on polarity classification.

As future work, we will work in close cooperation with Patient Opinion to further improve the accuracy of the system, both with the addition of further training material to refine the learning models, and by investigating which learning features could further improve the approach presented here. The end goal is to introduce the polarity classification system in Patient Opinion's work processes, delivering a crucial cost-cutting functionality.

References

1. Gindl, S., Liegl, J.: Evaluation of different sentiment detection methods for polarity classification on web-based reviews. In: Proceedings of the18th European Conference on Artificial Intelligence (ECAI 2008), ECAI Workshop on Computational Aspects of Affectual and Emotional Interaction, Patras, Greece (2008)
2. Abbasi, A., Chen, H., Salem, A.: Sentiment analysis in multiple languages: Feature selection for opinion classification in web forums. ACM Trans. Inf. Syst. 26(3) (2008)
3. Boiy, E., Hens, P., Deschacht, K., Moens, M.: Automatic sentiment analysis in on-line text. In: Chan, L., Martens, B. (eds.) ELPUB, pp. 349–360 (2007)
4. Gamon, M.: Sentiment classification on customer feedback data: noisy data, large feature vectors, and the role of linguistic analysis. In: COLING 2004: Proceedings of the 20th international conference on Computational Linguistics, Morristown, NJ, USA, p. 841. Association for Computational Linguistics (2004)
5. Niu, Y., Zhu, X., Li, J., Hirst, G.: Analysis of polarity information in medical text. In: AMIA Annu. Symp. Proc., pp. 570–574 (2005)
6. Pang, B., Lee, L., Vaithyanathan, S.: Thumbs up? Sentiment classification using machine learning techniques. In: Proceedings of the 2002 Conference on Empirical Methods in Natural Language Processing (EMNLP), pp. 79–86 (2002)
7. Hatzivassiloglou, V., McKeown, K.: Predicting the semantic orientation of adjectives. In: Proceedings of the eighth conference on European chapter of the Association for Computational Linguistics, Morristown, NJ, USA, pp. 174–181. Association for Computational Linguistics (1997)
8. Koller, D., Sahami, M.: Hierarchically classifying documents using very few words. In: Fisher, D. (ed.) ICML, pp. 170–178. Morgan Kaufmann, San Francisco (1997)
9. Dumais, S., Chen, H.: Hierarchical classification of web content. In: SIGIR, pp. 256–263 (2000)
10. Pang, B., Lee, L.: A sentimental education: Sentiment analysis using subjectivity summarization based on minimum cuts. In: Proceedings of the ACL, pp. 271–278 (2004)
11. Niu, Y., Zhu, X., Hirst, G.: Using outcome polarity in sentence extraction for medical question-answering. In: AMIA Annu. Symp. Proc., pp. 599–603 (2006)
12. Esuli, A., Sebastiani, F.: Sentiwordnet: A publicly available lexical resource for opinion mining. In: Proceedings of the 5th Conference on Language Resources and Evaluation (LREC 2006), pp. 417–422 (2006)
13. Efron, M.: Cultural orientation: Classifying subjective documents by cocitation analysis. In: Proceedings of the 2004 AAAI Fall Symposium on Style and Meaning in Language, Art, Music, and Design, pp. 41–48 (2004)
14. Agrawal, R., Rajagopalan, S., Srikant, R., Xu, Y.: Mining newsgroups using networks arising from social behavior. In: WWW 2003: Proceedings of the 12th international conference on World Wide Web, pp. 529–535. ACM, New York (2003)
15. Yu, H., V.H.: Towards answering opinion questions: separating facts from opinions and identifying the polarity of opinion sentences. In: Proceedings of the 2003 conference on Empirical methods in natural language processing, Morristown, NJ, USA, pp. 129–136. Association for Computational Linguistics (2003)
16. Mccallum, A., Nigam, K.: A comparison of event models for naive bayes text classification. In: AAAI Workshop on Learning for Text Categorization (1998)

Update Summarization Based on Latent Semantic Analysis

Josef Steinberger and Karel Ježek

University of West Bohemia, Univerzitní 22, 306 14 Plzeň
{jstein,jezek_ka}@kiv.zcu.cz

Abstract. This paper deals with our recent research in text summarization. We went from single-document summarization through multi-document summarization to update summarization. We describe the development of our summarizer which is based on latent semantic analysis (LSA) and propose the update summarization component which determines the redundancy and novelty of each topic discovered by LSA. The final part of this paper presents the results of our participation in the experiment of Text Analysis Conference 2008.

1 Introduction

Four years ago we started to develop a summarization method whose core was covered by latent semantic analysis (LSA – [3]). The proposed single-document method [4] modified the first summarization approach, which used LSA representation of a document [5]. From single-document summarization we went on to produce multi-document summaries [6].

The next step from multi-document summarization is update summarization which piloted in DUC2007[1] and represented the main track in TAC2008. The task of update summarization is to generate short (~100 words) fluent multi-document summaries of recent documents under the assumption that the user has already read a set of earlier documents. The purpose of each update summary is to inform the reader of new information about a particular topic.

When producing an update summary the system has to decide which information in the set of new documents is novel and which is redundant. (Redundant information is already contained in the set of earlier documents.) This decision is crucial for producing summaries with a high update value.

In this paper we present our update summarizer which participated in the TAC2008 evaluation. Our method follows what has been called a term-based approach [7]. In term-based summarization, the most important information in documents is found by identifying their main terms, and then extracting from the documents the most important information about these terms. However, LSA provides a way how to work with

---

[1] The National Institute of Standards and Technology (NIST) initiated the Document Understanding Conference (DUC) series [1] to evaluate automatic text summarization. Its goal is to further progress in summarization and enable researchers to participate in large-scale experiments. Since 2008 DUC has moved to TAC (Text Analysis Conference) [2] that follows the summarization evaluation roadmap with new or upgraded tracks.

V. Matoušek and P. Mautner (Eds.): TSD 2009, LNAI 5729, pp. 77–84, 2009.

topics of the documents instead of terms only [5]. The idea of our approach is to use latent semantic information for the creation of a set of 'topics' contained in the set of recent documents. Then, we specify their redundancy, novelty and significance. Sentences containing novel and significant topics are then selected for the summary. We use our new sentence selection approach which fights with inner summary redundancy, i.e., selecting similar sentences. The advantage of our method is that it works only with the word context and thus it is independent of the language. This is important for us because the summarizer can be plugged into a larger multilingual environment/system.

The rest of the paper is organized as follows: We first describe the LSA summarization model (Section 2). Section 3 contains the core of the paper. Our new sentence-extractive update summarization method is proposed. In Section 4 we discuss the TAC results. Finally, in the last Section, we conclude the paper and reveal our next point of focus in summarization research.

2   Latent Semantic Analysis Model for Summarization

Latent semantic analysis is a fully automatic mathematical/statistical technique which is able to extract and represent the meaning of words on the basis of their contextual usage. Its fundamental idea is based on the fact that mutual similarity among the meanings of words or phrases can be obtained from the accumulated contexts in which the word or the phrase occurs and in which it does not. LSA was applied to various tasks: e.g. information retrieval [8], text segmentation [9], or document categorization [10]. The first LSA application in text summarization was published in the year 2002 [5].

The heart of LSA-based summarization is a document representation developed in two steps. In the first step we construct the terms by sentences association matrix $\mathbf{A}$. Each element of $\mathbf{A}$ indicates the weighted frequency of a given term in a given sentence. Having m distinguished terms and n sentences in the document(s) under consideration the size of $\mathbf{A}$ is $m \times n$. Element a_{ij} of $\mathbf{A}$ represents the weighted frequency of term i in sentence j and is defined as:

$$a_{ij} = L(i,j) \cdot G(i), \tag{1}$$

where $L(i,j)$ is the local weight of term i in sentence j and $G(i)$ is the global weight of term i in the document. The weighting scheme we found to work best uses a binary local weight and an entropy-based global weight:

$$L(i,j) = 1 \text{ if term } i \text{ appears at least once in sentence } j; \\ \text{otherwise } L(i,j) = 0 \tag{2}$$

$$G(i) = 1 - \sum_j \frac{p_{ij} \log(p_{ij})}{log(n)}, \quad p_{ij} = \frac{t_{ij}}{g_i}, \tag{3}$$

where t_{ij} is the frequency of term i in sentence j, g_i is the total number of times that term i occurs in the whole document and n is the number of sentences in the document.

The next step is to apply the Singular Value Decomposition (SVD) to matrix $\mathbf{A}$. The SVD of an $m \times n$ matrix is defined as:

$$\mathbf{A} = \mathbf{U} \cdot \mathbf{S} \cdot \mathbf{V}^T, \tag{4}$$

where $\mathbf{U}$ ($m \times n$) is a column-orthonormal matrix, whose columns are called left singular vectors. The matrix contains representations of terms expressed in the newly created (latent) dimensions. $\mathbf{S}$ ($n \times n$) is a diagonal matrix, whose diagonal elements are non-negative singular values sorted in descending order. $\mathbf{V}^T$ ($n \times n$) is a row-orthonormal matrix which contains representations of sentences expressed in the latent dimensions. The dimensionality of the matrices is reduced to r most important dimensions and thus, we receive matrices $\mathbf{U}$' ($m \times r$), $\mathbf{S}$' ($r \times r$) a $\mathbf{V}$'^{T} ($r \times n$). The optimal value of r can be learned from the training data.

From the mathematical point of view SVD maps the m-dimensional space specified by matrix $\mathbf{A}$ to the r-dimensional singular space. From an NLP perspective, what SVD does is to derive the latent semantic structure of the document represented by matrix $\mathbf{A}$: i.e. a breakdown of the original document into r linearly-independent base vectors which express the main 'topics' of the document. SVD can capture interrelationships among terms, so that terms and sentences can be clustered on a 'semantic' basis rather than on the basis of words only. Furthermore, as demonstrated in [8], if a word combination pattern is salient and recurring in a document, this pattern will be captured and represented by one of the singular vectors. The magnitude of the corresponding singular value indicates the importance degree of this pattern within the document. Any sentences containing this word combination pattern will be projected along this singular vector, and the sentence that best represents this pattern will have the largest index value with this vector. Assuming that each particular word combination pattern describes a certain topic in the document, each singular vector can be viewed as representing such a topic [11], the magnitude of its singular value representing the degree of importance of this topic.

3   Update Summarization Based on the LSA Model

In update summarization, we assume the reader's prior knowledge of the topic. The input consists of a set of older documents C_{old}, which represents the prior knowledge, and a set of newer documents C_{new}, which is intended for the summarization itself. The first step is to obtain a set of topics of the prior knowledge (denoted as 'old' topics) and the set of new topics. Thus we perform the analysis of sets C_{old} and C_{new} separately: an input matrix is created for each set – $\mathbf{A}_{old}$, respectively $\mathbf{A}_{new}$. Experiments show that the best weighting system is the Boolean local weight and the entropy-based global weight computed separately for each document. The values that correspond to terms in the narrative do not get any advantage at this stage because sentence vectors must be normalized in order to be used as an input to SVD. As one of the results of applying the SVD to the input matrices, we get matrices $\mathbf{U}_{old}$ and $\mathbf{U}_{new}$, whose columns contain topics of the analyzed sets of documents expressed in linear combinations of original terms. For each 'new' topic t (a column of $\mathbf{U}_{new}$, in the following text t denotes the column number) the most similar 'old' topic is found (a column of matrix $\mathbf{U}_{old}$). The similarity value indicates the 'redundancy' of topic t, denoted as $red(t)$:

$$red(t) = \max_{i=1}^{r_{old}} \frac{\sum_{j=1}^{m} \mathbf{U}_{old}[j, i] \cdot \mathbf{U}_{new}[j, t]}{\sqrt{\sum_{j=1}^{m} \mathbf{U}_{old}[j, i]^2} \cdot \sqrt{\sum_{j=1}^{m} \mathbf{U}_{new}[j, t]^2}}, \quad (5)$$

where r_{old} is the number of old topics (the number of latent dimensions arising from the decomposition of $\mathbf{A}_{old}$).

The corresponding 'novelty' of topic t, $nov(t)$, can be then computed:

$$nov(t) = 1 - red(t). \tag{6}$$

Thus the topic redundancy will be large if a similar topic is found in the set of older documents, reversely topic novelty. The topic importance is represented by its corresponding singular value, $s(t)$. For each topic t we can compute 'topic update score' $tus(t)$:

$$tus(t) = nov(t) \cdot s(t), \tag{7}$$

From topic update scores we create diagonal matrix **TUS**, in which the diagonal consists of $tus(1), tus(2), \dots, tus(r_{new})$. Final matrix $\mathbf{F}$ can then be computed as $\mathbf{F} = \mathbf{TUS} \cdot \mathbf{V}^T$. In this matrix, both the importance and novelty of the new topics are taken into account.

Sentence selection starts with the sentence that has the longest vector[2] in matrix $\mathbf{F}$ (the vector, the column of $\mathbf{F}$, is denoted as $\mathbf{f}_{best}$). After placing it in the summary, the topic/sentence distribution is changed by subtracting the information contained in that sentence:

$$\mathbf{F}_{(i+1)} = \mathbf{F}_{(i)} - \frac{\mathbf{f}_{best} \cdot \mathbf{f}_{best}^T}{|\mathbf{f}_{best}|^2} \cdot \mathbf{F}_{(i)}. \tag{8}$$

The vector lengths of similar sentences are decreased, thus preventing inner summary redundancy. After the subtraction the process of selecting the sentence that has the longest vector in matrix $\mathbf{F}$ and subtracting its information from $\mathbf{F}$ is iteratively repeated until the required summary length is reached.

4　Experiments

TAC corpus in 2008 contained 48 topics and two sets of documents for each – A (older documents) and B (newer documents). Both sets contained 10 documents. The target summary length was again 100 words. Only the summary for set B could use the prior knowledge from older documents in A and thus it was the true update summary. The summary for set A was a simple multi-document summary (here denoted as 'basic summary'). In total, 58 summarizers participated in the manual evaluation[3].

NIST also created a baseline automatic summarizer, which selected the first few sentences of the most recent document in the relevant document set, such that their combined length did not exceed 100 words. In addition to automatic summaries, 4 human-written summaries (denoted as model summaries) were available.

The main evaluation approach was the Pyramid method [12], accompanied by manually annotated scale-based linguistic quality and overall responsiveness. For automatic

---

[2] We experimented with boosting the score of sentences that contain narrative terms. A slight improvement, but not statistically significant, was observed.

[3] In total, 71 summarizers participated in the large-scale evaluation. Each group could submit up to three runs ordered by priority. Only the first two priority runs were annotated, resulting in 58 manually evaluated runs.

Table 1. Separate TAC results of our basic summaries (set A)

Evaluation metric	Best Human	Average Human	Base- line	Best System	Average System	Our System
The Pyramid score	0.848	0.664	0.186	0.362	0.266	0.301 (19/58)
Linguistic quality	4.917	4.786	3.250	3.000	2.331	2.646 (12/58)
Overall responsiveness	4.792	4.620	2.292	2.792	2.324	2.667 (10/58)
ROUGE-2	0.131	0.118	0.058	0.111	0.079	0.089 (18/71)
ROUGE-SU4	0.170	0.154	0.093	0.143	0.115	0.124 (19/71)
BE	0.095	0.078	0.030	0.064	0.044	0.053 (14/71)

Table 2. Separate TAC results of our update summaries (set B)

Evaluation metric	Best Human	Average Human	Base- line	Best System	Average System	Our System
The Pyramid score	0.761	0.630	0.147	0.344	0.207	0.287 (7/58)
Linguistic quality	4.958	4.797	3.417	3.208	2.318	2.833 (5/58)
Overall responsiveness	4.875	4.625	1.854	2.604	2.027	2.292 (15/58)
ROUGE-2	0.132	0.117	0.060	0.101	0.067	0.081 (18/71)
ROUGE-SU4	0.167	0.150	0.094	0.137	0.106	0.121 (17/71)
BE	0.103	0.089	0.035	0.076	0.045	0.059 (13/71)

evaluation ROUGE-2, ROUGE-SU4 [13] and Basic Elements (BE – [14]) were used as in the previous DUC campaigns.

We show in Tables 1 and 2 separated results of basic and update summaries. We compare best and average performance of humans and automatic summarizers, baseline summarizer, and our two submitted runs[4].

First of all, we can see that there is a gap between the performance of humans and systems in all evaluation scores. Automatic scores (ROUGE scores and BE) could not distinguish between system and model summaries that well.

Pyramid evaluation: NIST assessors evaluated the content of each summary within the Pyramid evaluation framework developed at Columbia University [12]. In the Pyramid evaluation, assessors first extract all possible 'information nuggets' from the four model summaries on a given topic. Each nugget is assigned a weight in proportion to the number of model summaries in which it appears. Once the nuggets are harvested from the model summaries, assessors determine how many of these nuggets can be found

---

[4] There was no system that worked the best in all metrics. Thus, we report the best system result for each evelution measure. The best system differs from one measure to another.

in each of the automatic summaries. Repetitive information is not rewarded. The final Pyramid score for an automatic summary is its total nugget weigth available to a summary of average length (where the average length is determined by the mean nugget count of the model summaries for this topic). A Pyramid score was also calculated for the model summaries evaluating each of them against the remaining three models for a given topic. We can see that the results of our summarizer were promising. Our basic summaries were ranked 19th and update summaries were ranked 7th. Our system outperformed the baseline and the average systems. The difference between our system and the best system was not statistically significant (organizers measured statistical significance by a t-test on per-topic scores).

Overall responsiveness and linguistic quality: Overall responsiveness evaluated the degree to which a summary is responding to the information need contained in the topic statement, considering the summary's content as well as its linguistic quality. Linguistic quality score reflected the fluency and structure of the summary, independently of content, and was based on such aspects as gramaticality, non-redundancy, referential clarity, focus, structure, and coherence. Both overall responsiveness and linguistic quality were evaluated according to a five-point scale (1=very poor, 5=very good). There was a high correlation between overall responsiveness and the Pyramid score (Pearson's correlation: 0.637 for human summaries and 0.950 for automatic summaries). Overall responsiveness ranked our basic summaries 10th and update summaries were ranked 15th. If we take a look at the linguistic quality we can see that although we performed only a simple sentence extraction without any sentence modifications, the linguistic quality

---

ID: D0824E

Title: Breast Cancer Research

Narrative: Describe the causes and therapies being studied to help decrease breast cancer.

Basic Summary (set A)

Olive oil can help fight breast cancer, a discovery that researchers said in a study out Sunday could guide the development of related treatment. A new U.S. study will investigate genetic and environmental causes of breast cancer by enrolling 50,000 sisters of women already diagnosed with the disease. Analyzing cells from milk ducts doesn't reliably detect breast cancer, U.S. researchers reported Tuesday dashing hopes that the experimental technique could diagnose high-risk women sooner. Deficiencies in the ability of cells to repair damaged DNA are associated with an increased risk of breast cancer, researchers at Columbia University said Monday.

Update Summary (set B)

Women with benign breast lesions bear a higher risk of eventually developing breast cancer, according to a study released Wednesday in the United States. Women stricken with breast cancer can boost their chances of survival by 50 percent with some exercise, according to a study published Wednesday. A study examining the role childhood diet plays in breast cancer has found an association between eating French fries regularly during the preschool years and developing breast cancer as an adult. Breast cancer patients who follow diets low in fat may reduce the chance that their tumors will return, scientists reported Monday.

Fig. 1. Example summaries from TAC2008 corpus

was pretty good compared to other systems. Basic summaries were ranked 12th and update summaries were ranked 5th. As in the case of Pyramid evaluation the differences between our system and the best system were not statistically significant in both overall responsiveness and linguistic quality.

Automatic measures: Automatic scores were produced for all participated runs (71 in total). Organizers calculated two versions of ROUGE: ROUGE-2 and ROUGE-SU4, as well as BE-HM[5] recall. While ROUGE/BE correlations with manual assessment were universally high for automatic systems, when it comes to model summaries, only overall responsiveness is relatively well reflected in their scores. All the three metrics were evaluated as equally good at predicting manual scores. The best system for basic summaries evaluated by automatic measures obtained bad ranks by manual metrics, especially linguistic quality. Our system performed reasonably well in automatic measures. Most of the numbers are in top 20 from 71 systems and the best system is not better in the sense of statistical significance.

At last we show example summaries for one topic in TAC corpus – Figure 1. The basic summary describes older facts about breast cancer research and it is updated by more actual information in the update summary.

5 Conclusion

The work discussed here is novel in two respects. Firstly, we improved the method of sentence selection. In the past we simply selected the sentences with the greatest combined weights accross the important topics. The improved method uses an iterative process in which the information contained in the selected sentence is subtracted from the representation in each iteration. The second and the main contribution deals with the proposal of a topic-based determination of novel and redundant information in the context of the update summarization task. The advantage of the summarization method is that it works just with the context of terms and thus it is completely language independent. The participation in the TAC evaluation campaign showed promising results, however, a lot of work has to be done to make the summaries more fluent, starting with sentence compression.

Acknowledgement

This research was partly supported by project 2C06009 (COT-SEWing).

References

1. Document understanding conference (2007), http://duc.nist.gov/
2. Text analysis conference (2008),
   http://www.nist.gov/tac/tracks/2008/index.html
3. Landauer, T., Dumais, S.: A solution to platos problem: The latent semantic analysis theory of the acquisition, induction, and representation of knowledge. Psychological Review 104 (1997)

---

[5] HM corresponds to matching based on Head-Modifier criterion.

4. Steinberger, J., Ježek, K.: Text summarization and singular value decomposition. In: Yakhno, T. (ed.) ADVIS 2004. LNCS, vol. 3261, pp. 245–254. Springer, Heidelberg (2004)
5. Gong, Y., Liu, X.: Generic text summarization using relevance measure and latent semantic analysis. In: Proceedings of ACM SIGIR (2002)
6. Steinberger, J., Křišťan, M.: Lsa-based multi-document summarization. In: Proceedings of 8th International Workshop on Systems and Control (2007)
7. Hovy, E., Lin, C.: Automated text summarization in summarist. In: Proceedings of ACL/EACL workshop on intelligent scalable text summarization (1997)
8. Berry, M., Dumais, S., O'Brien, G.: Using linear algebra for intelligent ir. SIAM Review 37(4) (1995)
9. Choi, F., Wiemer-Hastings, P., Moore, J.: Latent semantic analysis for text segmentation. In: Proceedings of EMNLP (2001)
10. Lee, C.H., Yang, H.C., Ma, S.M.: A novel multilingual text categorization system using latent semantic indexing. In: Proceedings of the First International Conference on Innovative Computing, Information and Control. IEEE Computer Society, Los Alamitos (2006)
11. Ding, C.: A probabilistic model for latent semantic indexing. Journal of the American Society for Information Science and Technology 56(6) (2005)
12. Nenkova, A., Passonneau, R.: Evaluating content selection in summarization: The pyramid method. In: Document Understanding Conference (2005)
13. Lin, C.: Rouge: A package for automatic evaluation of summaries. In: Proceedings of the Workshop on Text Summarization Branches Out (2004)
14. Hovy, E., Lin, C.Y., Zhou, L.: Evaluating duc 2005 using basic elements. In: Proceedings of the Document Understanding Conference (2005)

WEBSOM Method – Word Categories in Czech Written Documents

Roman Mouček and Pavel Mautner

Department of Computer Science and Engineering, University of West Bohemia
Univerzitní 8, 306 14 Pilsen, Czech Republic
{moucek,mautner}@kiv.zcu.cz

Abstract. We applied well-known WEBSOM method (based on two layer architecture) to categorization of Czech written documents. Our research was focused on the syntactic and semantic relationship within word categories of word category map (WCM). The document classification system was tested on a subset of 100 documents (manual work was necessary) from the corpus of Czech News Agency documents. The result confirmed that WEBSOM method could be hardly evaluated because humans have problems with natural language semantics and determination of semantic domains from word categories.

1 Introduction

Finding relevant information from the vast material in the electronic form is a difficult and time consuming task. Therefore, an enormous scientific and commercial effort is paid to development of new methods and approaches, which help people to find and refer to (or extract) required information in accessible electronic sources. Some approaches try to involve as many aspects of natural language as possible whereas some of them are strictly limited by elaborated domain or processed language aspects.

However, the following question is rarely asked: which approaches are really practicable. We got used to enter key words using search engines and go through a set of returned documents to find the right one. Since entering key words does not limit or annoy people in general, scanning a large set of documents is a tiring and unpleasant work.

1.1 Semantic Web

Inability to find required information in documents properly led to idea of semantic web. Semantic web provides a common framework that allows data to be shared and reused across application, enterprise, and community boundaries [1]. This idea is based on common formats for interchange of data (not interchange of documents) from various sources. It supposes that documents are designed for humans to read, not for computer programs to manipulate them meaningfully.

V. Matoušek and P. Mautner (Eds.): TSD 2009, LNAI 5729, pp. 85–92, 2009.
© Springer-Verlag Berlin Heidelberg 2009

Searching for documents means to work with semantics of natural language. Natural language gives freedom to express a real word in various ways; to choose between synonyms, to use different styles, emphasis, different levels of abstractions, anaphoric and metaphoric expressions, etc. Then the idea of semantic web corresponds to the idea that there is no reliable way to fully process natural language semantics using a computer.

There are two necessary conditions to succeed in the next development of semantic web. However, acceptance of these conditions is very indeterminate, because they relate more to common human behavior than to technical solutions. The first condition is general agreement of people working in the elaborated domain because only widely accepted domain ontology can be respected and used. The second condition deals with the human ability and willingness to organize data respecting domain ontology.

1.2 Document Organization

The actual progress in the development of semantic web leads to the suggestion that a lot of people will prefer writing documents also in the future. Then there is a question if we can help people with document organization, eventually with parsing techniques, which extract relevant data from previously organized documents. We focus on the first step of this process: organization of a set of large documents.

We suppose a common scenario of searching for relevant documents. This scenario is based on asking a question (query including keywords from a domain area), and the following matching of keywords with a document content. One possibility to accelerate information retrieval in large document collections is a categorization of documents into classes with a similar content. Based on the keywords included in the query, we suggest that it is possible to estimate a document domain and to search only in documents from this domain. In this case, search time and the list of returned documents are strongly reduced.

In the past, some methods of document classification into domains were developed. These methods usually require a suitable representation of the stored documents. Documents are most often represented by the vector model [2]. The main problem of this representation is the large vocabulary of document collection and the high dimensionality of document vectors. Then methods for reducing this dimensionality have to be used.

Grouping similar items together is a technique used by methods based on word clustering. Documents are than represented as histograms of word clusters. One from various approaches to word clustering is the self-organizing map, which is based on distribution of words in their immediate context.

2 Self-organizing Feature Map and WEBSOM Method

Self-organizing feature map (SOFM) has been developed by T. Kohonen and it has been described in several research papers and books [3], [4]. The purpose of SOFM is to map a continuous high-dimensional space into a discrete space of lower dimension (usually 1 or 2). The map contains one layer of neurons, arranged to a two-dimensional grid, and two layers of connections. The aim of this network is to implement the winner-take-all strategy, i.e. only one neuron is selected and labeled as the best matching unit (BMU).

2.1 WEBSOM Architecture

WEBSOM method [5] is based on SOFM. This method was designed for automatic processing and categorization of arbitrary English and Finish written documents accessible on internet and the following information retrieval in these documents. Like WEBSOM, our classifier is based on two layer architecture.

The first layer processes the input feature vector representing the document words and creates the word category map (WCM). The second layer (document category map - DCM) processes the output from WCM and creates the clusters corresponding to document categories.

2.2 Document Preprocessing

Each document in a collection can be initially preprocessed using various techniques to reduce the computational load: lemmatization is done, non-textual information is removed, numerical expressions are replaced by textual forms, words occurring only a few times or common words not distinguishing document topics are removed.

2.3 Word Category Map

Word category map is supposed as "self-organizing semantic map" [6] because describes relation of words based on their averaged contexts. The word category map is trained by context vectors (input feature vector, which includes word context).

It is suggested that the words occurring in the similar context in the given document will have a similar representing vector v_i and they will also belong to the same word category.

In Fig. 1 we can see an example of the word category map trained by the words from the set of 100 documents. We can see that some map units respond to the words from certain syntactic categories (e.g. verbs, proper nouns etc.), whereas other units respond to the words from various syntactic categories.

2.4 Document Category Map

Document category map (DCM) classifies the input document to the given class. The size of input vector of DCM, i.e. the word category vector, is the same as the number of neurons in WCM. Each component of this vector represents a frequency of occurrence of the given word category in the input document. It is assumed that documents with the similar or the same content will have the similar word category vector. Based on this assumption, it is possible to use these vectors for training of DCM.

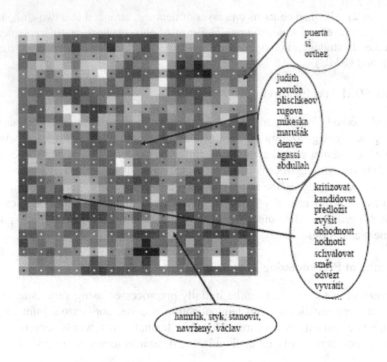

Fig. 1. Example of word category map

Since a Kohonen map is unsupervised learning paradigm, only the clusters of similar documents are created during the training. The given categories are assigned to these clusters afterwards.

3   Experiments and Results

3.1   Document Collection

The document classification system described in the previous sections was tested on the corpus of Czech News Agency documents. Generally, there were 7600 documents from 6 domains, containing 145766 words (stop words were removed). SOM-PAK [7], SOM toolbox [8] and own implementation of SOFM have been used for Kohonen map simulation. Both layers, WCM and DCM, were trained by the sequential training algorithm. The documents were classified by hand into 6 classes according to the document topics.

Our main task was to examine syntactic and semantic relationships within the word categories of WCM. Basic syntactic categories (nouns, adjectives, etc.) can be easily detected automatically, whereas semantic relationships have to be marked manually. Thus we worked only with a limited number of documents to manage this tiring and time consuming process of word categories evaluation. Finally we randomly selected

100 documents from the document collection. These documents contained 7421 different words after lemmatization and stop words removing.

3.2 Word Category Map

All words from the selected collection of 100 documents appeared in WCM; no threshold was applied to frequency of word occurrence, because a lot of words, which occurred only once, had impact on document semantics. The size of WCM was 437 neurons (19 x 23 grid), i.e. on average 17 words were placed into each category. The dimension of context vector was set to 60; ε was set to 0.2.

3.3 Syntactic Evaluation of Word Categories

The syntactic evaluation of word categories was done by the following process. Distribution of words into three basic word classes (nouns, adjectives, verbs) within word categories was observed. The fourth class named "others" was settled for all other word classes. Word categories contained in total 55.0% of nouns, 19.2% of adjectives and 13.5% of verbs (document collection contained a large number of geographical names and proper nouns). Fig. 2 represents distribution of word classes

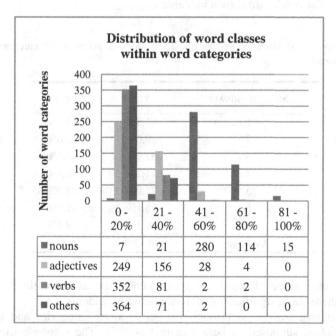

	0 - 20%	21 - 40%	41 - 60%	61 - 80%	81 - 100%
■ nouns	7	21	280	114	15
▦ adjectives	249	156	28	4	0
■ verbs	352	81	2	2	0
■ others	364	71	2	0	0

Fig. 2. Distribution of word classes (*nouns, adjectives, verbs*) within word categories, the percentage shares (*five groups*) of the word class within word category is presented on *X-axis*, each column indicates the number of word categories, in which the given word class appears with given percentage share, e.g. there are seven word categories in which the percentage share of nouns is up to 20%, otherwise there are 15 word categories, where the percentage share of nouns is greater than or equal to 81%.

within word categories. It is obvious that adjectives and verbs usually create up to 20% of words in the word category, while 280 word categories contain between 40% and 60% of nouns. Because the document collection contains a higher number of nouns, this word distribution corresponds to standard distribution of investigated word classes within text documents.

3.4 Semantic Evaluation of Word Categories

Semantic content of word categories can be hardly evaluated automatically. It is not possible to compare word categories e.g. to WordNet sets and expect some level of similarity. Thus semantic processing of word categories was done by hand. We used the following method: seven students were asked to go through 437 word categories three times in three weeks (the week break was necessary to ensure that students forgot the content of word categories from the previous task). Each round they got a different task concerning semantics of word categories. All tasks were time limited (one second of reading time for each five words in a word category). The response time was different according to task complexity.

The first task was to resolve if the given word category represents a semantic domain; the answer was simply yes or no. The response time was three seconds for each category. The results are shown in Table 1.

Table 1. Responses of students to the question: Does a given word category represent a semantic domain?

Student/ Answer	Yes	No
1	81,5%	18,5%
2	68,0%	32,0%
3	57,9%	42,1%
4	71,2%	28,8%
5	76,0%	24,0%
6	58,1%	41,9%
7	80,3%	19,7%
Average	**70,43%**	**29,57%**

The percentage share of word categories considered as semantic domains was 70.43%, but there was a significant difference between students.

The second task was to go through the set of word categories and name each category, which is supposed to be a semantic domain. The response time was six seconds for each category. The results are shown in Table 2.

There is obvious that students had problems to name a semantic category even in the case they marked it as a semantic domain in the previous task.

The third task was to classify a given word category to four predefined domains (sport, politics, legislation, and society). Students had a possibility to answer that

Table 2. Responses of students to the task: If a given word category represents a semantic domain, write its name. Answer 'Yes' means that word category was named, answer 'No' means none or senseless answer.

Student/Category name	Yes	No
1	55,1%	44,9%
2	35,2%	64,8%
3	29,3%	70,7%
4	25,9%	74,1%
5	28,1%	71,9%
6	36,2%	63,8%
7	40,0%	60,0%
Average	**35,69%**	**64,31%**

a given word category did not match any from the predefined set of domains. The response time was three seconds. The results are available in Table 3.

Students classified 63.16% of word categories into a domain selected from the predefined set of domains.

Table 3. Responses of students to the task: Classify a given word category to the predefined domains (*sport, politics, legislation, and society*). If a given word category does not match any from predefined domains, give no answer. Answer 'Yes' means classification in a domain from the predefined set of domains.

Student/ Predefined Category name	Yes	No
1	67,5%	32,5%
2	71,4%	28,6%
3	52,9%	47,1%
4	65,4%	34,6%
5	52,4%	47,6%
6	64,1%	35,9%
7	68,4%	31,6%
Average	**63,16%**	**36,84%**

4 Conclusion

The results obtained by application of WEBSOM method to the collection of Czech written documents confirmed a general problem connected with document semantics (i.e. with semantics of natural language) and document classification. Not only WEBSOM method but also humans had problems with classification of word categories into semantic domains. Moreover, there were significant differences between students undergoing the semantic experiments. However, an effort to interpret these differences would lead only to a speculative result. It is possible that the obtained results correspond to the idea that the semantics of natural language cannot be processed with computer in any reliable way.

Acknowledgment. This work was supported by Grant No. 2C06009 Cot-Sewing.

References

1. Semantic Web, http://www.w3.org/2001/sw
2. Manning, C.D., Raghavan, P., Schütze, H.: Introduction to Information Retrieval. Preliminary draft. Cambridge University Press, Cambridge (2007)
3. Kohonen, T.: Self-Organizing map. Springer, Heidelberg (2001)
4. Fausset, L.V.: Fundamentals of neural networks. Prentice Hall, Engelwood Cliffs (1994)
5. Kaski, S., Honkela, T., Lagus, K., Kohonen, T.: WEBSOM – Self-Organizing Maps of Document Collections. Neurocomputer, 101–117 (1998)
6. Ritter, H., Kohonen, T.: Self-organizing semantic maps. Biological Cybernetics 61, 241–254 (1989)
7. Kohonen, T., Hynninen, J., Kangas, J., Laaksonen, J.: SOM-PAK, The self-organizing map program package (1996)
8. Vesanto, J., Himberg, J., Alhoniemi, E., Parhankangas, J.: SOM Toolbox for Matlab (2000)

Opinion Target Network: A Two-Layer Directed Graph for Opinion Target Extraction

Yunqing Xia[1] and Boyi Hao[2]

[1] Tsinghua National Laboratory for Information Science and Technology,
Tsinghua University, Beijing 100084, China
yqxia@tsinghua.edu.cn
[2] Department of Computer Science and Technology, Tsinghua University
Beijing 100084, China
haoby@cslt.riit.tsinghua.edu.cn

Abstract. Unknown opinion targets lead to a low coverage in opinion mining. To deal with this, the previous opinion target extraction methods consider human-compiled opinion targets as seeds and adopt syntactic/statistic patterns to extract new opinion targets. Three problems are notable. 1) Manually compiled opinion targets are too large to be sound seeds. 2) Array that maintains seeds is less effective to represent relations between seeds. 3) Opinion target extraction can hardly achieve a satisfactory performance in merely one cycle. The opinion target network (OTN) is proposed in this paper to organize atom opinion targets of component and attribute in a two-layer directed graph. With multiple cycles of OTN construction, a higher coverage of opinion target extraction is achieved via generalization and propagation. Experiments on Chinese opinion target extraction show that the OTN is promising in handling the unknown opinion targets.

1 Introduction

With increasing number of online reviews, opinion mining nowadays attracts huge research passion within natural language processing community. In opinion mining systems, one of the trickiest issues is that the opinion targets unknown to the system lead to a low coverage. To resolve this problem, many research efforts have been made in opinion target extraction, in which the popular approaches are those combining lexicon and corpus statistics. The typical solution considers manually compiled opinion targets in opinion dictionary or review corpus as seeds and seeks for expansion algorithms to find the unknown opinion targets.

Hu and Liu (2004) proposed to find frequent opinion targets with association miner and infrequent opinion targets with syntactic patterns based on opinion words [1]. To improve coverage, Hu and Liu (2006) further adopted *WordNet* to locate synonyms for the known opinion targets [2]. Popescu and Etzioni (2005) considered opinion targets as concepts that form certain relationships with the product and identified the opinion targets through corresponding meronymy discriminators [3]. Ghani et al.

V. Matoušek and P. Mautner (Eds.): TSD 2009, LNAI 5729, pp. 93–100, 2009.

(2006) considered generic and domain-specific opinion targets as seeds and proposed to infer implicit and explicit opinion targets with attribute-value pairs using *co-EM* algorithm [4]. Kobayashi et al. (2007) adopted machine learning techniques to extract *aspect-of* relation from a blog corpus [5]. Xia et al. (2007) made use of collocations of opinion targets and sentiment keywords to find new opinion targets [6].

The previous works setup a fundamental framework for the opinion target extraction system. However, the following issues remain tricky. Firstly, the human-compiled opinion targets are too large. Using them as seeds to find unknown opinion targets will inevitably lead to low coverage. In addition, most large opinion targets cannot be found in any synset, rendering the synset-based expansion methods not applicable. Secondly, seeds and opinion targets are normally stored in independent arrays, which are unfortunately proved very difficult to model the relations. Thirdly, the seed-based opinion target expansion methods cannot achieve a satisfactory coverage in one cycle. It is deemed that the iterative mechanism should be helpful.

In this work, the opinion target network (OTN) is proposed to deal with the above issues. OTN is firstly a directed graph. In OTN, nodes represent synsets of atom opinion targets (AOT), edges disclose the inter-seed relations, and paths model compound opinion targets (COT), each of which is combination of AOT's with certain order. In this work, human-complied opinion targets are considered as COT's and AOT's are extracted from COT's automatically. As a smaller and more general seed, AOT works better with synset and pattern. OTN is secondly a two-layer directed graph. AOT's are classified into components and attributes from the perspective of ontology. So, two layers are created to maintain components and attributes, respectively. This paper presents a bootstrapping method to construct the OTN and to recognize the unknown opinion targets in multiple cycles. Experiments on Chinese opinion target extraction show that the method outperforms the baseline by 0.117 on f-1 score in the first cycle and by 0.239 in the 8th cycle.

The rest of this paper is organized as follows. In Section 2, principle of the opinion target network is presented. In Section 3 and Section 4, the generalization module and propagation module for construction of the opinion target network are described, respectively. We present preliminary evaluation in Section 5 and conclude the paper in Section 6.

2 Principle

2.1 Motivation

In this work, *atom opinion target* (AOT) is defined as the opinion target that is internally cohesive and externally flexible. That is, words within each AOT are statistically dependent to each other and it tends to combine with many other AOT's to form real opinion targets. *Compound opinion target* (COT) is defined as the opinion target that combines some AOT's with certain patterns in real reviews. For example, *brightness of image is a* COT while *brightness* and *image* are AOT's. In this work, the AOT's are further classified into components (COM) and attributes (ATT) based on the ontology theory.

The opinion target network (OTN) is designed to reflect the following intentions.

1) To model AOT, COT and their relations, graph is adopted.
2) To differentiate components and attributes, the graph is split into two layers.
3) To manage thousands of seeds, synsets are adopted.
4) To recognize unknown opinion targets with a satisfactory coverage, the generalization and propagation modules are developed.

2.2 Formalism

The opinion target network is a two-layer directed graph G^{OTN} defined as follows,

$$G^{OTN} = <V^{COM}, E^{COM}; V^{ATT}, E^{ATT}; E^{\Theta}>$$

in which V^{COM} and V^{ATT} represents component node set and attribute node set, respectively; E^{COM} and E^{ATT} denotes component edge set and attribute edge set, respectively; E^{Θ} denotes the set of edges that connect component nodes and attribute ones. Within OTN, paths normally contain a few edges, which actually represent the compound opinion targets. Note that nodes within OTN are synsets for AOT's and each of them represents a group of AOT's. An illustrative OTN given in Fig. 1 explains the formalism vividly. Note that the virtual edges on the attribute layer are used for categorization purpose. They are not used for COT constitution.

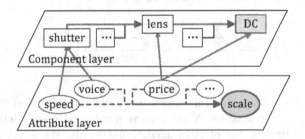

Fig. 1. An illustrative opinion target network for digital camera (DC) contains a component layer and an attribute layer

2.3 Construction

The opinion target network is constructed via generalization and propagation in a bootstrapping style. The workflow is given as in Fig. 2.

The annotation review corpus provides the initial COT set. The generalization module extracts AOT's from COT's, classifies the AOT's into components and attributes, assigns AOT's proper synset labels, and finally generates opinion target patterns (OTP). In the propagation module, unknown opinion targets are recognized with OTN and dependency relations. To achieve a wider coverage, the bootstrapping algorithm executes the generalization module and propagation module to upgrade the OTN in multiple cycles.

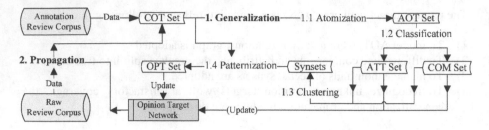

Fig. 2. Workflow for construction of the opinion target network

Note that in every cycle, language parsers for word segmentation and dependency parsing should be updated with newly obtained AOT's so that AOT's will be considered as normal words in lexical analysis and dependency parsing.

3 Generalization

3.1 Atomization

AOT's are extracted from COT's based on cohesion degree and flexibility degree. The cohesion degree is obtained by calculating the point-wise mutual information [3]. The flexibility degree of word W is calculated as follows,

$$F(W) = \frac{1}{2} \left(\frac{\sum_{W_i \in N^L} \frac{1}{N^R(W_i)}}{N^L(W)} + \frac{\sum_{W_i \in N^R} \frac{1}{N^L(W_i)}}{N^R(W)} \right), \tag{1}$$

in which N^L denotes the set of neighboring words to the left, N^R the set of neighboring words to the right. The function $N^L(x)$ returns number of unique left-neighboring words, and $N^R(x)$ returns that of right-neighboring words. We select the words as AOT's if cohesion degree and flexibility degree both satisfy empirical thresholds.

3.2 Classification

A probabilistic classifier is designed to recognize components (COM) and attributes (ATT) considering the following two features.

(1) Average Edit Distance (d^{AVG})

Average edit distance measures string similarity which is calculated as follows,

$$d^{AVG}(t \mid X) = \frac{1}{|X|} \sum_{x_i \in X} d(t, x_i), \tag{2}$$

in which t denotes the to-be-classified AOT, $X=\{x_i\}$ represents AOT set of component or attribute, $|X|$ represents size of set X and $d(t, x_i)$ is the function to measure edit distance between t and x_i. With Equation (2), we are able to calculate how likely an AOT belongs to COM set C or ATT set A.

(2) Overall Position Tendency (t^{OVA}).

Overall position tendency measures how likely an AOT is COM or ATT according to position heuristics. In certain language, the attributes tend to appear at the end of COT's. So, the overall position tendency is calculated as follows,

$$t^{OVA}(t) = \frac{count(t, A)}{count(C, t)},$$ (3)

in which $count(t, A)$ returns number of COT's in which t appears before an attribute, and $count(C,t)$ returns number of COT's in which t appears at the end of them.

Note that the initial component set and attribute set are extracted from the annotation review corpus. To improve coverage, we extract attribute words from *WordNet* and *HowNet*.

3.3 Clustering

To assign synset labels to every new AOT's, we first apply k-means clustering algorithm to group the AOT set into a few clusters. Then we adjust parameters to find a cluster that satisfies two conditions: (i) The cluster contains more than three AOT's carrying same synset label; (ii) The cluster contains at least one new AOT's.

Once such a cluster is found, the new AOT's in the cluster are assigned the synset label held by the known AOT's. We repeat the clustering process until no AOT's can be assigned any synset label. Two features are used in AOT clustering, i.e., (1) opinion words neighboring the AOT's in the raw reviews and (2) edit distance between new AOTs and known ones. Some new AOT's might remain unknown to any synset. We then run the clustering algorithm merely on the remaining new AOT's and attempt to find new synsets. A new synset is created if one cluster satisfies two conditions: (1) containing more than three AOT's and (2) AOT occurring for over three times in the review corpus. We choose the AOT with most occurrences as label for the new synset. There might be still some new AOT's without synset labels. We put them aside and allow the system to handle them in the next cycle.

3.4 Patternization

The opinion target patterns (OTP) comply with regular expressions like

$$\{A_c\}^*\{string\{B_c\}^*\}^*,$$

in which A_c and B_c represent synset labels for the AOT's; and *string* is a constant in the pattern. As an example, the pattern *<color> of <image>* combines the synset labeled by *image* and the synset labeled by *color*, in which *of* is the pattern *string*.

The patterns are learned from the COT set. COT may contain many words while some of them are not AOT. We consider the non-AOT words as *string* candidates. With thousands of COT's, we extract the pattern *string*s such as *of* based on statistics.

When new AOT's are extracted, we put them in OTN. New edges should also be drawn if new OTP's are generated.

4 Propagation

The propagation module aims to extract new opinion targets from the raw reviews with the opinion target network (OTN). The OTN may naturally handle this job. In addition, the dependency relations are adopted to locate opinion targets with new synsets. In this process, the raw review corpus is used to filter the false candidates.

4.1 OTN-Based Propagation

OTN is capable of inferring new opinion targets with AOT's and OTP's. In other words, if an edge exists between AOT synset A and B, combinations between AOT's in synset A and that in synset B probably exist. Based on this assumption, a large number of opinion target candidates can be inferred. The assumption might lead to false candidates. To filter them, we apply the sequence confidence measure to estimate how likely AOT A appears before AOT B. Given that a candidate X contains N AOT's, i.e. $\{A_i\}_{i=1,\ldots,N}$, the sequence confidence (SC) is calculated as follows,

$$SC(X) = \frac{1}{C_N^2} \sum_{i<j} count(A_i, A_j) \; , \qquad (4)$$

in which $count(A_i, A_j)$ denotes number of occurrences that A_i, appears before A_j. An empirical threshold for sequence confidence is set in our experiments.

4.2 Dependency-Based Propagation

To locate opinion targets with new synsets, we study the dependency relations to find new AOT's that the patterns fail to cover. To make the dependency-based propagation reliable, we setup three constraints: (1) Only four dependency relations, i.e. ATT (modifying), COO (coordinating), QUN (numbering) and DE (DE structure[1]), are condisered. (2) AOT candidate should adjoin the known AOT accept that one conjunction or 的(de0, of) appears in the middle. (3) AOT candidate is not adjective or pronoun.

5 Evaluation

5.1 Experiment Setup

Two corpora are used in this work. Opinmine corpus [7] contains 8,990 human-judged opinions on digital camera. The raw review corpus contains 6,000 reviews in the same domain. We divide Opinmine corpus into training set and test set evenly and adopt precision (p), recall (r) and f-1 score (f) in this evaluation.

We apply HIT Language Technology Platform [8] for Chinese word segmentation and dependency paring. We also extract attribute synsets from *HowNet* [9] manually to expand the initial set of attribute AOT's.

---

[1] In Chinese DE structure refers to the clause that contains 的(de0, of).

5.2 Results

The baseline method is defined as the one that uses human-judged opinion targets as seeds. To find more opinion targets, dependency relations and patterns are also used. The intention is to prove necessity and effectiveness of the opinion target network (OTN) for the opinion target extraction. Our method is the one that uses AOT's as seeds and executes the generalization module and propagation module in multiple cycles. The thresholds in our method are empirically configured, i.e., threshold for cohesion degree is set as 0.001, that for flexibility degree is 0.333, and that for sequence confidence is 0.8. Experimental results are shown in Figure 3.

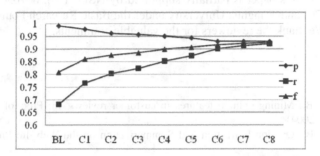

Fig. 3. Experimental results on precision (p), recall (r) and f-1 measure (f) of the baseline method (BL) and our method in eight cycles

5.3 Discussions

In the 1st cycle our method outperforms the baseline by 0.051 on f-1 score, where recall is improved by 0.085 with precision loss of 0.014. In the 8th cycle, our method outperforms the baseline by 0.117 on f-1 score, where recall is improved by 0.239 with precision loss of 0.063. After the cycle C6, f-1 score starts to converge. Two conclusions can thus be drawn: (1) OTN improves coverage significantly without significant loss of precision; (2) bootstrapping process makes significant contribution.

To understand significance of the bootstrapping process, we look into COM set, ATT set, COT set and OTP set that are produced in every cycle. Statistics show that in eight cycles, size of COM set climbs from 177 to 1,291, from 67 to 254 for ATT set, from 978 to 51,724 for COT set, and from 294 to 9,077 for OTP set. Interestingly, size of OTP set starts to converge after the cycle C6. This proves our claim on the opinion target network.

6 Conclusions and Future Works

To improve coverage in opinion target extraction, the opinion target network is proposed to manage atom opinion targets, compound opinion targets and their relations in a two-layer directed graph. The network is constructed by generalizing compound opinion targets into atom opinion targets and patterns. The network is in turn used in the propagation algorithm to find unknown opinion targets. In a bootstrapping style, the opinion target network helps to recognize unknown opinion targets in multiple cycles. Experiments on Chinese opinion target extraction show the OTN is promising in handling the unknown opinion targets.

This work is still preliminary and some future works are planned. For example, we will conduct more experiments to evaluate how our method fits into reviews for other domains such as mobile phone, movie and hotel.

Acknowledgements

Research work in this paper is partially supported by NSFC (No. 60703051), MOST (2009DFA12970) and Tsinghua University under the Basic Research Foundation (No. JC2007049). We thank the reviewers for the valuable comments.

References

1. Hu, M., Liu, B.: Mining opinion features in customer reviews. In: Proc. of AAAI 2004, pp. 755–760 (2004)
2. Hu, M., Liu, B.: Opinion Extraction and Summarization on the Web. In: Proc. of AAAI 2006 (2006)
3. Popescu, A., Etzioni, O.: Extracting product features and opinions from reviews. In: Proc. of HLT-EMNLP 2005, pp. 339–346 (2005)
4. Ghani, R., Probst, K., Liu, Y., Krema, M., Fano, A.: Text mining for product attribute extraction. SIGKDD Explorations Newsletter 8(1), 41–48 (2006)
5. Kobayashi, N., Inui, K., Matsumoto, Y.: Extracting Aspect-Evaluation and Aspect-of Relations in Opinion Mining. In: Proc. of EMNLP-CoNLL 2007, pp. 1065–1074 (2007)
6. Xia, Y., Xu, R., Wong, K.F., Zheng, F.: The Unified Collocation Framework for Opinion Mining. In: Proc. of ICMLC 2007, vol. 2, pp. 844–850 (2007)
7. Xu, R., Xia, Y., Wong, K.F.: Opinion Annotation in On-line Chinese Product Reviews. In: Proc. of LREC 2008 (2008)
8. Ma, J., Zhang, Y., Liu, T., Li, S.: A statistical dependency parser of Chinese under small training data. In: Proc. of Workshop: Beyond shallow analyses-formalisms and statistical modeling for deep analyses, IJCNLP 2004, pp. 1–5 (2004)
9. Dong, Z., Dong, Q.: HowNet and the Computation of Meaning. World Scientific Publishing, Singapore (2006)

The Czech Broadcast Conversation Corpus

Jáchym Kolář and Jan Švec

Department of Cybernetics, Faculty of Applied Sciences,
University of West Bohemia, Univerzitní 8, CZ-306 14 Plzeň, Czech Republic
{jachym,honzas}@kky.zcu.cz

Abstract. This paper presents the final version of the Czech Broadcast Conversation Corpus released at the Linguistic Data Consortium (LDC). The corpus contains 72 recordings of a radio discussion program, which yield about 33 hours of transcribed conversational speech from 128 speakers. The release not only includes verbatim transcripts and speaker information, but also structural metadata (MDE) annotation that involves labeling of sentence-like unit boundaries, marking of non-content words like filled pauses and discourse markers, and annotation of speech disfluencies. The annotation is based on the LDC's MDE annotation standard for English, with changes applied to accommodate phenomena that are specific for Czech. In addition to its importance to speech recognition, speaker diarization, and structural metadata extraction research, the corpus is also useful for linguistic analysis of conversational Czech.

1 Introduction

Spoken language corpora are important for training and testing automatic speech recognition and understanding systems. For widespread languages such as English and Mandarin, many spoken language resources from various domains are publicly available, but for smaller languages, such as Czech, the speech resource availability to researchers is limited. Although the two major language resource publishers, the Linguistic Data Consortium (LDC) at the University of Pennsylvania and the European Language Resources Association (ELRA) offer Czech broadcast news [1,2] and prompted speech corpora [3,4] in their catalogs, no Czech conversational speech resources have been publicly available.

This has been a significant handicap for Czech researchers since the problem of automatically processing conversational speech is without a doubt one of the most important tasks in the field. Hence, in order to support broader research on the problem of conversational Czech, we have decided to create and publish a new speech corpus of broadcast conversations. The broadcast conversation genre was selected because of the easy data acquisition as well as for its increasing popularity in the speech processing community – for example, some current large research projects (such as GALE) include automatic translation of broadcast conversations [5].

The Czech Broadcast Conversation Corpus not only contains audio recordings and standard transcripts; the important additional value of this corpus lies in its "structural metadata" annotation. This annotation involves partitioning verbatim transcripts into sentence-like units (SUs) that function to express a complete idea; and identifying fillers and edit disfluencies. The structural information is critical to both increasing

V. Matoušek and P. Mautner (Eds.): TSD 2009, LNAI 5729, pp. 101–108, 2009.
© Springer-Verlag Berlin Heidelberg 2009

human readability of the transcripts and allowing application of downstream NLP methods (e.g., machine translation, summarization, parsing), which are typically trained on fluent and formatted text.

The corpus will be released by the LDC as catalog numbers LDC2009S02 (audio) and LDC2009T20 (transcripts and structural metadata annotations) in summer 2009. The remainder of this paper describing the final version of the corpus is organized as follows. Section 2 describes the audio data, Section 3 presents details about speech transcription, Section 4 is devoted to structural metadata annotation, and Section 5 provides a summary and conclusions.

2   Audio Data

The broadcast conversation speech database contains recordings of a radio discussion program called *Radioforum*, which is broadcast by Czech Radio 1 (CRo1) every weekday evening. *Radioforum* is a live talk show where invited guests (most often politicians but also journalists, economists, teachers, soldiers, crime victims, and so on) spontaneously answer topical questions asked by one or two interviewers. The number of interviewees in a single program ranges from one to three. Most frequently, one interviewer and two interviewees appear in a single show. The material includes passages of interactive dialog, but longer stretches of monolog-like speech slightly prevail. Because of the scope of the talk show, all speakers are adults. Although the corpus was recorded from a public radio where standard (literary) Czech would be expected, many speakers, especially those not used to talking on the radio, use colloquial language as well. Literary and colloquial word forms are often mixed in a single sentence. The usage of colloquial language, however, is not as frequent as in unconstrained informal conversations.

The number of transcribed shows has been increased over the last year. The final release of the corpus contains 72 recordings acquired over the air during the period from February 12, 2003 through June 26, 2003. The signal is single channel, sampled at 22.05 kHz with 16-bit resolution. Typical duration of a single discussion is 33–35 minutes (shortened to 26–29 minutes after removing compact segments of telephonic questions asked by radio listeners, which were not transcribed). In total, the duration of the audio data is 40 hours, which yield approximately 33 hours of pure transcribed speech. The total number of speakers in the whole corpus is 128; male speakers are more frequent than females (108 males, 20 females).

3   Transcripts

The goal of the transcription phase was to produce precise time-aligned verbatim transcripts of the broadcast recordings. The data were manually transcribed in the Transcriber tool [6] using the careful transcription approach. The transcripts were created by a large number of annotators. To keep them maximally correct and consistent, all submitted transcripts were manually checked by a senior annotator.

The transcripts contain speaker turn labels – time stamps and speaker IDs were recorded at each speaker change. Overlapping speech regions were also labeled; within

Table 1. Corpus size (* – 'lexemes' include words, special interjections, and filled pauses; 'tokens' include lexemes plus speaker and background noises)

Number of shows	72	Number of unique words	30.5k
Total duration	40.0h	Total number of speakers	128
Duration of transcribed speech	33.0h	— males	108
Total number of tokens	306.6k	— females	20
Total number of lexemes*	292.6k	Number of speaker turns	8.0k

these regions, each speaker's speech was transcribed separately (if intelligible). Regions of unintelligible speech were marked with a special symbol. To break up long turns, breakpoints roughly corresponding to "sentence" boundaries within a speaker's turn were added. The transcripts contain standard punctuation, but acceptable marks were limited to periods and question marks at the end of a sentence, and commas within a sentence. Capitalization was used for proper names, but not for the beginnings of sentences (unless they start with a proper name). Word fragments and mispronounced words were also tagged. In addition to words and punctuation, the transcripts contain special tags marking speaker noises (BREATH, COUGH, LAUGH, and LIP-SMACK), other noises (MUSIC, BACKGROUND-SPEECH, and unspecified NOISE), and "inarticulate" interjections expressing agreement (HM) and disagreement (MH).

Special attention was paid to transcription of filled pauses (FPs). FPs are hesitation sounds used by speakers to indicate uncertainty or to keep control of a conversation while thinking what to say next. In order to support maximal annotation consistency, we distinguished only two types of Czech FPs: *EE* (most typical example of EE is an FP similar to long Czech vowel *é*, but this group also includes all hesitation sounds that are phonetically closer to vowels), and *MM* (all FPs that are phonetically more similar to consonants or mumble-like sounds, typically pronounced with a closed mouth).

The overall size of the corpus in terms of a number of different measures is presented in Table 1. Among others, note that the number of distinct words in the vocabulary created from the corpus transcripts is quite large given the size of the corpus. Czech, same as other Slavic languages, is highly inflectional, and thus uses an extremely large number of distinct word forms.

4   Structural Metadata (MDE) Annotation

4.1   Annotation Approach

The structural metadata (MDE) annotation can be viewed as a post-processing step applied to the standard transcription. Structural information is critical to both increasing human readability of the transcripts and allowing application of downstream NLP methods, which typically require a fluent and formatted input. Because spontaneous utterances are not as well-structured as read speech and written text, annotating structure by simply making reference to standard punctuation is inadequate. Hence, several different schemes have been proposed for annotation of typical spontaneous speech phenomena. Earliest efforts include the manual for disfluency tagging of the Switchboard corpus [7],

Heeman's annotation scheme for the Trains dialog corpus [8], and a syntactic-prosodic labeling system for spontaneous speech called "M" presented in [9]. Recently, Fitzgerald and Jelinek [10] presented a new annotation scheme for spontaneous speech reconstruction.

For our work on conversational Czech, we have decided to adopt the "Simple Metadata Annotation" approach introduced by the LDC as part of the DARPA EARS program [11]. Originally, this standard was defined only for English. Later, the authors proposed to extend the guidelines for use with Mandarin and Arabic [12], but because of the premature termination of the EARS project, these efforts ended up as early as during the pilot annotation tests and no reasonably-sized corpora have been created to this day. When developing particular annotation rules for Czech structural metadata annotation, the LDC's annotation guidelines for English were taken as the starting point, with changes applied to accommodate specific phenomena of Czech. In addition to the necessary language-dependent modifications, we further proposed and applied some language-independent modifications. The annotation involves insertion of sentence-like unit breakpoints (SUs) to the flow of speech and identification of a range of spontaneous speech phenomena (fillers and disfluencies). The following subsections briefly describe individual annotation subtasks. A more detailed description is given in [13].

SUs: Dividing the stream of words into sentence-like units is a crucial component of the MDE annotation. The goal of this part of annotation is to improve transcript readability and processability by presenting it in small coherent chunks rather than long unstructured turns. Because speakers often tend to use long continuous compound sentences in spontaneous speech, it is nearly impossible to identify the end-of-sentence boundaries with consistency using only a vague notion of a "conversational equivalent" of the "written sentence" definition; strict segmentation rules are necessary. One possible solution is to divide the flow of speech into some "minimal meaningful units" functioning to express one complete idea on the speaker's part. These utterance units are called SUs (Sentence-like/Syntactic/Semantic Units) within the MDE task. The SU symbols are the following:

/. – Statement break – end of a complete SU functioning as a declarative statement
(*Theresa loves irises /.*)
/? – Question break – end of an interrogative
(*Do you like irises /?*)
/, – Clausal break – identifies non-sentence clauses joined by subordination
(*If it happens again /, I'll go home /.*)
/& – Coordination break – identifies coordination of either two dependent clauses or two main clauses that cannot stand alone
(*Not only she is beautiful /& but also she is kind /.*)
/- – Incomplete (arbitrarily abandoned) SU
(*Because my father was born there /, I know a lot about the /- They must try it /.*)
/~ – Incomplete SU interrupted by another speaker
(*A: Tell me about /~   B: Just a moment /.*)

The SU symbols may be divided into two categories: sentence-internal (/& and /,) and sentence-external (others). Sentence-external breaks are used to indicate the presence of a main (independent) clause. These independent main clauses can stand alone as a sentence and do not depend directly on the surrounding clauses for their meaning. Sentence-internal breaks are secondary and have mainly been introduced to support inter-annotator agreement. They delimit units (clauses) that cannot stand alone as a complete sentence.

We did not use the identical set of SU symbols as originally defined in [11] but introduced two significant modifications. First, the original set contains only one symbol for incomplete SUs, but we decided to distinguish two types of incomplete SUs: /– indicating that the speaker abandoned the SU arbitrarily; and /~ indicating that the speaker was interrupted by another speaker. This distinction of incompletes is useful since their patterns differ significantly in prosody, semantics, and syntax. Second, in order to identify some "core boundaries", we introduced two new symbols: //. and //? — the double slashes indicate a strong prosodic marking on the SU boundary, i.e. pause, final lengthening, and/or strong pitch fall/rise.

Other modifications in the SU annotation pertain to differences between Czech and English. One example of a difference affecting the SU annotation is the possibility of subject dropping in Czech. In English, subject dropping is only allowed in the second clause of a compound sentence when both clauses share the same subject. In Czech, the subject (pronoun) can be dropped every time it is "understood" from the context and/or from the form of a conjugated predicate (verb). Since the conjugation of the verb includes both person and number of the subject, it is possible to say for instance *Běžím /.*, lit. *(I am) running /.* As a result, subject dropping in the coordinated clause does not imply the use of the coordinating break (/&), as is the case for English. Instead, we separate the coordinated clauses with an SU-external break, even if the subject is present in the first clause and dropped in the second (*Robert do práce šel pěšky /. ale domů jel vlakem /.*, lit. *Robert walked to work /. but (he) took the train home /.*).

Fillers: Four types of fillers are labeled: filled pauses (FPs, already described in Section 3), discourse markers (DMs), explicit editing terms (EETs) and asides/parentheticals (A/Ps). Annotating fillers consists of identifying the filler word(s) and assigning them an appropriate label.

DMs are words or phrases that function primarily as structuring units of spoken language. They do not carry separate meaning, but signal such activities as a change of speaker, taking or holding control of the floor, giving up the floor or the beginning of a new topic. Frequent examples of Czech DMs are *tak* (lit. *so*) or *no* (*well*). Unlike English, DMs containing a verb (such as *you know*) are less frequent. We also labeled a DM subtype – Discourse Response (DR). DRs are DMs that are employed to express an active response to what another speaker said, in addition to mark the discourse structure. For instance, the speaker may also initiate his/her attempt to take the floor. DRs typically occur turn-initially.

EETs are fillers only occurring within the context of an edit disfluency. These are explicit expressions by which speakers signal that they are aware of the existence of a disfluency on their part. EETs are quite rare. In our corpus, by far the most frequent one is *nebo* (lit. *or*). The further filler type, A/P, occurs when a speaker utters a short side comment and then returns to the original sentence pattern (e.g., *And then that last question {it was a funny question} came up /.*). Strictly speaking, A/Ps are not fillers, but because as with other filler types, annotators must identify the full span of text

Table 2. Structural metadata statistics

Total number of SUs	21.7k	Tokens in DelRegs	2.9%
Average length of complete SUs	13.7	DelRegs having correction	88.2%
- statements	13.9	Tokens in A/Ps	1.5%
- questions	11.6	Tokens annotated as DMs	1.5%
Average length of incomplete SUs	10.2	Tokens annotated as EETs	0.1%

functioning as an A/P, they are included with fillers in the MDE definition. Some very common words or short phrases, that can be denoted as "lexicalized parentheticals" (e.g. *řekněme*, lit. *say*) are not annotated as A/Ps.

Edit disfluencies: Edit disfluencies are portions of speech in which a speaker's utterance is not complete and fluent. Instead, the speaker corrects or alters the utterance, or abandons it entirely and starts over. In MDE, edit disfluencies consist of the deletable region (DelReg, speaker's initial attempt to formulate an utterance that later gets corrected), interruption point (IP, the point at which the speaker breaks off the DelReg with an EET, repetition, revision or restart), optional EET (an overt statement from the speaker recognizing the existence of a disfluency), and correction (portion of speech in which the speaker corrects or alters the DelReg). Whereas corrections had not been explicitly tagged within the MDE project for English, we decided to label them in order to obtain relevant data for further research of disfluencies. An example of a disfluency follows (* denotes IP, DelReg is displayed within square brackets, EET is typed in boldface, and correction is underlined):

Naše děti milují [kočku] EE **nebo** psa pana Bergera /.*
lit. *Our children love [the cat]* uh **or** the dog of Mr. Berger /.*

Table 2 displays some interesting numbers relating to the metadata annotation of the corpus. Among others, the numbers indicate that statement SUs are on average longer than interrogative SUs, and that, as expected, complete SUs are slightly longer than incomplete SUs. Furthermore, note that the number of DelRegs that were corrected is quite high, indicating that false starts (the disfluencies that do not have a correction) are not that frequent. The numbers also show that EETs are very rare, while DMs and A/Ps are more frequent.

4.2 Annotation Tool and Formats

When creating MDE annotations, annotators not only work with the verbatim transcripts, but also listen to audio and use prosody to resolve potential syntactic ambiguities. To ease the annotation process, a special software tool called QAn (Quick Annotator) was developed. It allows annotators to highlight relevant spans of text, play corresponding audio segments, and then record annotation decisions. A screenshot of the tool is displayed in Figure 1. The tool can be freely downloaded from the project website http://www.mde.zcu.cz/.

In the corpus release, the structural annotations are provided in two different formats. The first is the format used by QAn. It is based on the XML-based Transcriber format,

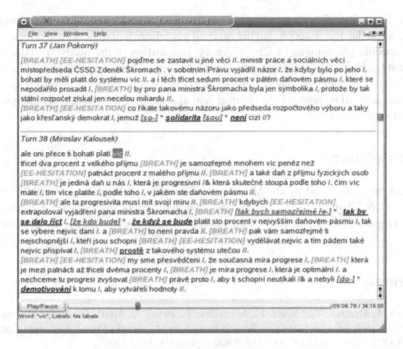

Fig. 1. QAn – the MDE annotation tool

which is extended by special tags representing structural metadata information. The format uses two types of metadata tags: SUs associated with interword boundaries, and Labels spanning over one or more words (i.e., these are begin/end pairs).

The second format is RTTM. It is based on the RTTM-format-v13 specification that was used for storing MDE annotations in the EARS project. The format uses object-oriented representation of the rich text data. There are four general object categories to be represented: speech-to-text objects, MDE objects, source (speaker) objects, and structural objects. Except for the speaker information object, each object exhibits a temporal extent with a beginning time and duration. Note that the duration of interruption points and clausal boundaries is zero by definition. The objects are represented individually, one object per record, using a flat record format with object attributes stored in white-space separated fields.

5   Conclusion

In this paper, we have presented the final version of the Czech Broadcast Conversation Corpus that will be released by the LDC in summer 2009. The corpus was created in order to support broader research on the problem of conversational Czech. It contains 72 recordings of a broadcast discussion program, which yields about 33 hours of pure transcribed speech from 128 adult speakers. The annotations not only include verbatim transcripts and speaker information, but the additional value of this corpus is that it also contains structural metadata annotations capturing important information about sentence-like units, fillers, and edit disfluencies. The metadata annotation is based on the LDC's standard for English, but the original guidelines had to be adjusted to accommodate specific phenomena of Czech syntax. In addition to the necessary language-dependent modifications, we further proposed and applied some language-independent modifications. We believe that besides its importance to speech recognition, speaker diarization, and MDE research, the corpus will also be useful for linguistic analysis of conversational Czech.

Acknowledgments

This work was supported by the Ministry of Education of the Czech Republic under projects 2C06020 and ME909. The authors also thank Stephanie Strassel, Christopher Walker, Dagmar Kozlíková, and Vlasta Radová for their help with the creation of the corpus.

References

1. Psutka, J., Radová, V., Müller, L., Matoušek, J., Ircing, P., Graff, D.: Voice of America (VOA) Czech broadcast news audio and transcripts. Linguistic Data Consortium Catalog No. LDC2000S89 and LDC2000T53, Philadelphia, PA, USA (2000)
2. Radová, V., Psutka, J., Müller, L., Byrne, W., Psutka, J.V., Ircing, P., Matoušek, J.: Czech Broadcast News Speech and Transcripts. Linguistic Data Consortium Catalog No. LDC2004S01 and LDC2004T01, Philadelphia, PA, USA (2004)
3. ELRA: Czech SpeechDat(E) database. Catalog Reference ELRA-S0094 (2001)
4. ELRA: GlobalPhone Czech. Catalog Reference ELRA-S0196 (2006)
5. Zheng, J., Wang, W., Ayan, N.F.: Development of SRI's translation systems for broadcast news and broadcast conversations. In: Proc. Interspeech 2008, Brisbane, Australia (2008)
6. Boudahmane, K., Manta, M., Antoine, F., Galliano, S., Barras, C.: Transcriber: A tool for segmenting, labeling and transcribing speech, http://trans.sourceforge.net
7. Meeter, M.: Dysfluency annotation stylebook for the Switchboard corpus (1995), ftp://ftp.cis.upenn.edu/pub/treebank/swbd/doc/DFL-book.ps
8. Heeman, P.: Speech Repairs, Intonational Boundaries and Discourse Markers: Modeling Speakers' Utterances in Spoken Dialogs. PhD thesis, University of Rochester, NY, USA (1997)
9. Batliner, A., Kompe, R., Kiessling, A., Mast, M., Niemann, H., Nöth, E.: M = Syntax + Prosody: A syntactic–prosodic labelling scheme for large spontaneous speech databases. Speech Communication 25, 193–222 (1998)
10. Fitzgerald, E., Jelinek, F.: Linguistic resources for reconstructing spontaneous speech text. In: Proc. LREC 2008, Marrakech, Morocco (2008)
11. Strassel, S.: Simple metadata annotation specification V6.2 (2004), http://www.ldc.upenn.edu/Projects/MDE/Guidelines/SimpleMDE_V6.2.pdf
12. Strassel, S., Kolář, J., Song, Z., Barclay, L., Glenn, M.: Structural metadata annotation: Moving beyond English. In: Proc. Interspeech 2005, Lisbon, Portugal (2005)
13. Kolář, J.: Automatic Segmentation of Speech into Sentence-like Units. PhD thesis, University of West Bohemia, Pilsen, Czech Republic (2008)

Vector-Based Unsupervised Word Sense Disambiguation for Large Number of Contexts

Gyula Papp

Pázmány Péter Catholic University, Faculty of Information Technology,
Interdisciplinary Technical Sciences Doctoral School
Práter utca 50/a, 1083 Budapest, Hungary
gyupa@digitus.itk.ppke.hu

Abstract. This paper presents a possible improvement of unsupervised word sense disambiguation (WSD) systems by extending the number of contexts applied by the discrimination algorithms. We carried out an experiment for several WSD algorithms based on the vector space model with the help of the SenseClusters ([1]) toolkit. Performances of algorithms were evaluated on a standard benchmark, on the nouns of the Senseval-3 English lexical-sample task ([2]). Paragraphs from the British National Corpus were added to the contexts of Senseval-3 data in order to increase the number of contexts used by the discrimination algorithms. After parameter optimization on Senseval-2 English lexical sample data performance measures show slight improvement, and the optimized algorithm is competitive with the best unsupervised WSD systems evaluated on the same data, such as [3].

1 Introduction

Word sense disambiguation (WSD) is a widely researched topic in natural language processing. There are two main approaches applied by WSD researchers: supervised or unsupervised machine learning.

Supervised WSD methods need hand-tagged data to learn how the actual word sense of an ambiguous word (usually called *target word*) in a specific context can be decided. Unsupervised WSD algorithms don't need any tagged data. They don't even try to decide from a previously given list of senses the actual sense of the target word. Their goal is to separate the different uses of the ambiguous word. (The term *word use* is preferred to *word sense* in cases of unsupervised WSD approaches.) They use clustering techniques to group similar contexts together. Hopefully the induced clusters represent the different uses of the ambiguous word. Each new context of the target word will be compared to the clusters and the most similar cluster will be the selected word use.

One of the advantages of unsupervised WSD over supervised WSD is that it doesn't require any sense-tagged corpora. We tried to exploit this advantage by using a lot of contexts to discriminate the different word uses of a target word. Our assumption was that the more contexts the learning algorithm sees, the better its sense discrimination ability becomes. It is easy to find untagged contexts that contain the target word, for example collecting paragraphs containing the target word from a large text corpus is a common way.

V. Matoušek and P. Mautner (Eds.): TSD 2009, LNAI 5729, pp. 109–115, 2009.

A large number of unsupervised WSD systems are based on the vector space model which was proposed by [4]. However, very good results have been achieved by graph-based systems as well, e.g. by the HyperLex algorithm [5] and its optimized version [3].

Our goal was to test how the number of contexts affects the performance of unsupervised WSD algorithms. We carried out an experiment with vector-based algorithms to decide if seeing more contexts causes better discrimination ability. In the future some graph-based methods (e.g. HyperLex) could be tested as well.

The following section shortly describes the basics of vector-based unsupervised WSD methods. Section 3 deals with the analyzed WSD system. Section 4 presents the experiment. Performance results of the experiment are described in section 5. Finally, conclusions of the experiment are summarized in section 6.

2   Vector-Based Unsupervised Word Sense Disambiguation

Vector-based unsupervised WSD systems usually need a text corpus for each target word. Each corpus consists of contexts in which its target word occurs. A context is a short unit of text, usually a sentence, a paragraph or a window of k words centered on the ambiguous word. These corpora don't have to be tagged with the senses of the target word.

In this section the usual process of unsupervised WSD algorithms for each corpus is presented.

2.1   Feature Selection

As the first step, vector-based systems extract features from the corpus. (Features can be collected from separate held out data as well, although this isn't the common way of feature selection.) Features can be unigrams, bigrams or co-occurrences. Unigram features are words that occur more than a given number of times in the corpus. Bigrams are frequent ordered word pairs inside a small word window (usually between 2-4 words). Co-occurrences are unordered word pairs mostly inside a larger window, e.g. a sentence or a paragraph. Beside minimal frequency numbers statistical tests can be used to select features as well. Feature extraction is a critical step of algorithms because the collected features serve as the dimensions of the context representation vectors.

2.2   Context Representation

Feature selection extracts a list of features from the corpus belonging to a target word. The context representation step computes a vector for each context of the corpus. These vectors can be first order or second order context representations. The ith element of a first order vector is the number of occurrences of the ith feature in the context.

Second order context vectors (proposed by [4]) are computed in a more complex way. They can be used in cases of bigram or co-occurrence features. At first, a word by word matrix is created for the features. The first words of the features are assigned to the row labels, while the second words become column labels in the matrix. Then, each word in a context that is a row label in the matrix is replaced with its row in the matrix. The mean of these replacement vectors is the computed context representation.

[6] compared first and second order context representations. He showed that first order representation performs better if lots of data are available but he proposed second order representation in cases when less data can be used for learning.

2.3 Dimensionality Reduction

Usually the context representation vectors are very sparse. In some cases the number of dimensions is too high for the clustering algorithms. This is why [4] used the method singular value decomposition (SVD) to smooth the word by word second order matrix and reduce its size. When using first order representations the context by feature matrix can be transformed with SVD as well. SVD can improve the performance of WSD systems. (Other methods can also be used for dimensionality reduction.)

2.4 Clustering

When the contexts vectors have been created any clustering algorithms can be used to discriminate the senses of the target word. Usually, clustering algorithms require the number of clusters as an input parameter. [7] proposed several criterion functions to determine the "right" number of senses before clustering.

2.5 Evaluation

There are several methods to evaluate unsupervised WSD systems. These are summarized in [3]. One alternative is to decide the correctness of an algorithm manually. Another possibility is to measure the performance of a WSD system in an application. Evaluation on publicly available hand-tagged corpora can be performed as well. Another possibility is to compare the generated clusters with gold standard clusters.

3   The Examined WSD System

The analyzed unsupervised WSD system consists of modules of the publicly available SenseClusters ([1]) toolkit and a separate evaluation module. This section shortly describes the above mentioned parts of the system and briefly presents its parameters.

3.1 SenseClusters Modules

As an input SenseClusters requires a set of contexts for each word to be disambiguated. Its feature selection module is able to extract unigram, bigram and co-occurrence features from the input corpus. The set of collected features depends on either a minimal occurrence number parameter or on a statistical measure.

After feature selection the context representation module computes context vectors for the contexts. This module can produce first order context representation for unigram features and both first and second order representation in cases of bigram and co-occurrence features.

Dimensionality reduction can be performed by SenseClusters with SVD transformation. (This module doesn't have to be used.)

The SenseClusters toolkit uses the CLUTO ([8]) program for clustering. It supports many clustering techniques, such as agglomerative, partitional and hybrid methods. These clustering algorithms require the number of clusters to create as an input parameter. The clustering module of SenseClusters is able to predict this number with the help of the PK1, PK2, PK3 measures ([7]) and the Gap Statistic ([9]). These measures are called cluster stopping criteria.

3.2   Evaluation Module

In order to be able to evaluate an unsupervised system as [3] did, sense-tagged contexts are needed. These contexts are split into a training and a test set. The evaluation process is a supervised learning task: the cluster-sense mapping has to be learnt on the training set and its performance is measured on the test set.

Although SenseClusters supports some evaluation methods, these couldn't be used for the earlier described way of evaluation. This is the reason why we had to write an own evaluation module.

Although this module makes the system a mixture of unsupervised and supervised, it is important to notice that the way of creating clusters is completely unsupervised, only the mapping is supervised.

3.3   Parameters

The SenseClusters toolkit was examined as an unsupervised WSD system. Its different configuration settings described in subsection 3.1 (e.g. the order of context representation or the applied clustering algorithm), were regarded as free parameters to be tuned.

Certainly, parameter optimization has to be performed on different data from the data which performance is measured on. Then the final performance results can be obtained by running the system with the best set of parameters on the test data.

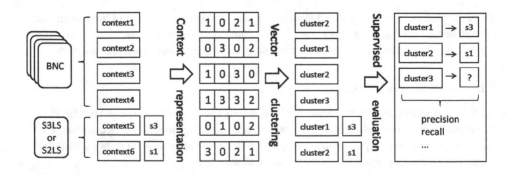

Fig. 1. Process of the WSD system

4   Experiment

In order to be able to evaluate the experiment in a standard benchmark, the toolkit was applied on the 20 nouns in the Senseval-3 English lexical-sample (S3LS) task. The standard training-test split was used during the experiment. Since the evaluation was performed on this dataset, performance measures of the experiment can be compared to other systems' performances.

Tuning of the parameters was executed on the nouns of the Senseval-2 English lexical sample task (S2LS), as in [3].

The process of the experiment can be seen on figure 1. At first, corpora for each target word were created. For each ambiguous noun, paragraphs from the British National Corpus (BNC) were collected which contain the target word. The sense-tagged contexts from the Senseval dataset were added to the appropriate corpora. (Senseval dataset means S2LS data while parameter tuning and S3LS data while testing.) In this way about 2000-3000 context were collected for each target word. After that the SenseClusters modules were executed. Finally, the evaluation was performed.

The whole experiment was executed again for only the Senseval contexts. This is why the effect of additional contexts in the data to be clustered could be examined.

4.1   Optimal Parameters

As mentioned in section 4, parameter tuning was performed on the Senseval-2 data. In case of SenseClusters executed over extended number of contexts the best performance was achieved with first order co-occurrence features, which corresponds with [6]. A partitional clustering approach called globally optimized *Repeated Bisection* ([8]) algorithm proved to be optimal over the Senseval-2 data. Using SVD transformation didn't improve the results.

The best set of parameters of the system executed over only the Senseval contexts was almost the same, only the type of the features were different. In this case using unigram features produced better results.

5   Results

Table 1 shows the performance results of the algorithms. The most frequent sense heuristic serves as a baseline method. Started with the optimized parameters the experiment run over many contexts slightly outperformed the experiment carried out only over the S3LS data. Both method significantly outperformed the baseline method.

The achieved results are competitive with other systems evaluated on the Senseval-3 data, only the optimized HyperLex algorithm of [3] performs better, as it can be seen on table 2. SCBNC and SCS3 correspond to the analyzed systems. (Performance scores of systems apart from SCS3 and SCS3BNC were adopted from [3].)

Although these systems were evaluated on the same data it is still difficult to compare them because of the different training methods they use. Some of them have access to the most frequent sense information (MFS-Sc if counted over SemCor, MFS-S3 if counted over S3LS data), some use 10% of the S3LS training part for mapping, some use the full amount of S3LS training data for mapping (S3LS). [3]

Table 1. Performance results of the experiment over the nouns of the S3LS data. The first and second columns show the target words and the most frequent sense. The last two columns correspond to the precision values of the optimized SenseClusters algorithm executed on only the S3LS dataset and on corpora extended with contexts from BNC. (Coverage is in all cases 100%.)

Word	*MFS*	*SCS3*	*SCBNC*
argument	51.4	48.6	51.4
arm	82.0	85.0	85.7
atmosphere	66.7	72.8	71.6
audience	67.0	70.0	76.0
bank	67.4	72.7	72.0
degree	60.9	67.2	68.8
difference	40.4	48.2	43.0
difficulty	17.4	47.8	26.1
disc	38.0	71.0	66.0
image	36.5	60.8	60.8
interest	41.9	59.1	66.7
judgment	28.1	40.6	40.6
organization	73.2	73.2	69.6
paper	25.6	44.4	52.1
party	62.1	64.7	65.5
performance	32.2	42.5	46.0
plan	82.1	78.6	77.4
shelter	44.9	42.9	48.0
sort	65.6	65.6	65.6
source	65.6	50.0	50.0
Average:	54.5	61.9	62.9
(Over S2LS)	51.9	59.0	59.8

Table 2. Comparison among unsupervised WSD systems evaluated over the S3LS data

System	*Type*	*Precision*	*Coverage*
HyperLex	S3LS	64.6	1.0
SCBNC	S3LS	62.9	1.0
SCS3	S3LS	62.0	1.0
Cymfony	10%-S3LS	57.9	1.0
Prob0	MFS-S3	55.0	0.98
MFS	-	54.5	1.0
Ciaosenso	MFS-Sc	53.95	0.90
clr04	MFS-Sc	48.86	1.0
duluth-senserelate	-	47.48	1.0

6 Conclusions

This paper has presented a possible improvement of unsupervised WSD systems by applying the WSD algorithm to large number of contexts containing the target word.

Although this approach has computational disadvantages, performance of the tested algorithm rose slightly.

After parameter optimization the vector-based algorithm is competitive with the best similar systems, although it is difficult to compare these systems because of the different methods of mapping.

References

1. Purandare, A., Pedersen, T.: SenseClusters - finding clusters that represent word senses. In: Proceedings of the Nineteenth National Conference on Artificial Intelligence (AAAI 2004), San Jose, pp. 1030–3031 (2004)
2. Mihalcea, R., Chklovski, T., Kilgarriff, A.: The Senseval-3 English lexical sample task. In: Senseval-3 proceedings, pp. 25–28 (2004)
3. Agirre, E., Martínez, D., de Lacalle, O.L., Soroa, A.: Evaluating and optimizing the parameters of an unsupervised graph-based wsd algorithm. In: Proceedings of the TextGraphs Workshop: Graph-based algorithms for Natural Language Processing, New York, pp. 89–96 (2006)
4. Schütze, H.: Automatic word sense discrimination. Computational Linguistics 24(1), 97–123 (1998)
5. Véronis, J.: HyperLex: lexical cartography for information retrieval. Computer Speech & Language 18(3), 223–252 (2004)
6. Purandare, A., Pedersen, T.: Word sense discrimination by clustering contexts in vector and similarity spaces. In: Proceedings of the Eighth Conference on Computational Natural Language Learning (CoNLL), Boston, pp. 41–48 (2004)
7. Pedersen, T., Kulkarni, A.: Selecting the "right" number of senses based on clustering criterion functions. In: Proceedings of the Posters and Demo Program of the Eleventh Conference of the European Chapter of the Association for Computational Linguistics, Trento, pp. 111–114 (2006)
8. Zhao, Y., Karypis, G.: Evaluation of hierarchical clustering algorithms for document datasets. In: Proceedings of the 11th Conference of Information and Knowledge Management (CIKM), McLean, USA, pp. 515–524 (2002)
9. Tibshirani, R., Walther, G., Hastie, T.: Estimating the number of clusters in a dataset via the Gap statistic. Journal of the Royal Statistics Society (Series B) 63(2), 411–423 (2001)

Chinese Pinyin-Text Conversion on Segmented Text

Wei Liu and Louise Guthrie

Department of Computer Science, University of Sheffield
{w.liu,l.guthrie}@dcs.shef.ac.uk

Abstract. Most current research and applications on Pinyin to Chinese word conversion employs a hidden Markov model (HMMs) which in turn uses a character-based language model. The reason is because Chinese texts are written without word boundaries. However in some tasks that involve the Pinyin to Chinese conversion, such as Chinese text proofreading, the original Chinese text is known. This enables us to extract the words and a word-based language model can be developed. In this paper we compare the two models and come to a conclusion that using word-based bi-gram language model achieve higher conversion accuracy than character-based bi-gram language model.

1 Introduction

Pinyin is the most widely adopted Romanisation system for Mandarin Chinese. It was introduced in 1956. The main objectives of introducing Pinyin was to help both native and non-native speakers to learn to pronounce Chinese characters. With the introduction of computing, Pinyin has also become the dominant method to type in Chinese characters or words with a Roman alphabet keyboard.

Nevertheless Pinyin and Chinese characters have a two-way ambiguity. For example, many Chinese characters are heteronyms, meaning each of these characters can be pronounced in many different ways. Also many Chinese characters are homonyms, i.e. many different characters (with different meanings) have identical pronunciation and therefore identical Pinyin letters.

As a result converting a Pinyin sequence to its correct Chinese text is difficult. Many researches [1] and [2] use a hidden Markov models approach to tackle this problem. In the HMM state transition probablity matrix a character based n-gram is commonly used. In the following sections we will examine this phenomenon, then we will build a word-based hidden Markov model and compare the conversion accuracies to the character-based language model.

2 Chinese Pinyin and Word

2.1 Heteronyms and Homonyms

Heteronyms are defined as words with identical spellings but different pronunciations and meanings. In Chinese that means characters with identical writing (same ideogram) that can be pronounced differently (with more than one Pinyins). It is generally believed

V. Matoušek and P. Mautner (Eds.): TSD 2009, LNAI 5729, pp. 116–123, 2009.
© Springer-Verlag Berlin Heidelberg 2009

that one-tenth of the Chinese characters are heteronyms. To find out, we use the Chinese Gigaword corpus and a character-Pinyin dictionary to gather some statistics. The Chinese Gigaword corpus is a large Chinese news-wire text collection of more than one billion Chinese characters. The character-Pinyin dictionary [1] lists every Chinese character along with its all possible pronunciations. With them we can have the following table 1 and figure 1 statistics.

Table 1. Gigaword heteronyms statistics

	No. of distinct characters	with 2 Pinyins	with 3 Pinyins	with 4 Pinyins	with 5+ Pinyins
Top 5000 common characters	5000	505	51	5	1
Total distinct characters	9114	702	65	6	1

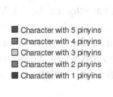

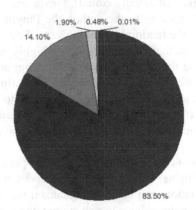

Fig. 1. Gigaword heteronyms distribution

Table 1 shows that out of the 9114 unique characters found in the Gigaword corpus, 774 are heteronyms, and the majority of them contains two possible pronunciations. Furthermore, about 73% of the heteronyms are common characters (they appeared in the top 5000 frequent characters that make up the Gigaword).

In terms of token-wise distribution, of the one billion characters in the Gigaword corpus, heteronyms make up 16.5% of the population (figure 1. It is interesting to note that although only 83.5% of the Gigaword characters are ambiguous in terms of pronunciation, a simple algorithm we wrote that assigns the most frequent pronunciation to each character has an accuracy of over 98%. This suggests that a large proportion of the heteronyms have one prevalent pronunciation.

On the other hand, homonyms present a more ambiguous situation: there are 406 distinct Pinyins [1] and 9114 distinct characters in the Gigaword corpus. This means that

---

[1] From the pinyin4j software package.

on average, each Pinyin corresponds to 22 different characters. This is the main reason why "translating" from a Pinyin sequence to the correct Chinese character sequence is difficult.

2.2 Converting Chinese Text to Pinyin

Automatic Chinese character to Pinyin convertion is considered to be a solved problem mostly because even using a character to Pinyin lookup table can achieve 95% accuracy [3]. Errors are caused by homonyms. In most applications homonyms are simply assigned the most frequently used pronunciation from a corpus.

We also performed character-to-pinyin conversion experiments to validate this claim. We use the publicly available Pinyin4j[2] package to do the conversion. Pinyin4j uses a character-Pinyin hash table to produce all valid Pinyins for a given character. For test documents we gathered The Lancaster Corpus of Mandarin Chinese (LCMC). The LCMC is a balanced Mandarin Chinese text corpus including news-wire, academic papers and fictions. It contains over 1 million Chinese characters. The LCMC corpus is POS tagged (manually edited). Pinyin transcription of the whole corpus are also provided. According to the author, the Pinyin transcription is automaticly generated by machine. We randomly examine a small portion of the text (about 2000 characters) and can confirm that the transcription is highly accurate.

We use the Pinyin4j package to generate Pinyin for each character in the LCMC corpus. We use the LCMC Pinyin transcription as our reference. For homonyms the most frequently used Pinyin is assigned to that character. The result can be seen in table 2. It should be noted that we do not include tonal information when we convert Chinese texts into Pinyins. This simple conversion approach can achieve high accuracy (98%).

To boost conversion accuracy, we propose a word based approach to convert Chinese text to Pinyins. This approach takes advantage of the fact that many homonym character pronunciations can be disambiguated if we know the "words" they are in.

In this approach we use an automatic segmenter to first segment the text before converting them into Pinyins, and then we assign Pinyin based on a word-Pinyin hash table.

Table 2. Results for Chinese text to Pinyin conversion

Method	Accuracy
Character to Pinyin	98.06%
Word to Pinyin	98.74%

3   Hidden Markov Model in Chinese Pinyin-Text Conversion

A hidden Markov model is a statistical model that is used to model a system who satisfies the Markov process, and to determine the hidden parameters from observable parameters. A Markov process is a random process whose future probabilities are only

---

[2] http://pinyin4j.sourceforge.net/

determined by its most recent history values. More specifically, a process x_t is Markovian if for every n and $t_1 < t_2 ... < t_n$ we have:

$$P(x_{t_n} \leq x_n | x_{t_{n-1}}, ..., x_{t_1}) = P(x_{t_n} \leq x_n | x_{t_{n-1}})$$

This assumes that random variables x are not independent, but the current value of $x(t_n)$ depends on previous value $x(t_{n-1})$, and we can ignore the values of all the past random variables in the sequence.

For example, word n-gram models are Markov models. In a bi-gram language model, the probability of a sentence is estimated by the consecutive bi-grams:

$$P(w_n | w_{n-1}, w_{n-2}, ... w_1) = P(w_n | w_{n-1}) * P(w_{n-1} | w_{n-2}) * ... * P(w_2 | w_1)$$

so that the probability of the current word only depends on the previous word (most recent history).

In a Markov model, the state sequence we are modelling is visible to us, i.e. in the above bi-gram model, we know how to estimate the probability of a word given we have seen the previous word. On the other hand, the hidden Markov model operates on a higher level of abstraction, and this "hidden" structure allows us to look at the order of categories in the state sequence.

Given the prevalence of HMM implementation in speech recognition systems [4,5,6], several researchers have noticed the similarities between speech recognition and Pinyin-to-text conversion: in a speech recognition task, we want to produce words but what we have is a sequence of acoustic signals. Converting Pinyin to words is similar - we want to produce a Chinese sentence from a sequence of Pinyins [1] [2].

Details of the hidden Markov modelling is thoroughly described in the classical paper [5] and will not be repeated here. Instead we will be focusing on how we implement the HMM system to convert Pinyins into Chinese characters.

In HMM Pinyin-to-text conversion, our goal is to find out the most likely character sequence given a Pinyin sequence, using Bayesian formula this can be written as:

$$C = \text{argmax } p(C|P) = \text{argmax } p(P|C)p(C)/p(P)$$
$$\propto \text{argmax } p(P|C)p(C)$$

C denotes character sequence and P denotes Pinyin sequence. The first part $p(P|C)$ is typically referred to as the Pinyin conversion model, assuming that the Pinyin(s) of a character does not depend on any other characters (which is almost true), we can break it down so that

$$p(P|C) = \prod_{i=1} p(P_i|C_i) \tag{1}$$

$p(P_i|C_i)$ is the probability of the a Pinyin is P_i produced by the character C_i. This can be easily estimated by using a Pinyin tagged Chinese corpus.

The second part $p(C)$ is the language model and can be estimated by bi-gram language model.

$$p(C) = \prod_{i=1} p(C_i|C_{i-1}) \tag{2}$$

So the final equation for determining the optimal character sequence is:

$$C = \text{argmax}_C \prod_{i=1}^{n} p(P_i|C_i)p(C_i|C_{i-1}) \tag{3}$$

Figure 2 demonstrates an example of a hidden Markov model for Pinyin-to-text conversion. In this figure $o_1...o_n$ are Pinyin observations, $s_1...s_n$ are hidden states, here the hidden states are character sequence we need to recover. Blue dotted lines indicate the transaction from one state to another through time. To effectively calculate the optimal sequence, Viterbi algorithm is used to search for the optimal solution. A detailed description can be found in [7].

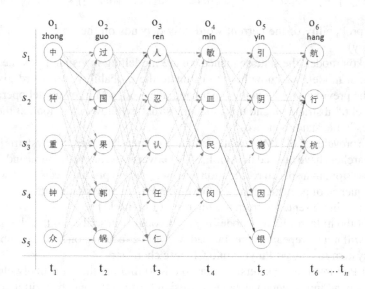

Fig. 2. An example of HMM model to Pinyin-to-word conversion

3.1 Language Model

Most HMM Pinyin-text conversion approaches utilise character N-grams as their language models. The reason is quite obvious - a Pinyin sequence is usually written without word boundaries. In other words, looking at a string of Pinyins it is impossible to tell how many "words" they represent. Let's consider a Pinyin sequence: "**Zhong Guo Ren Min Yin Hang**" and want to find out the most likely character sequence from that. If we are using a character N-gram language model, then we can tell that there are six observations here "Zhong", "Guo", "Ren", "Min", "Yin" and "Hang", and the corresponding character sequence will also contain six characters. However when we want to switch to word states, we have to decide how many observation symbols there are. Without a deeper knowledge we do not know if "Zhong" is a word, or "Zhong Guo" is a word or the whole string "Zhong Guo Ren Min Yin Hang" is just one word.

Fortunately, tasks such as Chinese text proofreading we are interested in finding out the difference between the original Chinese text and the HMM produced text. This means that we know the actual characters in the first place so we can use an automatic segmenter to break down the text into words. Once we know the words we can convert them into Pinyins reliably, therefore we can then build a HMM model with a segmented and Pinyin tagged Chinese corpus to estimate the emission probabilities as well as state transition probabilities.

3.2 Using Bi-gram Word Language Model

We assume that we know the original text that we are going to regenerate. We will use the following example to illustrate our steps to adopt words as the language model. Suppose we have a short phrase ”美国银行” ("Bank of America"). First we use a greedy match segmenter to segment the phrase into words. In this example, the result words will be ”美国” (the United States) and ”银行” (bank as a financial institution).

Second, we assign the most likely Pinyin to this phrase using a word-based Pinyin conversion approach. It is worth pointing out that some heteronyms, It is worth pointing out that the pronounciation of some heteronyms becomes unambiguous when they are used as part of a word. For example the common character ”行” have two pronunciations: "xing" when it is used as a verb, or "hang" when it is used as a noun, but if we see the word ”银行” (bank, as a financial institution), then the character ”行” will always be pronounced "hang". After this process we will have the Pinyin sequence: "Mei Guo", "Yin Hang".

Third, we build a HMM using word bi-gram language model. We use the Chinese Gigaword (version 2) as our training corpus. This corpus is segmented and each word is assigned its most likely Pinyin using the same segmenter and method in our first step, From the corpus we can derive and estimate the probability of a certain word being pronounced in different ways. A Tri structure on disk containing word bi-grams and their counts is also used to build the transition matrix in the HMM model.

Lastly we take the Pinyin sequence as our input and use Viterbi algorithm to search for the most probable state sequence, and compare to our original phrase/sentence.

3.3 Memory/Time Complication

We also want to point out some technical details of the implementation for the word-state HMM model because problems arise when words are used instead of characters in the HMM model. In most implementations of a HMM model and its related algorithms, a two dimensional array is commonly used to represent the transition matrix. For a HMM model that adopts character bi-gram as transitions, using the most common 5,000 character states, a transition matrix will require 5000^2 entries, consuming approximately 190 megabyte[3] of memory storage. However, since a moderate Chinese dictionary contains over 60 thousand entries (words)[4], our smaller word list used in the segmenter also contains over 11 thousand words (including single character words). To fully accommodate the word states transition matrix using array structure we will need 11000^2 entries, this amounts to approximately 968 megabytes.

---

[3] Assuming that each entry requires a 64bit double precision storage.

[4] According to the most prestigious "The Contemporary Chinese Dictionary 2002 Edition".

However we find out that there are only about 23 million distinct bi-grams in our corpus. This means that there are only about 23 million possible word transitions. Therefore we decide to use the Compressed Row Matrix (CRM) [8] structure to store these 23 million numbers in memory.

4   Experiment and Discussion

To compare the performance of the word-based HMM approach to the character-based HMM, we carry out the exact experiment, using the same training and testing data, on both word bi-gram language model and character bi-gram language in our HMM. We use the first 91 sentences from The LCMC as our testing data, for the training data we use the segmented and Pinyin tagged Gigaword Chinese (ver.2).

Table 3. HMM Pinyin-to-text accuracy on 91 sentences

	No. of	Char. bi-gram HMM		Word bi-gram HMM	
	Sentences	Character accuracy	Sentence accuracy	Character accuracy	Sentence accuracy
Short sen. (1-10 chars.)	11		36.36%		45.45%
Mid sen. (11-30 chars.)	45		11.11%		2.2%
Long sen. (30+ chars.)	35		0%		0%
Overall	**91**	**78.86%**	**9.89%**	**81.51%**	**6.59%**

5   Conclusion

In this paper we demonstrate that using word-based HMM model we can achieve higher character recognition accuracy on Chinese Pinyin to text conversion (table 3). In addition, [1] reported improved results over HMM model on Pinyin to character conversion with a Segment-based HMM model and interpolated smoothing technique. This means that a proper semantic unit should be used instead of isolated characters. This can be explained by the fact that once a multi-character word (a word that contains more than one character) is determined its Pinyin is unambiguous and the Pinyin can only map to this word only. By adopting words as states we elimiate a large propotion of ambiguity with the conversion process. We believe that to adopt state-of-the-art smoothing, such as Kneaser-Ney smoothing [9], [10] one can further improve the Pinyin-text conversion accuracy.

References

1. Zhou, X., Hu, X., Zhang, X., Shen, X.: A segment-based hidden markov model for real-setting pinyin-to-chinese conversion. In: CIKM 20007: Proceedings of the sixteenth ACM conference on Conference on information and knowledge management, pp. 1027–1030. ACM, New York (2007)

2. Chen, Z., Lee, K.F.: A new statistical approach to chinese pinyin input. In: Proceedings of the 38th Annual Meeting on Association for Computational Linguistics, Hong Kong, pp. 241–247 (2000)
3. Sen, Z., Laprie, Y.: Mandarin text-to-pinyin conversion based on context knowledge and d-tree. In: Natural Language Processing and Knowledge Engineering, pp. 227–230 (2003)
4. Poritz, A.B.: Hidden markov models: a guided tour. In: International Conference on Acoustics, Speech, and Signal Processing, ICASSP 1988, pp. 7–13 (1988)
5. Rabiner, L.R.: A tutorial on hidden markov models and selected applications in speech recognition, 267–296 (1990)
6. Gales, M., Young, S.: The application of hidden markov models in speech recognition. Found. Trends Signal Process. 1(3), 195–304 (2007)
7. Manning, C.D., Schütze, H.: Foundations of Statistical Natural Language Processing. MIT Press, Cambridge (1999)
8. Smailbegovic, F., Georgi, N., Gaydadjiev, S.V.: Sparse matrix storage format. In: Proceedings of the 16th Annual Workshop on Circuits, Systems and Signal Processing, pp. 445–448 (2005)
9. Goodman, J.T.: A bit of progress in language modeling, extended version. Technical report, Machine Learning and Applied Statistics Group, Microsoft Research (2001)
10. James, F.: Modified kneser-ney smoothing of n-gram models. Technical report (2000)

Mining Phrases from Syntactic Analysis

Miloš Jakubíček, Aleš Horák, and Vojtěch Kovář

NLP Centre, Faculty of Informatics, Masaryk University,
Botanická 68a, 602 00 Brno, Czech Republic

Abstract. In this paper we describe the exploitation of the syntactic parser synt to obtain information about syntactic structures (such as noun or verb phrases) of common sentences in Czech. These phrases/structures are from the analysis point of view usually identical to nonterminals in the grammar used by the parser to find possible valid derivations of the given sentence. The parser has been extended in such a way that enables its highly ambiguous output to be used for mining those phrases *unambiguously* and offers several ways how to identify them. To achieve this, some previously unused results of syntactic analysis have been evolved leading to more precise morphological analysis and hence also to deeper distinction among various syntactic (sub)structures. Finally, an application for shallow valency extraction and punctuation correction is presented.

1 Introduction

Usually a derivation tree is presented as the main output of syntactic analysis of natural languages, but currently most of the syntactic analysers for Czech lack precision, i.e. there is large amount (actually, in some cases up to billions) of trees given on the output. However, there are many situations in which it is not necessary and sometimes even not desirable to have such derivation trees, may it be basic information extraction and retrieval, sentence segmentation, transformation of sentences into a predicate-arguments structure, shallow valency extraction or even punctuation correction. In such cases we rather need to mine whole phrases in the given sentence, especially noun, prepositional and verb phrases, numerals or clauses, each of them in a specific way. Moreover, so as not to end up with the same problems as with the standard parser output, we need to identify the structures *unambiguously* so that they are directly usable for further post-processing.

Therefore we modified the Czech parser synt so that it would be possible to obtain syntactic structures corresponding to the given nonterminal in a number of ways according to the user's choice. To improve the structure detection we also employed the results of contextual actions used in synt as described in Section 4, which increased the precision of morphological analysis by almost 30 %. We also present results of the mining from sample sentences as well as two example applications, first, the shallow extraction of verb valencies from annotated corpora and, second, the punctuation correction based on the described mining.

2 Syntactic Parser synt

Syntactic parser synt [1] has been developed for several years in the in the Natural Language Processing Centre at Faculty of Informatics, Masaryk University. It performs

V. Matoušek and P. Mautner (Eds.): TSD 2009, LNAI 5729, pp. 124–130, 2009.

a chart-type syntactic analysis based on the provided context-free head-driven grammar for Czech. For easy maintenance this grammar is edited in form of a metagrammar (having about 200 rules) from which the full grammar can be automatically derived (having almost 4,000 rules). Contextual phenomena (such as case-number-gender agreement) are covered using the per-rule defined contextual actions.

In recent evaluation [2, p. 77] it has been shown that synt accomplishes a very good recall (above 90 %) but the analysis is highly ambiguous: for some sentences even billions of output syntactic trees can occur. There are two main strategies developed to fight such ambiguity: first, the grammar rules are divided into different priority levels which are used to prune the resulting set of output trees. Second, every grammar rule has a ranking value assigned from which the ranking for the whole tree can be efficiently computed to sort the trees on the output accordingly.

For the purpose of the mining process the internal parsing structure of synt is used, the so called *chart*, a multigraph which is built up during the analysis holding all the resulting trees. What is important about chart is its polynomial size [3, p. 133] implying that it is a structure suitable for further effective processing – as the number of output trees can be up to exponential to the length of the input sentence, processing of each tree separately would be otherwise computationally infeasible. By processing of the chart we refer to the result of the syntactic analysis, i.e. to the state of the chart after the analysis.

3   Mining Phrases

Several ways how to identify the phrases have been developed respecting the reality that the syntactic structures behind the phrases differ a lot and thus no universal procedure can be used for all of them. Since we want the output of the mining process to be unambiguous, we cover all possible structures found in the resulting chart data structure used during the syntactic analysis.

There are two straightforward approaches for structure detection which consist in extracting the biggest or smallest found structure, however, to achieve high quality results, more sophisticated methods have to be employed for each structure/nonterminal specifically. By speaking about biggest or smallest we mean that with regard to the fact that many of the rules in the grammar used by synt are recursive and thus we have to decide what recursion level should be extracted – e.g. in a simple sentence: *Vidím starý velký dům*[1] we may extract both *velký dům* and *starý velký dům* as noun phrases. This decision is, however, not only nonterminal-specific, but depends also heavily on the particular application of the mining. Sample mining results for various nonterminals are listed in Examples 1–4.

- *Example 1.* – clause (nested)
  Input:
  Muž, který stojí u cesty, vede kolo.
  (A man who stands at the road leads a bike.)
  Output:
  [0-9): Muž , , vede kolo *(a man leads a bike)*
  [2-6): který stojí u cesty *(who stands at the road)*

---
[1] In English: *I see an old big house.*

- *Example 2.* – verb phrase
  Input:
  Kdybych to byl býval věděl, byl bych sem nechodil.
  (If I had known it, I would not have come here.)
  Output:
  [0-5]^2 : byl býval věděl *(had known)*
  [6-10]: byl bych nechodil *(would not have come)*

- *Example 3.* – clause (sequence)
  Input:
  Vidím ženu, jež drží růži, která je červená.
  (I see a woman who holds a flower which is red.)
  Output:
  [0-3): Vidím ženu , *(I see a woman)*
  [3-7): jež drží růži , *(who holds a flower)*
  [7-10): která je červená *(which is red)*

- *Example 4.* – noun phrase
  Input:
  Tyto normy se však odlišují nejen v rámci různých národů a států, ale i v rámci sociálních skupin, a tak považuji dřívější pojetí za dosti široké a nedostačující.
  (But these standards differ not only within the scope of various nations and countries but also within the scope of social groups and hence I consider the former conception to be wide and insufficient.)
  Output:
  [0-2): Tyto normy *(These standards)*
  [6-12): v rámci různých národů a států *(within the scope of various nations and countries)*
  [15-19): v rámci sociálních skupin *(within various social groups)*
  [23-30): dřívější pojetí za dosti široké a nedostačující *(former conception for wide and insufficient)*

4   Morphological Refinement

In order to further divide large structures into separate meaningful segments it is possible to segment them according to the morphological agreement – i.e. in such a way that words in each structure agree in case, number and gender. To improve this technique some previously unused results of the syntactic analysis have been exploited, namely the contextual actions used by the parser to handle the case-number-gender agreement. In each analysis step, the results of the contextual actions are propagated bottom-up so that they can be used in the next step to prune possible derivations.

So far these results in form of morphological values have not been used in any other way. Our enhancement backpropagates these values after the analysis top-down to the

---

^{2} The numbering denotes a (left inclusive, right exclusive) range of the structure in the input sentence (i.e. words indices).

Table 1. A comparison of morphological tagging before and after the refinement. The whole sentence in English was: *There was a modern shiny car standing on a beautiful long street.* Note that for readability purpose we abbreviate the tags so that k7{c4,c6} stands for k7c4, k7c6.

word	before	after
Na *(on)*	k7{c4, c6}	k7c6
krásné *(beautiful)*	k2eA{gFnPc1d1, gFnPc4d1, gFnPc5d1, gFnSc2d1, gFnSc3d1, gFnSc6d1, gInPc1d1, gInPc4d1, gInPc5d1, gInSc1d1wH, gInSc4d1wH, gInSc5d1wH, gMnPc4d1, gMnSc1d1wH, gMnSc5d1wH, gNnSc1d1, gNnSc4d1, gNnSc5d1}	k2eAgFnSc6d1
dlouhé *(long)*	k2eA{gFnPc1d1, gFnPc4d1, gFnPc5d1, gFnSc2d1, gFnSc3d1, gFnSc6d1, gInPc1d1, gInPc4d1, gInPc5d1, gInSc1d1wH, gInSc4d1wH, gInSc5d1wH, gMnPc4d1, gMnSc1d1wH, gMnSc5d1wH, gNnSc1d1, gNnSc4d1, gNnSc5d1}	k2eAgFnSc6d1
ulici *(street)*	k1gFnSc3, k1gFnSc4, k1gFnSc6	k1gFnSc6
stálo *(stand)*	k5eAaImAgNnSaIrD	k5eApNnStMmPaI[3]
moderní *(modern)*	k2eA{gFnPc1d1, gFnPc4d1, gFnPc5d1, gFnSc1d1, gFnSc2d1, gFnSc3d1, gFnSc4d1, gFnSc5d1, gFnSc6d1, gFnSc7d1, gInPc1d1, gInPc4d1, gInPc5d1, gInSc1d1, gInSc4d1, gInSc5d1, gMnPc1d1, gMnPc4d1, gMnPc5d1, gMnSc1d1, gMnSc5d1, gNnPc1d1, gNnPc4d1, gNnPc5d1, gNnSc1d1, gNnSc4d1, gNnSc5d1}	k2eAgNnSc1d1, k2eAgNnSc4d1, k2eAgNnSc5d1
nablýskané *(shiny)*	k2eA{gFnPc1d1rD, gFnPc4d1rD, gFnPc5d1rD, gFnSc2d1rD, gFnSc3d1rD, gFnSc6d1rD, gInPc1d1rD, gInPc4d1rD, gInPc5d1rD, gInSc1d1wHrD, gInSc4d1wHrD, gInSc5d1wHrD, gMnPc4d1rD, gMnSc1d1wHrD, gMnSc5d1wHrD, gNnSc1d1rD, gNnSc4d1rD, gNnSc5d1rD}	k2eAgNnSc1d1, k2eAgNnSc4d1, k2eAgNnSc5d1
auto *(car)*	k1gNnSc1, k1gNnSc4, k1gNnSc5	k1gNnSc1, k1gNnSc4, k1gNnSc5

chart nodes, i.e. input words, and prunes their original morphological tagging. This leads to more precise morphological analysis and hence it also enables more exact distinction between substructures according to grammar agreement. A detailed example of the impact of morphological refinement on particular sentence is provided in Table 1.

Testing on nearly 30,000 sentences from Czech annotated corpus DESAM [4] has shown that it is possible to increase the number of unambiguously analysed words by almost 30 % using this method while the number of errors introduced consequently remains very low, as shown in Table 2.

Segmenting structures according to their grammatical agreement is useful, for example, when mining noun or prepositional phrases, as can be seen in Example 5 (compare with Example 4 where the same sentence is extracted without morphological segmenting).

---

[3] The inconsistence in tagging on this row has purely technical background – the tag set has been changed.

Table 2. Morphological refinement results on the DESAM corpus

value	before	after
average unambiguous words	20.733 %	46.1172 %
average pruned word tags	38.3716 %	
error rate [4]	< 1.46 %	
number of sentences	29,604	

Example 5
Input:
Tyto normy se však odlišují nejen v rámci různých národů a států, ale i v rámci sociálních skupin, a tak považuji dřívější pojetí za dosti široké a nedostačující.

(These standards, however, differ not only within the scope of various nations and countries but also within the scope of social groups and hence I consider the former conception to be wide and insufficient.)
Output:
[0-4): Tyto normy se však
(But these standards)
[6-8): v rámci
(within the scope)
[8-12): různých národů a států
(various nations and countries)
[13-17): ale i v rámci
(but also within the scope)
[17-19): sociálních skupin
(social groups)
[23-25): dřívější pojetí
(former conception)
[25-30): za dosti široké a nedostačující
(for wide and insufficient)

Specific modifications how to extract nonterminals with important semantical representation (noun, prepositional and verb phrases, clauses, conjunctions etc.) have been developed. Furthermore, these modifications can be extended to other (possibly new) nonterminals easily as they are available as command-line parameters.

5 Applications

5.1 Shallow Valency Extraction

Currently, a new verb valencies lexicon for Czech, called *Verbalex* [5], is being developed in the NLP Centre. Since building of such a lexicon is a very time-consuming

---

[4] As an error we consider a situation when the correct tag has been removed during the refinement process. Actually, the error rate is even lower since many of the results marked as wrong were caused by an incorrect tag in the corpus.

long-term task for linguists, it is extremely important to use any possibilities to make this process easier. Therefore, we extended the mining of phrases so that it performs a shallow valency extraction from annotated corpora. The main idea is as follows: first, we mine separate clauses, then we identify individual verbs or verb phrases and finally we find noun and prepositional phrases within each clause. Sample output in BRIEF format [6][5] is provided in Example 6. Moreover, such basic extraction might be used for approximating the coverage of the valency lexicon by finding verbs that are not included as well as for finding suitable examples when building a valency dictionary.

Example 6
```
; extracted from sentence: Nenadálou finanční krizi musela
podnikatelka řešit jiným způsobem .
řešit <v>hPc4,hTc4,hPc7,hTc7
```
*(The businessman had to **solve** the sudden financial crisis in another way.)*
```
; extracted from sentence: Hlavní pomoc ale nacházela
v dalších obchodních aktivitách .
nacházet <v>hPc4,hTc4,hPc6r{v},hTc6r{v}
```
*(However she **found** the main help in further business activities.)*
```
; extracted from sentence: U výpočetní techniky se pohybuje
v rozmezí od 8000 Kč do 16000 Kč .
pohybovat <v>hPc2r{u},hTc2r{u},hPc6{v},hTc6{v}
```
*(By IT [it] **ranges** between 8,000 Kč and 16,000 Kč.)*

5.2 Punctuation Correction

Punctuation correction represents one of the possible usages of the mined phrases. Our preliminary results show that if we have the necessary syntactic knowledge it is possible to solve many of the corner-cases which cannot be covered by a simple rule-based punctuation correction system, such as inserted sentences (and similar structures which are enclosed with punctuation) or coordinations. The basic approach to cover these phenomena is to introduce a special nonterminal marking the punctuation into the grammar and adjust the grammar rules so that they describe proper punctuation usage. The punctuation nonterminal is then extracted together with clauses, relative clauses and other syntactic elements which enables us to fill the punctuation in the sentence.

6 Conclusions

We presented recent improvements in the Czech parser synt that can be used for mining various phrases (equivalent to syntactic substructures). We have also shown practical usage of syntactic analysis for refining morphological tagging as well as examples using the resulting tagging for structures distinguishing. Furthermore, we have presented an application of the proposed method, namely shallow extraction of valencies.

---

[5] The BRIEF format is primarily a sequence of *attribute-value* pairs. In the given example, *hP/hT* stands for animate/inanimate object, *c[1-7]* for the case and *r{preposition}* for a prepositional valency.

In the future there will be further work on the development of the mining process. We would like to compare the results of morphological refinement with similar oriented methods (e.g. with morphological disambiguation as described in [7]) as well as perform more detailed experiments with the shallow valency extraction on big corpora. This method allows us also to add basic syntactic annotation to existing corpora easily, e.g. to mark noun or prepositional phrases for future usage.

Acknowledgements

This work has been partly supported by the Academy of Sciences of Czech Republic under the project 1ET100300419 and by the Ministry of Education of CR within the Center of basic research LC536 and in the National Research Programme II project 2C06009.

References

1. Kadlec, V., Horák, A.: New meta-grammar constructs in czech language parser synt. In: Matoušek, V., Mautner, P., Pavelka, T. (eds.) TSD 2005. LNCS (LNAI), vol. 3658, pp. 85–92. Springer, Heidelberg (2005)
2. Kadlec, V.: Syntactic analysis of natural languages based on context-free grammar backbone. PhD thesis, Faculty of Informatics, Masaryk University, Brno (2007)
3. Horák, A.: The Normal Translation Algorithm in Transparent Intensional Logic for Czech. PhD thesis, Faculty of Informatics, Masaryk University, Brno (2001)
4. Pala, K., Rychlý, P., Smrž, P.: DESAM – Annotated Corpus for Czech. In: Jeffery, K. (ed.) SOFSEM 1997. LNCS, vol. 1338, pp. 523–530. Springer, Heidelberg (1997)
5. Hlaváčková, D., Horák, A., Kadlec, V.: Exploitation of the verbaLex verb valency lexicon in the syntactic analysis of czech. In: Sojka, P., Kopeček, I., Pala, K. (eds.) TSD 2006. LNCS (LNAI), vol. 4188, pp. 79–85. Springer, Heidelberg (2006)
6. Pala, K., Ševeček, P.: The valence of czech words. In: Sborník prací FFBU, Brno, Masarykova univerzita, pp. 41–54 (1997)
7. Šmerk, P.: Unsupervised learning of rules for morphological disambiguation. LNCS. Springer, Heidelberg (2004)

Problems with Pruning in Automatic Creation of Semantic Valence Dictionary for Polish

Elżbieta Hajnicz

Institute of Computer Science, Polish Academy of Sciences

Abstract. In this paper we present the first step towards the automatic creation of semantic valence dictionary of Polish verbs. First, resources used in the process are listed. Second, the way of gathering corpus-based observations into a semantic valence dictionary and pruning them is discussed. Finally, an experiment in the application of the method is presented and evaluated.

1 Introduction

The primary task of our research is to create a semantic valence dictionary in an automatic way. To accomplish this goal, the valence dictionary of Polish verbs is supplemented with semantic information, provided by wordnet's semantic categories of nouns. In the present work we focus on NPs and PPs, whose semantic heads are nouns. In the current phase of work we dispose of two data resources:

- a purely syntactic valence dictionary,
- a syntactically and semantically annotated corpus.

In this paper the way of gathering corpus-based observations into a semantic valence dictionary is described. Such a task was often performed for the sake of a syntactic valence dictionary creation, often referred to as *subcategorisation acquisition* (e.g., [1,2]). Methods used there are adapted here. The main difference between these two tasks is that semantic dictionary has more complicated structure: semantic frames are assigned to syntactic schemata which are assigned to verbs. Moreover, syntactic frames are treated in the syntactic valence extraction task as indivisible entities, whereas it seems reasonable to treat semantic frames as a list of semantically interpreted slots.

2 Data Resources

Experiments were performed on a set of 32 manually chosen verbs [3]. The criteria of choice were: the need for maximisation of the variability of syntactic schemata[1] (in particular, diathesis alternations), polysemy of verbs within a single syntactic frame (schema), and the verb's frequency. This set of verbs is referred to as CHVLIST.

---

[1] We use the term syntactic *schema* instead of very popular syntactic *frame* in order to distinguish it from the term *semantic frame*.

V. Matoušek and P. Mautner (Eds.): TSD 2009, LNAI 5729, pp. 131–138, 2009.

2.1 Syntactic Valence Dictionary

In the experiments, an extensive valence dictionary was used, based on Świdziński's [4] valence dictionary containing 1064 verbs. It was specially prepared for the task. Świdziński's dictionary was supplemented with 1000 verb entries from the dictionary automatically obtained by Dębowski and Woliński [5] to increase the coverage of used dictionary on SEMKIPI. The most carefully elaborated part of the valence dictionary concerns verbs from CHVLIST [6].

A syntactic dictionary is a list of entries representing sets of schemata for every verb considered. The format encoding the syntactic dictionary used through the paper was applied for the first time in the process of automatic creation of the syntactic valence dictionary mentioned above. Every entry in the dictionary is composed of a verb lemma and a list of its arguments. The list of arguments can include: adjective phrases AdjP, adverb phrases AdvP, infinitive phrases InfP, noun phrases NP, prepositional-adjective phrases PrepAdjP, prepositional-nominative phrases PrepNP, clauses SentP and reflexive marker *się*. Arguments can be parametrised. AdvP has no parameters, the only parameter of AdjP and NP is their case, the only parameter of InfP is its aspect. PrepAdjP and PrepNP have two parameters: a preposition and the case of its AdjP or NP complement, respectively. SentP has one parameter showing the type of a clause.

Below we list syntactic dictionary entries for verb *interesować* (*to interest*).

(1) interesować   :np:acc: :np:nom:
     interesować   :np:inst: :np:nom: :sie:
     interesować   :np:nom: sentp:wh: :sie:

Formally, the syntactic valence dictionary $\mathcal{D}$ is a set of pairs $\langle v, g \rangle$, where $v \in V$ is a verb and $g \in G$ is a syntactic frame.

2.2 Słowosieć—Polish Wordnet

Semantic annotation is based on the Polish WordNet [7], [8], [9], called *Słowosieć*, which was modelled on the Princeton WordNet and wordnets constructed in the EuroWordNet project. In the present work we do not use the whole structure of the net, but the set S of 26 predefined semantic categories (see Table 1), which are assign to each lexical unit. The categories correspond to synsets positioned at the top of the actual hierarchy.

2.3 Corpus

The main resource was the IPI PAN Corpus of Polish written texts [10]. A small subcorpus was selected from it, referred to as SEMKIPI, containing 195 042 sentences. It is composed of sentences which contained: one or more verbs from CHVLIST and at most three verbs at all. SEMKIPI was parsed with the *Świgra* parser [11] based on the metamorphosis grammar GFJP [12] provided with the valence dictionary presented in section 2.1 above.

In order to reduce data sparseness, for the present experiment we considered only the top-most phrases being the actual arguments of a verb (i.e., the subject and complements

Table 1. Predefined set of general semantic categories in Polish WordNet

name	acr	name	acr	name	acr	name	acr
Tops	T	communication	cm	object	ob	relation	rl
act	ac	event	ev	person	pn	shape	sh
animal	an	feeling	fl	phenomenon	ph	state	st
artifact	ar	food	fd	plant	pl	substance	sb
attribute	at	group	gr	possession	ps	time	tm
body	bd	location	lc	process	pr		
cognition	cg	motive	mt	quantity	qn		

included in its valence schemata). This means that each obtained parse was reduced to its flat form identifying only these top-most phrases (called *reduced parses*, cf. [13]).

Świgra tends to produce large parse tree forests. The number of reduced parses of a sentence is much smaller than the number of entire parse trees, but it is still considerable. The most probable reduced parse was chosen by means of an EM algorithm called *EM selection algorithm* proposed by Dębowski [13].

Adopted versions of this algorithm were used to disambiguate semantic categories of nouns being semantic heads of NPs/PPs in the reduced parse of a sentence selected in the previous step. The whole process is described in detail in [6].

The EM selection algorithms from [6] work on the simplified version of the syntactic-semantic representation of a clause, namely it uses corresponding semantic frames. An exemplary sentence from the SEMKIPI corpus together with its syntactic schema (in <> brackets) and semantic frame selected subsequently during the process is shown in (2).

(2) % 'Mecz rozpoczął się od ataków Lotnika.'
    (The match started from attacks of Lotnik)
    <rozpocząć :np:nom: :prepnp:od:gen: :sie:>
    nom: **event**, od_gen: **action**

For the goal of collecting a semantic valence dictionary we can interpret SEMKIPI as a set of *observations* $\mathcal{O} = \langle s, v, g, F_s \rangle$, where $s \in S$ is a clause, $\langle v, g \rangle \in \mathcal{D}$ is a verb heading the clause with a corresponding syntactic frame, and $F_s \subseteq F$ is a set of semantic frames chosen for the clause. We cannot just speak about a single f, since the semantic annotation could be ambiguous. Let also $F_g \subseteq F$ be a set of all semantic frames corresponding to a syntactic schema g and $\mathcal{O}_g \subseteq \mathcal{O}$ be a set of all observations with syntactic schema g attached.

3 Semantic Dictionary

3.1 Protodictionary

Obtaining a semantic valence protodictionary from a set of observations $\mathcal{O}$ seems to be simple: semantic frames need to be counted w.r.t. verbs and syntactic schemata that co-occur with them. However, two difficulties complicate the process. First, clauses having more than one semantic frame assigned should be considered. Such frames are counted then in proportion to their number in a clause. The second difficulty needs more thought. Due to the ellipsis phenomenon, some observations are connected with syntactic schemata not belonging to $\mathcal{D}$, but being only the subschemata of elements of $\mathcal{D}$. Such frames should not be considered in the semantic dictionary. One of possible solutions is to ignore them. However, they carry an important piece of information that we do not want to lose. Since they are implicit realisations of the corresponding superschemata, the idea is to transfer this information to semantic frames connected with direct superschemata. This goal can be achieved in several ways, assuming that semantic frames are compatible on slots belonging to a subschema. First, an additional slot can be "distributed" between all senses with a small frequency (e.g., $1/(2 \cdot 26)$). However, this would mainly increase noise. Second, if compatible superframes exist, then subframes can be "unified" with them, by increasing their frequency.

Let us present this process on an example. In (3) a subset of the set of frames connected with the schema :np:acc: :np:dat: :np:nom: of the verb *proponować* (*to propose*) is shown. First, all frames connected with schemata from outside $\mathcal{D}$ are moved to all their direct superschemata, with an additional slot having artificial semantic category UNDEF (on the left). Next, such an additional frame is unified with all other frames of a particular schema that are consistent with it. This operation does not create new frames, but increases a frequency of the frames that match (proportionally to their number). The resultant frequencies are listed on the right, while the deletion of artificial UNDEF-containing frames used in the unification is marked with '—'.

(3) proponować

:np:acc: :np:dat: :np:nom:	573	
acc: ac; dat: UNDEF; nom: gr	47.25	—
acc: ac; dat: UNDEF; nom: qn	0.5	0.5
acc: ac; dat: UNDEF; nom: pn	107.75	—
acc: ac; dat: fl ; nom: gr	1	4.63
acc: ac; dat: ac; nom: gr	5	8.63
acc: ac; dat: ac; nom: pn	23	31.29
acc: ac; dat: gr; nom: gr	13	16.63
acc: ac; dat: gr; nom: pn	42	50.35
acc: ac; dat: gr; nom: ev	1	1
acc: ac; dat: qn; nom: gr	2	5.63
acc: ac; dat: lc; nom: pn	1	9.29
acc: ac; dat: pn; nom: gr	19	22.63
acc: ac; dat: pn; nom: pn	43	51.36

Formally, a semantic dictionary $\mathfrak{D}$ is a set of tuples $\langle\langle v, g, f\rangle, n, m\rangle$, where $\langle v, g\rangle \in \mathcal{D}$ is a schema of a verb, $f \in F_g$ is one of its frames, n is the frequency of $\langle v, g\rangle$ and m is the frequency of $\langle v, g, f\rangle$.

3.2 Pruning

The resulting set of collected observations is very noisy, the more so as the statistical EM selection algorithm [13] was applied twice: first for a syntactic schema selection for each clause and next for semantic frame selection for the clause with the syntactic schema already assigned. Thus, in order to make the resultant semantic valence dictionary more reliable we needed to perform some pruning.

Using a syntactic valence dictionary in parsing ensures that all schemata are properly attached to verbs, even if they were improperly selected for some clauses. Thus, there is no reason for pruning here. We only could filter some schemata that were too rare to guarantee the good quality of semantic frames connected with them. Therefore, we decided to focus only on pruning the semantic valence dictionary.

We decided to adapt here the *Binomial Hypothesis Test* (BHT). It is the most frequently employed statistical test for syntactic subcategorisation acquisition, originally introduced for the purpose by Brent [1]. Instead of counting syntactic schemata g against verbs v, we count semantic frames f against syntactic schemata $\langle v, g \rangle$ of a particular verb, treated as a whole.

The method is based on the assumption that every erroneous observation occurs in $\mathcal{O}$ with some small probability p^e, and that this probability has the binomial distribution. Thus, the probability of m occurrences of frame f within n occurrences of schema g is $B(m, n, p^e)$. The main idea is that if the probability that g occurs just because of a noise more than m time is large enough, then it is not an actual frame of g. Thus, the probability

$$B^+(m, n, p^e) = \sum_{i=m}^{n} B(i, n, p^e) \tag{4}$$

is beeing counted. If $B^+(m, n, p^e) < \zeta$, where threshold ζ is traditionally set to 0.05, then f is assumed to be a frame of g.

Very important part of the method is establishing p^e value. Brent himself accomplished this experimentally w.r.t. precision and recall being achieved. One p^e was chosen for all verbs.

Briscoe and Carroll[2] estimate p^e as follows (after adapting it for the semantic level):

$$p_f^e = \left(1 - \frac{|H_{\langle g,f \rangle}|}{|H_g|}\right) \frac{|\mathcal{O}_g|}{|\mathcal{O}|} \tag{5}$$

where $H_{\langle g,f \rangle}$ is a set of frames of a schema in a manually-prepared dictionary HD (see below) and H_g is a set of schemata in this dictionary.

4 Experiments

Experiments were performed on a protodictionary collected from SEMKIPI. Its two versions were considered: SD obtained from the source dictionary by deleting all sub-schemata (together with corresponding frames) and UD obtained after the unification.

The enormous amount of semantic frames causes a significant sparseness of dictionaries. In particular, most of frames occur in HD at most once. As a result, p_f^e estimated w.r.t equation (5) is very small (about 10^{-5}) hence the threshold should be proportionally small. Our experiments were conducted with $p_f^e = 0.05$ and $p_f^e = 0.01$.

The very characteristic feature of semantic frames is their length (understood as the number of NPs/PPs). Thus, we decided to estimate the error probability of all frames with the same length:

$$p_l^e = \left(1 - \frac{|H_l|}{|S|^l}\right) \frac{|O_l|}{|O|}, \tag{6}$$

where $H_l = \sum_{|g|=l} H_g$ and $O_l = \sum_{|g|=l} O_g$. This estimation is rather coarse, hence it needs a proportionally large threshold. For our experiments we choose $p_l^e = 0.25$ and $p_l^e = 0.05$.

Both dictionaries were pruned w.r.t. two estimations of p^e, which led to the dictionaries P_f-t-SD, P_l-t-SD, P_f-t-UD and P_l-t-UD, respectively, where t denotes the threshold used in pruning.

The results were evaluated against a small manually-prepared dictionary HD composed of all syntactic schemata and corresponding semantic frames for 5 verbs: *interesować* (*to interest*), *minąć* (*to pass*), *proponować* (*propose*), *rozpocząć* (*to begin*) and *widzieć* (*to see*).

Below, some exemplary manually-prepared dictionary entries for *minąć* (*to pass*) are presented. HD differs from the other dictionaries presented in this paper, as it was already elaborated in the aggregated form, collecting all NP/PP senses connected with the same meaning of the verb. Senses (here: semantic categories) are separated with commas and the whole slots are separated with semicolons.

(7) minąć   :np:acc: :np:nom:
             acc: ar,mt;   nom: ac
             acc: mt;   nom: ar
             acc: an,ar,gr,lc,pl,pn; nom: an,ar,gr,pn
             acc: gr,pn;   nom: ac,cg,cm,mt,ps,rl

    minąć   :np:dat: :np:nom:
             dat: gr,pn;   nom: at
             dat: gr,pn;   nom: ac,ev,pr,tm
             dat: ps;   nom: cg,fl

    minąć   :np:nom: :prepnp:od:gen:
             nom: tm; od_gen: ac,ev,st,tm
             nom: tm; od_gen: ps
             nom: tm; od_gen: cg

In table 2 the results of evaluation of all dictionaries discussed above w.r.t. HD dictionary are presented. We added information about the percentage of deleted schemata (column *s-dels*) and frames (column *f-dels*).

It is easy to notice that the unification helps a bit. But results are confusing. Good accuracy is the effect of large amount of true negatives (26^l). Poor recall is an obvious consequence of data sparseness. However, we hope that it is going to increase during an aggregation. The most problematic is precision. Only pruning with p_l^e estimate of p^e increases it (significantly), but the deletion of at least 8% of schemata and 83% of

Table 2. Evaluation of semantic valence dictionaries before and after pruning

	s-dels	f-dels	acc	prec	rec	F
SD	—	—	98.841	46.023	23.549	31.156
P_f-.01-SD	2.60	22.7	98.841	45.055	19.172	26.898
P_f-.05-SD	0.24	12.3	98.839	45.455	21.681	29.359
P_l-.05-SD	34.04	91.6	98.817	63.492	8.351	14.760
P_l-.25-SD	9.93	83.4	98.938	63.084	9.165	16.005
UD	—	—	98.859	47.460	25.482	33.160
P_f-.01-UD	1.76	21.2	98.853	46.381	20.892	28.807
P_f-.05-UD	0.29	11.2	98.849	46.361	22.888	30.646
P_l-.05-UD	26.69	91.5	98.819	63.682	8.895	15.610
P_l-.25-UD	8.21	84.4	98.931	60.731	9.029	15.721

frames seems to be too drastic. Please observe that after pruning F-measure always decreases.

5 Conclusions and Future Works

In this paper the method for an automatic creation of a semantic valence dictionary of Polish verbs has been presented, meaning augmenting syntactic valence schema with semantic senses of arguments.

The task is similar to syntactic valence dictionary creation, but it is more complicated due to much larger data sparseness. A very popular method of pruning data was applied, but it failed. The reasons, in spite of obvious data sparseness, are some systematic errors which appeared when SEMKIPI was created, which contradicts the assumption of random distribution of errors in it.

Future works will develop in three directions. First, other pruning methods will be checked, including BHT with p^e estimated with an additional assumption of independence of slots. Second, we want to employ an aggregation to the collected semantic dictionary and check its influence on the obtained results. Third, we plan to perform the whole procedure on the entire structure of *Słowosieć* as well.

Acknowledgement

This paper is a scientific work supported within the Ministry of Science and Education project No. N N516 0165 33.

References

1. Brent, M.R.: From grammar to lexicon: unsupervised learning of lexical syntax. Computational Linguistics 19(2), 243–262 (1993)
2. Briscoe, T., Carrol, J.: Automatic extraction of subcategorization from corpora. In: Proceedings of the 5th ACL Conference on Applied Natural Language Processing, Washington, DC, pp. 356–363 (1997)

3. Hajnicz, E.: Dobór czasowników do badań przy tworzeniu słownika semantycznego czasowników polskich. Technical Report 1003, Institute of Computer Science, Polish Academy of Sciences, Warsaw (2007)
4. Świdziński, M.: Syntactic Dictionary of Polish Verbs. Uniwersytet Warszawski / Universiteit van Amsterdam (1994)
5. Dębowski, Ł., Woliński, M.: Argument co-occurrence matrix as a description of verb valence. In: Vetulani, Z. (ed.) Proceedings of the 3rd Language & Technology Conference, Poznań, Poland, pp. 260–264 (2007)
6. Hajnicz, E.: Semantic annotation of verb arguments in shallow parsed Polish sentences by means of EM selection algorithm. In: Marciniak, M., Mykowiecka, A. (eds.) Aspects of Natural Language Processing. LNCS, vol. 5070. Springer, Heidelberg (2009)
7. Derwojedowa, M., Piasecki, M., Szpakowicz, S., Zawisławska, M.: Polish WordNet on a shoestring. In: Data Structures for Linguistic Resources and Applications: Proceedings of the GLDV 2007 Biannual Conference of the Society for Computational Linguistics and Language Technology, Universität Tübingen, Tübingen, Germany, pp. 169–178 (2007)
8. Derwojedowa, M., Piasecki, M., Szpakowicz, S., Zawisławska, M., Broda, B.: Words, concepts and relations in the construction of Polish WordNet. In: Tanacs, A., Csendes, D., Vincze, V., Fellbaum, C., Vossen, P. (eds.) Proceedings of the Global WordNet Conference, Seged, Hungary, pp. 162–177 (2008)
9. Derwojedowa, M., Szpakowicz, S., Zawisławska, M., Piasecki, M.: Lexical units as the centrepiece of a wordnet. In: Kłopotek, M.A., Przepiórkowski, A., Wierzchoń, S.T. (eds.) Proceedings of the Intelligent Information Systems XVI (IIS 2008). Challenging Problems in Science: Computer Science. Academic Publishing House Exit, Zakopane (2008)
10. Przepiórkowski, A.: The IPI PAN corpus. Preliminary version. Institute of Computer Science, Polish Academy of Sciences, Warsaw (2004)
11. Woliński, M.: Komputerowa weryfikacja gramatyki Świdzińskiego. PhD thesis, Institute of Computer Science, Polish Academy of Sciences, Warsaw (2004)
12. Świdziński, M.: Gramatyka formalna języka polskiego. Rozprawy Uniwersytetu Warszawskiego. Wydawnictwa Uniwersytetu Warszawskiego, Warsaw (1992)
13. Dębowski, Ł.: Valence extraction using the EM selection and co-occurrence matrices. arXiv (2007)

Disambiguating Tags in Blogs

Xiance Si and Maosong Sun

Tsinghua University, Beijing, China
sxc@mails.tsinghua.edu.cn,
sms@tsinghua.edu.cn

Abstract. Blog users enjoy tagging for better document organization, while ambiguity in tags leads to inaccuracy in tag-based applications, such as retrieval, visualization or trend discovery. The dynamic nature of tag meanings makes current word sense disambiguation(WSD) methods not applicable. In this paper, we propose an unsupervised method for disambiguating tags in blogs. We first cluster the tags by their context words using Spectral Clustering. Then we compare a tag with these clusters to find the most suitable meaning. We use Normalized Google Distance to measure word similarity, which can be computed by querying search engines, thus reflects the up-to-date meaning of words. No human labeling efforts or dictionary needed in our method. Evaluation using crawled blog data showed a promising micro average precision of 0.842.

1 Introduction

Tagging becomes increasingly popular among web applications, it allows user to describe a document with one or more free-chosen words. Ambiguity in tags leads to inaccurate results in tag-based services. For example, in tag-based document browsing, a.k.a Tag Cloud, one cannot distinguish the documents about the apple fruit or the Apple electronics, since they share the same tag "apple". Solving the ambiguity in tags can help improving the accuracy of related applications, including but not limited to retrieval, browsing and trend discovery.

Disambiguating tags is not the same as disambiguating words. Word disambiguation methods mainly serve the needs of machine translation. They either rely on a human edited dictionary, or need sufficient training samples. Unfortunately, the meaning of tags changes fast, new named entities and new use of old words cause most of the ambiguity. Changing fast means maintaining the meanings in dictionary or annotating training samples are not feasible. We need an unsupervised method to find and distinguish the meanings of tags. Several previous literatures explored the unsupervised word sense disambiguation methods. They either need parallel corpus of different languages, or use labeled examples as seed to extend the training set. It is hard to find aligned multilingual documents with tags. Again labeled examples for tags are expensive to obtain.

In this paper, we propose an unsupervised method to disambiguate tags, specifically tags for text documents. Our method need no dictionary or labeled training data, it consists of two phases. In the first phase, we collect the context words of a tag in all documents, and cluster them with Spectral Clustering, each cluster represents a meaning of the tag. In the second phase, we compare a tag's context words in a document to

V. Matoušek and P. Mautner (Eds.): TSD 2009, LNAI 5729, pp. 139–146, 2009.

each of the clusters, then choose the most similar cluster as the meaning of the tag in this document. Both phases need to compare the semantic similarity between two words. We employ Normalized Google Distance(NGD) [3], one from the family of information distance, to compute the word similarity. NGD uses the results of querying search engines (like Google) as its input, so the similarity is always up-to-date. We evaluate our method with a large blog data set crawled from the web. Experiments show that the method is effective.

To summarize, the main contributions of this paper are 1) proposed an unsupervised method for tag disambiguation, specifically for tags of documents, 2) Evaluated the method on a real-world blog data set. The rest of this paper is organized as follows: Section 2 summarizes the related work. Section 3 describes our method. Section 4 introduces the experiment settings and the results. Section 5 concludes the paper.

2   Related Work

Word Sense Disambiguation(WSD) has been an important topic in NLP for decades. Currently supervised methods have the best performance. They use several local features like surrounding words and part-of-speech tags to train a SVM or Maximum Entropy model, and achieve precision and recall as high as 88%/88% in recent SemEval-2007 Coarse-grained English all words open evaluation [10]. Unsupervised WSD methods also draw a lot of attention since they do not need expensive human labeled data. Parallel multilingual corpora were used to find training samples for supervised learning automatically [2] . Lu et al [6] use Equivalent Pseudowords to extend the training set of supervised method, and got better results than using just the given training set. McCarthy et al [9] combined WordNet similarity and thesaurus learned from raw text to find the predominant sense of a word. This is useful since a baseline method which just assigns the predominant sense to all word tokens has better performance than most of the unsupervised method [10]. Yet for the tag disambiguation task, there are several barriers that make these WSD methods hard to apply. The cost of human labeling for supervised methods is high due to the dynamic nature of tags, and there is no proper parallel corpus, thesaurus or dictionary for unsupervised methods.

Han et al [4] presented a way to disambiguate names in paper citation using k spectral clustering. Like Han et al, we used the spectral method to disambiguate name entities. But our method incorporates Normalized Google Distance as the similarity measure, whereas Han et al use direct word overlap to express the relations. Normalized Google Distance is computed from large corpus so it is better able to overcome the word sparseness problem.

A Yeung et al [1] presented a method to disambiguate collaborative tags. They first built a tag graph, with "co-tagged the same object" relationship as edges between tags, then a community discovery algorithm was applied to the tag graph to find different meanings represented by different communities. Experiments on the del.icio.us online bookmark data showed promising results. Their method requires that tags can cover most aspects of an object, so the graph can be dense enough. In the blog tag situation, where a document have 2-3 tags usually, this requirement is not fulfilled.

3 Methodology

Our method consists of two steps. The first step is using Spectral Clustering to cluster context words of a tag. The second step is comparing a document with a tag's clusters to decide the meaning of the tag. Both clustering and comparing involve with how to measure the semantic similarity between words. In this section, we first introduce the similarity distance we used, along with the reason of choice. Then we describe the process of clustering context words. Finally, we introduce how to determine the meaning of a tag in a document.

3.1 Normalized Google Distance

We use Normalized Google Distance [3] as our similarity distance measure. It computes as follows.

$$NGD(x,y) = \frac{\max\{\log f(x), \log f(y)\} - \log f(x,y)}{\log N - \min\{\log f(x), \log f(y)\}} \tag{1}$$

where $f(x)$ is the approximated number of relevant documents returned by a search engine when searching for word x (for example by querying Google), and N is the approximated total number of documents indexed. Cilibrasi et al [3] derived this measure based on information distance and Kolmogorov complexity. Since the computation of NGD need no human labeling or pre-constructed dictionary, we can get it by indexing a large amount of plain text, which can be done at low cost.

3.2 Cluster Context Words of a Tag

In order to resolve the ambiguity of a tag T, we have to know the context of every sense. A widely accepted assumption is *one sense per context* [7], which means a sense is determined by its context, and only one sense of a word exists within the same context. So our basic idea is to find different dense clusters of T's context words.

We first collect the possible context words C_T of T. We issue a query for T to the search engine, then locate the positions of T in the returned documents. After that, we select every sentence contains T and two neighbor sentences, extract nouns and verbs in these sentences, and sort these context words by their NGD to T.

Then we build a $N \times N$ affinity matrix A to represent the relationship between top N context words, where $A_{i,j} = NGD(w_i, w_j), w_i, w_j \in C_T$. Finding different senses of T can be seen as finding clusters in the matrix A, under the assumption that context words from the same sense have closer semantic distance, while words from the different senses tend to stay apart from each other. This assumption is generally accepted in the WSD literature [12] [16] [5]. There are many clustering methods, such like K-Means, Gaussian Mixture Model, Spectral Clustering and clique-based methods. Among these, Spectral Clustering can find a global optimal solution of k-way clustering, it allows real-valued distances between all objects, which fits well in the k main senses discovery task. We use the algorithm described by Ng et al [11]. It starts by calculating the k largest eigenvalues and corresponding eigenvectors of the following matrix L.

$$L = D^{-\frac{1}{2}} A D^{-\frac{1}{2}} \tag{2}$$

where D is a diagonal matrix, $D_{i,j} = \frac{1}{N} \sum_{k=0}^{N} A_{i,k}$. The k orthogonal eigenvectors $x_1, x_2, \ldots, x_k$ corresponding to the k largest distinct eigenvalues are selected to form a new $N \times k$ matrix X. Then the row vectors of X are normalized to form matrix Y with all N row vectors as unit length vectors. The i-th row vector represents the position of context word w_i in the new lower dimensioned space after spectral projection. Normally words from k main senses of tag T are in nearly orthogonal positions to one another on the unit hypersphere in the new space.

After spectral projection, we pick up the context words C_k for each of the k senses by finding k cluster centriods that are furthest from each other in the new space. Here we use cosine similarity between word vectors as the distance measure. Then, the words belonging to different clusters are sorted by $\frac{p}{d}$, where p is the distance between the word and the cluster centriod, and d is the NGD between the word and T. We use top 10 words as final context words for a sense. To avoid totally irrelevant words, we set a threshold to filter out the words too far away from T under NGD.

In general, it is hard to decide the number of clusters k automatically in unsupervised clustering. A common strategy is to choose a k first, say 2, have a look at the extracted context words, then adjust k to fit the data better. Real world applications may need some human interaction to get fine-tuned results.

3.3 Disambiguation

We determine a tag's meaning by comparing the document with the tag's context words. For a document d with tag T, we use the same method described in Section 3.2 to extract the context words of T in d, and pick n nearest words to the tag measured by NGD as the local context C_{local}. We then compare the average distance between $C_i, 0 < i < k$ and C_{local}, and pick the nearest cluster as the disambiguated meaning $S(T, d)$ of the tag. This can be described as

$$S(T, d) = argmax_i \sum_{w \in C_i} \sum_{w_l \in C_{local}} NGD(w, w_l) \tag{3}$$

Because NGD provides a smoothed similarity between words, the size of local context and per-sense context can be small. We use 10 words as context words.

4 Experiments

4.1 Data Set

We use a collection of real world Chinese blog posts. The data set is crawled from the web, it contains 386,012 posts, all with at least one tag. Tags in the data set are chosen by the author of the post. There are 136,961 distinct tags in total, 2.5 tags for one post in average, the average length of title and content are 8.3 and 796 Chinese characters respectively. All the 386,012 blog posts were segmented and indexed using the open source indexing software Lucene for performing NGD queries.

4.2 Evaluation

Labeling the correct senses of tags by human is a resource consuming job. To make normal scale evaluation possible, Schütze [13] introduced a automatic test set construction method called pseudoword, which is widely accepted in WSD domain [7][14]. Pseudoword combines two or more non-ambiguous words to form a pseudo-ambiguous word, then we know the exact meaning of every occurrence of a pseudo-ambiguous word.

Pseudoword is not the perfect solution for evaluating sense disambiguation, but is a reasonable choice here. Gaustad [15] argued that using the same data set and algorithm, the performance on pseudowords and real words differs, the main reason is pseudoword didn't always model the correlations between senses well. We use pseudo-tags here because it's a rather new application, pseudo-tags provided a possible way for evaluation when we had no annotated corpus. Besides pseudo-tags, we've also run the algorithm on a real world ambiguity tag, and will show the result as an example. In future work, we'll consider evaluation with human efforts.

To construct our test set with ground truth, we select k non-ambiguous tags, replace all occurrences of them with a same pseudo-tag. We build a test set with 27,327 blog posts, in which 36 non-ambiguous tags are combined into 16 pseudo-tags.

4.3 Context Clustering

First we perform the context clustering algorithm on all pseudo-tags. We collect $N = 200 \times k$ context words for every pseudo-tag. If a word appears only in the context of one sense s_i, we consider it a context word of s_i. If a word appears in the context of more than one senses, we consider it a common word. We illustrate the results by plotting all non-common context words of a tag in the k-dimension space. Figure 1a showed a good case of the projected data points, we can see that three original senses of the pesudo-tag (football, lose-weight, stock) are well separated. A bad case is

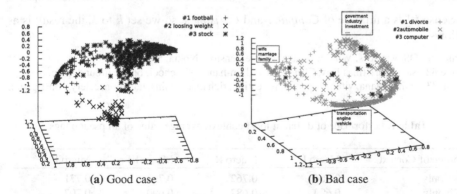

(a) Good case   (b) Bad case

Fig. 1. Good and bad cases of tag sense discovery. In (a), the algorithm puts 3 main different senses of the pseudotag to 3 orthogonal corners respectively. In (b), insufficient context words for sense *computer* leads to wrong clusters. *automobile* and *divorce* are separated, while words for *computer* scattered over the whole area.

illustrated in Figure 1b, the pseudo-tag is (divorce,automobile,computer). In this case there are too few context words for *computer*, the algorithm splits the context words for *automobile* into *industry* and *vehicle*. Overall we clustered 87.5% of the pseudo-tags' meanings correctly.

4.4 Disambiguation

Next we show the result of disambiguation. We see the task as a classification problem, each pseudo-tag is classified into one of its meanings. Thus, we can use evaluation metrics for classification to evaluate the results. Specifically, we use Precision, Recall and F1 to measure the effectiveness of our method.

For comparison, we use the predominant meaning of a tag as the disambiguated meaning. Considering that most of the ambiguous words have skewed senses, the predominant sense method outperforms many unsupervised methods in recent SemEval contest [10] and it is accepted as a baseline in WSD [7].

We compared different sources of context words. Context words can be extracted from the title, co-occurred tags or the content. If a tag T does not appear in the content, we use $TF \times IDF$ to select the top n keywords as context words. When combining different sources of context, we first extract and sort the words from the sources separately, then use a round-robin strategy to pick words from every source in turn until enough words are collected. We use 20 context words. The average macro precision/recall/F1 measure and micro precision of all pseudo-tags using different context sources are shown in Table 1. It is clear that combining tag+title+content has the best performance.

Pseudo-tags provide a test data set with groundtruth, but they are still "pseudo" tags. To know how the algorithm performs on real world data, we chose a typical ambiguity tag, *Apple* to show the result of context clustering. We first chose $k = 2$ as the number of senses, the clusters of context words are as follows, all words are translated to English:

Sense #1 OS, Christmas, Peace, Windows, wireless, java, vista, bluetooth
Sense #2 fruit, banana, vitamin, pectin, for-eating, rich-in, contains, nutrition, celery, constipation

Sense #1 is a mixture of *Computer* and *Christmas*, so we set k to 3, the result is as follows,

Sense #1 OS, wireless, Windows, java, vista,extension, Nokia, body, user
Sense #2 Safe, Christmas Day, Christmas, red, in-hand, off school, on the ground, give, in street
Sense #3 fruit, banana, vitamin, pectin, for eating, rich in, contains, nutrition, celery, constipation

Table 1. Performace of different local context, average value of all pseudo tags

Source of Context	Macro P	Macro R	Macro F1	Micro P
tags only	0.764	0.762	0.762	0.771
title only	0.699	0.682	0.690	0.707
tags + title	0.788	0.786	0.787	0.795
title + content	0.837	0.811	0.824	0.831
tags + title + content	**0.847**	**0.825**	**0.836**	**0.842**
baseline	0.275	0.459	0.341	0.592

In Chinese, 苹果*(apple) and* 平安(safe) are homophones, people occasionally use "apple" to represent wishes for peace in holidays like Christmas. In this example, our method successfully discovered the different senses of tag 苹果(apple).

5 Conclusion

In this paper, we present an unsupervised method for tag disambiguation, and showed its effectiveness on Chinese blog tags. Our method is based on context clustering. Specifically we use Normalized Google Distance as the word similarity measure. The main advantage of the method is that it needs no pre-constructed knowledge such as parallel corpus, dictionary or thesaurus. Quantitive evaluation using pseudo-tags showed that our method can effectively disambiguate tags. We also provided an example on real ambiguity tag.

The method can help to improve tag-based applications, like trend discovery or information retrieval in folksonomy. Future work includes the automatic detection of number of coarse-gained sense, and incorporating other features to enrich the local context. Further evaluation with sufficient human labeled real ambiguity tags is also necessary.

Acknowledgements

This work is supported by the National 863 Project under Grant No. 2007AA01Z148 and the National Science Foundation of China under Grant No. 60873174.

References

1. Yeung, C.A., Gibbins, N., Shadbolt, N.: Tag Meaning Disambiguation through Analysis of Tripartite Structure of Folksonomies. In: Proceedings of IEEE/WIC/ACM International Conferences on Web Intelligence and Intelligent Agent Technology Workshops, pp. 3–6 (2007)
2. Chan, Y.S., Ng, H.T., Zhong, Z.: NUS-PT: Exploiting Parallel Texts for Word Sense Disambiguation in the English All-Words Tasks. In: Proceedings of the Fourth International Workshop on Semantic Evaluations (SemEval 2007), pp. 253–256 (2007)
3. Cilibrasi, R., Vitányi, P.: The Google Similarity Distance. IEEE transactions on Knowledge and Data Engineering 19(3) (2007)
4. Han, H., Zha, H., Giles, L.C.: Name disambiguation in author citations using a K-way spectral clustering method. In: Proceedings of the 5th ACM/IEEE-CS joint conference on Digital libraries (2005)
5. Lin, D., Pantel, P.: Concept Discovery from Text. In: Proceedings of COLING 2002, pp. 577–583 (2002)
6. Lu, Z., Wang, H., Yao, J., Liu, T., Li, S.: An Equivalent Pseudoword Solution to Chinese Word Sense Disambiguation. In: Proceedings of ACL 2007 (2007)
7. Manning, C., Schutze, H.: Foundations of Statistical Natural Language Processing. MIT Press, Cambridge (2001)
8. Marlow, C., Naaman, M., Boyd, D., Davis, M.: HT06, Tagging Paper, Taxonomy, Flickr, Academic Article, To Read. In: Proceedings of the Seventeenth Conference on Hypertext and Hypermedia (2006)

9. McCarthy, D., Koeling, R., Weeds, J., Carroll, J.: Unsupervised Acquisition of Predominant Word Senses. Computational Linguistics 33(4), 553–590 (2007)
10. Navigli, R., Litkowski, K.C., Hargraves, O.: SemEval-2007 Task 07: Coarse-Grained English All-Words Task. In: Proceedings of the Fourth International Workshop on Semantic Evaluations (SemEval 2007), pp. 30–35 (2007)
11. Ng, A.Y., Jordan, M.I., Weiss, Y.: On Spectral Clustering Analysis and an algorithm. In: Proceedings of NIPS (2002)
12. Pereira, F.C.N., Tishby, N., Lee, L.: Distributional Clustering of English Words. In: Proceedings of ACL 1993, pp. 183–193 (1993)
13. Schütze, H.: Context Space. In: Working Notes of the AAAI Fall Symposium on Probabilistic Approaches to Natural Language, pp. 113–120. AAAI Press, Menlo Park (1992)
14. Bordag, S.: Word Sense Induction: Triplet-Based Clustering and Automatic Evaluation. In: Proceedings of the EACL 2006, pp. 137–144 (2006)
15. Gaustad, T.: Statistical Corpus-Based Word Sense Disambiguation: Pseudowords vs. Real Ambiguous Words. In: Proceedings of ACL 2001 Student Research Workshop (2001)
16. Widdows, D., Dorow, B.: A Graph Model for Unsupervised Lexical Acquisition. In: Proceedings of COLING 2002, pp. 1093–1099 (2002)

Intraclausal Coordination and Clause Detection as a Preprocessing Step to Dependency Parsing

Domen Marinčič, Matjaž Gams, and Tomaž Šef

Jozef Stefan Institute, Jamova cesta 39, 1000 Ljubljana, Slovenia
{domen.marincic,matjaz.gams,tomaz.sef}@ijs.si

Abstract. The impact of clause and intraclausal coordination detection to dependency parsing of Slovene is examined. New methods based on machine learning and heuristic rules are proposed for clause and intraclausal coordination detection. They were included in a new dependency parsing algorithm, PACID. For evaluation, Slovene dependency treebank was used. At parsing, 6.4% and 9.2 % relative error reduction was achieved, compared to the dependency parsers MSTP and Malt, respectively.

1 Introduction

Long sentences are complex and hard to parse. By identifying smaller units and parsing them separately, one could decompose a sentence into more easily manageable parts. Clauses and intraclausal coordinations are an example of such units. The first systems for clause detection were only able to retrieve non-embedded, simplex clauses and were based on hand-crafted rules [1,2]. With more recent systems, embedded clauses can be found as well [3]. These algorithms use machine learning (ML) techniques. The research for intraclausal coordination detection is scarcer. The algorithm described in [4] is limited to nominal coordinations since it is based on the semantic similarity of conjoined nouns.

Clause identification has been mainly treated as a separate task – with some exceptions. In [1], a parsing system using a clause filter to delimit non-embedded clauses before parsing is described. In [5], a system for incremental, clause-by-clause parsing of spoken monologue is presented. In [6], we can find a short description of a rule-based parser where clause identification is included in the parsing process.

We describe a new Algorithm for *PA*rsing with *C*lause and *I*ntraclausal coordination *D*etection – PACID, which joins clause and intraclausal coordination detection with dependency parsing. Slovene Dependency Treebank [7] was used as training data and gold standard for evaluation.

2 The PACID Algorithm

The algorithm consists of two phases:

- Detection and reduction of clauses and intraclausal coordinations
- Dependency tree construction

V. Matoušek and P. Mautner (Eds.): TSD 2009, LNAI 5729, pp. 147–153, 2009.

The first phase is a loop, which begins by splitting the sentence into segments as proposed by [8]. Punctuation tokens and conjunctions represent delimiters between the segments. Then, intraclausal coordinations are detected and reduced. In the next step, the sentence is split into segments again. At the end, clause detection and reduction is performed. The loop iterates until no more units can be retrieved or only one segment remains. Several iterations enable the algorithm to retrieve embedded units.

In the second phase, the dependency tree is built by parsing the reduced units and merging them together with the sentence tree. In Figure 1, an example how the algorithm processes a sentence is presented.

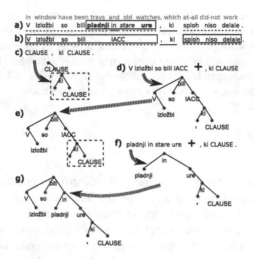

Fig. 1. The PACID algorithm. Segments and delimiters are underlined with dashed and solid lines, respectively. The reduced units are marked with rectangles. The head words are emphasized. The English word-by-word translation of the sentence in Slovene is provided in the top of the figure.

2.1 Intraclausal Coordination Detection, the First Phase

PACID works on prepositional, nominal and adjectival coordinations. First, candidate groups of head words of conjunct phrases ordered by their appearance in the sentence, are formed. A group is identified with the following heuristic rules:

- All head words must have the same POS and case.
- Between each pair of head words, there has to be a comma or a coordinating conjunction, where one or both surrounding segments contain no finite verbs.
- No colons, dashes, semicolons, brackets, finite verbs, relative pronouns and subordinating conjunctions are allowed between the head words.

Then, the the set of candidate groups of head words is filtered by ML classifiers. Each candidate group $(w_1,..., w_n)$ is converted to pairs of head words (w_i, w_{i+1}), $0 \geq i < n$. Each pair is classified by an AdaBoostM1 classifier with the J48 decision tree as the core classifier [9]. Separate classifiers were trained for prepositions, nouns and adjectives. The instances for training the classifiers were extracted from SDT. To describe a head word pair as a ML instance, two sections containing the tokens between the head words are formed. The section A consists of the tokens between the first head word and the delimiter. The section B consists of the tokens between the delimiter and the second head word. The attributes describing one section are the following:

- the presence of a preposition/an adverb in the section (two attributes, binary values),
- the presence of a noun/an adjective matching/non-matching with the head word in case, number and gender in the section (four attributes, binary values),
- the number of words in the section (one attribute, values: 0, 1, 2, >2),

In the instance description, the attributes appear two times, once for each section. To complete the description, the class attribute (binary values) is added. If all head word pairs are classified positively, the tokens spanning from the leftmost to the rightmost head word are reduced to a meta node named 'IACC'. In Figure 1a there is a group of two nouns, 'ure' and 'pladnji', the section A is empty, the section B is italicized. The tokens are reduced to the meta node 'IACC' in Figure 1b.

2.2   Clause Detection, the First Phase

At clause detection, the segments containing finite verbs are the candidates for reduction. In Figure 1b, there are two candidates, underlined with dashed lines.

First, the following heuristic rule is applied to a candidate: if the four neighboring segments (two to the left, two to the right) contain finite verbs, the segment is reduced to a meta node.

Otherwise, the segment is classified by a ML classifier. Two classifiers (both AdaBoost algorithm with J48 decision trees as base classifiers) were used: the first one when both neighboring segments are verb segments and the second one when at least one neighbor is a non-verb segment. To describe the segment as a ML instance, the delimiter preceding it, two segments left and two right with the preceding delimiters are included. For each pair delimiter/segment the following attribute set is used:

- The presence of a coordinating/subordinating conjunction or a pronoun (three attributes, binary values).
- The presence of a punctuation token (values: 'none', 'comma', 'colon_or_semicolon', 'other').
- Auxiliary verb before participle (values: 'yes', 'no', 'not_def').
- Segment type (values: 'verb', 'non_verb', 'not_def').
- Possible existence of a crossing intraclausal coordination, where some head words lie in the described segment and the others lie in the neighboring segments (binary values)

Accordingly, in the attribute-value vector, each of the attributes listed above appears five times and the class attribute (binary values) is added to get the complete description of the instance. The positively classified segments are reduced to meta nodes named 'CLAUSE' (Figure 1c). The punctuation tokens and the conjunctions immediately in front of the 'CLAUSE' meta nodes do not play the role of delimiters any more in the next iterations of the first stage.

2.3 The Second Phase

In this phase, three base parsers and the rule-based parser are used to create the sentence tree. For base parsers, we can use any parser that can be trained on a dependency treebank.

At the beginning, the sequence of tokens that remains unreduced is parsed, producing the initial version of the sentence tree (Figure 1c). The base parser for sentence trees is used. If the tree consists of meta nodes and punctuation tokens only, certain errors can be detected, e.g. the meta node of a subordinate clause is the root of the initial tree. In such cases, the initial sentence tree is built from scratch by a simple, newly developed, rule-based parser.

The phase continues with a loop processing the meta nodes inside the sentence tree. In Figure 1, two iterations of the loop are shown. The steps d and e correspond to the first iteration, while the steps g and h correspond to the second iteration. In each iteration, first, the paths from the root of the sentence tree to the meta nodes are examined. If a path contains only one meta node – the one at the end of the path, this meta node is processed. In the first iteration in Figure 1, this is the upper 'CLAUSE' meta node of the initial sentence tree (Figure 1c), while in the second iteration, this is the 'IACC' meta node.

The units corresponding to the meta nodes are joined with the tokens of the meta node subtree, excluding the meta node itself (marked with the dashed rectangles in Figure 1). The new sequences are then parsed with the clause or intraclausal coordination base parser (Figure 1, steps d and f). At the end of the iteration, the meta node subtrees in the sentence tree are replaced with the new subtrees (Figure 1, steps e and g). When there are no more meta nodes in the sentence tree, the iteration finishes and the final dependency tree of the sentence is created (the third and final iteration is missing in Figure 1).

3   Evaluation

Experiments were performed to evaluate (i) parsing and (ii) clause and intraclausal coordination retrieval, using 10-fold cross validation. Slovene dependency treebank (SDT) served as the train and test set. The text in SDT (55,208 tokens) originates from the Orwell's novel "1984" (38,646 tokens) and from the SVEZ-IJS corpus [10] (16,562 tokens). Each experiment was carried out either with the Maximum spanning tree parser (MSTP) [11] or the Malt parser [12] used as base parsers.

3.1   Parsing

Two sets of experiments were performed. As the measure of parsing accuracy, unlabeled attachment score (UAS) was used, which equals the quotient between scoring tokens assigned the correct head and all scoring tokens. All words were scoring tokens. In addition, all the punctuation tokens that are roots of coordination subtrees in the gold standard were scoring tokens as well. As the baseline results, the accuracies of the plain MSTP and Malt parsers were chosen.

In the first set of experiments, various versions of the PACID algorithm were compared. For these experiments, the portion of SDT from the novel "1984" was used.

The results are presented in Table 1. First, the accuracy of the baseline parsers without clause and intraclausal coordination detection was measured. Second, PACID including intraclausal coordination detection, without clause detection was used. In the third experiment, both clauses and intraclausal coordinations were retrieved; the rule-based parser was switched off and all candidates for clauses and intraclausal coordinations were reduced, e.g. no filtering by ML classifiers. In the fourth version, ML classifiers were turned on while the rule-based parser was still not used. Finally, the accuracy of the full version was measured. The results marked with * are statistically significantly different (95% confidence level, resampled t-test [9]) from the baseline result, comparing the experiments with the same type of base parsers used. Furthermore, the results using clause detection (last tree rows) are statistically significantly different from the results using intraclausal coordination detection only (not explicitly marked in the table). However, there is unfortunately no statistically significant difference between the results of the versions with clause detection.

Table 1. Parsing accuracy

Parsing algorithm	Malt	MSTP
Baseline	73,28 %	80,24 %
PACID, intraclausal coordination detection only	*72,42 %	80,57 %
PACID, no ML classifiers, no rule-based parser	*74,63 %	*81,05 %
PACID, with ML classifiers, no rule-based parser	*74,83 %	*81,34 %
PACID, full version	*75,19 %	*81,51 %

The full version achieved the highest accuracy, 6.4% and 9.2% relative decrease of error in comparison with the baseline result, using the MSTP and Malt base parsers, respectively. By comparing the version without the rule-based parser to the full version, we can conclude that the use of the rule-based parser increases the accuracy. In [13], the accuracy of the plain MSTP and Malt parsers on sentences of various complexity was examined. The lowest accuracy was measured on the sentences without verbs, which are normally the shortest ones. Since the target of the rule-based parser are short sequences of meta nodes and punctuation tokens, the rule-based parser might compensate for the inability of the base parsers to deal with very simple sentences effectively. Using the PACID algorithm without ML classifiers, the errors of retrieving false positives among clauses and intraclausal coordinations caused the algorithm to perform worse than the previously described versions. The version without clause detection using the Malt base parsers shows worse results than baseline.

When using the MSTP base parsers, better results are achieved – for the price of larger time consumption. The complexity of the MSTP parser is $O(n^2)$ compared to the complexity $O(n)$ of the Malt parser, n being the number of tokens in a sentence.

The motivation to do the experiments only on the part of SDT originating from the novel "1984" was the relatively unfavorable clausal structure of the second part coming from the corpus SVEZ-IJS. In the "1984" part (1999 sentences), we can find 3718 clauses in sentences with two or more clauses, while in the SVEZ-IJS part (820 sentences) only 764 such clauses can be found. This means that in the SVEZ-IJS part, there are substantially less sentences, where the PACID algorithm can contribute to improving parsing accuracy.

The second set of experiments was done on the whole SDT. The baseline parsers MSTP and Malt parser achieved the accuracies of 80.34% and 73.91% respectively, while the full version of the PACID algorithm yielded the accuracy of 81.06% and 75.52%, using the MSTP and Malt base parsers, respectively. Measured on the whole SDT, the improvement of accuracy by the use of the PACID algorithm is smaller than the improvement measured on the first part of SDT only.

3.2   Clause and Intraclausal Coordination Retrieval

In SDT, clauses and intraclausal coordinations are represented as subtrees. If the subtree's root is a finite verb or a subordinating conjunction with a finite verb as a child, the nodes of the subtree constitute a clause. If the incoming edge of the subtree's root has the label 'Coord', and none of its childs is a finite verb, the subtree represents an intraclausal coordination. Retrieval of clauses and intraclausal coordinations was evaluated by comparing the subtrees in the parsed trees to the subtrees in the gold standard. The "1984" part of SDT was used in experiments described here.

For clause retrieval, only the sentences containing two or more clauses were considered (3718 clauses in the gold standard). Clause subtrees were compared node by node, excluding punctuation token nodes. In the second column of Table 2, there is the number of gold standard clause subtrees which were found identical in the parsed trees.

The gold standard contains 615 intraclausal coordinations of prepositions, nouns and adjectives. At the evaluation, groups of head words were compared extracted from parsed and gold trees. In the third column of Table 2, there is the number of gold head word groups which were found identical in the parsed trees.

As expected, the results show the same tendency as in Table 1. The PACID algorithm using the MSTP base parsers achieves better results than the plain parsers. When using the Malt base parsers, PACID is less successful in intraclausal coordination retrieval as the plain Malt parser, but still better at retrieving clauses.

Table 2. Clause and intraclausal coordination retrieval

Parsing algorithm	Clauses	Intraclausal coordinations
Plain MSTP	1961	220
PACID, full version (MSTP base parsers)	2233	427
Plain Malt	860	419
PACID, full version (Malt base parsers)	1401	388

4   Conclusion

The experiments show that by dividing complex sentences to smaller, more easily manageable units one can improve parsing accuracy. For existing parsers, some tasks like determining the clausal structure of a sentence seem to be hard to discover even if trained on text annotated with dependency trees. The parsers were upgraded with the methods based on machine learning and heuristic rules which include additional information provided by the richly inflected languages. The upgrade enables to carry out such tasks more accurately.

The time complexity of clause and intraclausal coordination detection methods is $O(n)$, where n is the length of the sentence in tokens, meaning that additional computational time required by the upgrade is acceptable.

There are further ways how to improve the PACID algorithm. The set of attributes used in ML classifiers could be extended. The rule-based parser could be more elaborated. Maybe the most important improvement would be to change the rigid treatment of the reduced units, which are either declared to be reduced or not. It would be worth trying to encode the knowledge about clause and intraclausal structure as additional features of sentence tokens. As such, a parser would be able to combine this knowledge with other information about the parsed text more smoothly.

References

1. Abney, S.P.: Rapid Incremental Parsing with Repair. In: Proceedings of the 6th New OED Conference, pp. 1–9 (1990)
2. Ejerhed, E.I.: Finding clauses in unrestricted text by finitary and stochastic methods. In: Proceedings of the second conference on Applied natural language processing, pp. 219–227 (1988)
3. Tjong Kim Sang, E.F.: Memory-Based Shallow Parsing. Journal of Machine Learning Research 2, 559–594 (2002)
4. Hogan, D.: Empirical measurements of lexical similarity in noun phrase conjuncts. In: Proceedings of the 45th Annual Meeting of the Association for Computational Linguistics (ACL), pp. 149–152 (2007)
5. Ohno, T., Matsubara, S., Kashioka, H., Maruyama, T., Inagaki, Y.: Incremental Dependency Parsing of Japanese Spoken Monologue Based on Clause Boundaries. In: Proceedings of the 21st International Conference on Computational Linguistics and the 44th annual meeting of the Association for Computational Linguistics (ACL), pp. 169–176 (2006)
6. Holán, T., Žabokrtský, Z.: Combining czech dependency parsers. In: Sojka, P., Kopeček, I., Pala, K. (eds.) TSD 2006. LNCS (LNAI), vol. 4188, pp. 95–102. Springer, Heidelberg (2006)
7. Džeroski, S., Erjavec, T., Ledinek, N., Pajas, P., Žabokrtský, Z., Žele, A.: Towards a Slovene Dependency Treebank. In: Proceedings of the 5th International Conference on Language Resources and Evaluation (LREC), pp. 1388–1391 (2006)
8. Kuboň, V., Lopatková, M., Plátek, M., Pognan, P.: Segmentation of complex sentences. In: Sojka, P., Kopeček, I., Pala, K. (eds.) TSD 2006. LNCS (LNAI), vol. 4188, pp. 151–158. Springer, Heidelberg (2006)
9. Witten, I.H., Frank, E.: Data Mining: Practical Machine Learning Tools and Techniques, 2nd edn. Morgan Kaufmann Publishers, San Francisco (2005)
10. Erjavec, T.: The English-Slovene ACQUIS Corpus. In: Proceedings of the 5th International Conference on Language Resources and Evaluation (LREC), pp. 2138–2141 (2006)
11. McDonald, R., Pereira, F., Ribarov, K., Hajič, J.: Non-projective Dependency Parsing Using Spanning Tree Algorithms. In: Proceedings of the Joint Conference on Human Language Technology and Empirical Methods in Natural Language Processing (HLT-EMNLP), pp. 523–530 (2005)
12. Nivre, J.: Inductive Dependency Parsing. Springer, Heidelberg (2006)
13. Marinčič, D., Gams, M., Šef, T.: How much can clause identification help to improve dependency parsing? In: Proceedings of the 10th International Multiconference Information Society (IS), pp. 92–94 (2007)

Transcription of Catalan Broadcast Conversation

Henrik Schulz[1], José A. R. Fonollosa[1], and David Rybach[2]

[1] Technical University of Catalunya (UPC)
TALP Research Center
08034 Barcelona, Spain
{hschulz,adrian}@gps.tsc.upc.edu
[2] RWTH Aachen University
Human Language Technology and Pattern Recognition
52056 Aachen, Germany
rybach@i6.informatik.rwth-aachen.de

Abstract. The paper describes aspects, methods and results of the development of an automatic transcription system for Catalan broadcast conversation by means of speech recognition. Emphasis is given to Catalan language, acoustic and language modelling methods and recognition. Results are discussed in context of phenomena and challenges in spontaneous speech, in particular regarding phoneme duration and feature space reduction.

1 Introduction

The transcription of spontaneous speech still poses a challenge to state-of-the-art methods in automatic speech recognition. Spontaneous speech exhibits a significant increase in intra-speaker variation, in speaking style and speaking rate during its term. It involves phenomena such as repetition, repair, hesitation, incompleteness and disfluencies. Furthermore the reduction in spectral or feature space respectively, and in duration. The paper focuses on aspects of the development of a transcription system for Catalan broadcast conversations by means of automatic speech recognition carried out in the framework of the TECHNOPARLA project [1].

Catalan, mainly spoken in Catalonia - a north-eastern region of Spain - and Andorra, is a Romance language. As its geographic proximity suggests, Catalan shares several acoustic phonetic features and lexical properties with its neighbouring Romance languages such as French, Italian, Occitan and Spanish. Nevertheless there are fundamental differences to all of them. Substantial dialectual differences devide the language into an eastern and western group on the basis of phonology as well as verb morphology. The eastern dialect includes Northern Catalan (French Catalonia), Central Catalan (the eastern part of Catalonia), Balearic, and Alguerès limited to Alghero (Sardinia). The western dialect includes North-western Catalan and Valencian (south-west Catalonia). Catalan shares many common lexical properties with the languages of Occitan, French, and Italian which are not shared with Spanish or Portuguese. In comparison with Spanish that has a faint vowel reduction in unstressed positions, Catalan exposes vowel reduction in various varieties - in particular with the presence or absence of the neutral vowel "schwa" /@/. More specifically, the appearance of a neutral vowel in reduced

V. Matoušek and P. Mautner (Eds.): TSD 2009, LNAI 5729, pp. 154–161, 2009.

position in eastern Catalan is regarded as a fundamental distinction to western Catalan. Among the eastern dialects, Balearic allows the neutral vowel in stressed position unlike Central Catalan and the western dialects [2]. The voiced labiodental fricative /v/ is confined to Balearic and northern Valencian, while in the remaining dialects the sound converges as bilabial /B/ [3]. In Eastern Catalan, the Nasals /m/ (bilabial), /n/ (alveolar), /J/ (palatal), and /N/ (velar) appear in final position. /m/, /n/, and /J/ also appears intervocalically. /N/ is only found word internally preceding /k/ [4]. The voiced alveolar liquid /rr/ in word final position only appears to be pronounced in Valencian. Furthermore, a word final voiceless dental stop /t/ is omitted in the Eastern and Northern dialectual region.

The subsequent sections deal with major aspects of feature extraction, acoustic and language modelling, and recognition. Finally, the recognition results are discussed and put into context by examining phenomena of spontaneous speech, assessing feature distribution, duration and disfluencies of speech in broadcast conversation.

The ASR acoustic model (AM) training and decoding subsystem has been developed in the RWTH Open Source ASR framework [5].

2   Broadcast Conversational Speech

The broadcast conversational speech data used during these studies originate from 29 hours of transcribed Catalan television debates (known as Àgora), 16% interferred with background music, 4% with overlapping speech and 3% originating from replayed telephony speech. The debates exhibit sporadic applause, rustle, laughing, or harrumph of the participants. Segments containing background music, speaker overlap, and telephony speech have been excluded at this stage, and are subject of separate studies. Short term events of the same remained in the data, since a removal of affected words may fragment the recordings. Speakers intermittently also tend to use Spanish words in conversations due to their virtual bilinguality. Also Spanish proper names remain as such. The gender distributes to $1/3$ female, $2/3$ male respectively. The speaking style features 95% spontaneous speech, the remainder planned speech. Most speakers are not considered professional.

3   Acoustic Model

An initial Catalan acoustic model (AM) was derived from a Spanish AM that was developed during the project TC-STAR [6]. While carrying out the first alignment iteration, Catalan allophones that extend the original set of Spanish allophones borrow the appropriate models from the original AM instead of following the approach of using monophone context independent models to bootstrap context dependent models.

The original feature space comprises 16 Mel frequency cepstral coefficients (MFCC) extended by a voicedness feature, whereas the cepstral coefficiants are subject to mean and variance normalisation. Vocal tract length normalization (VTLN) is applied to the filterbank. The temporal context is preserved by concatenating the features of 9 consecutive frames. Subsequently a linear transformation reduces the dimensionality.

A training phase is carried out by several steps: Prior to the AM estimation, a linear discriminative analysis estimates a feature space projection matrix (LDA). Furthermore, a new phonetic classification and regression tree (CART) is grown followed by Gaussian mixture estimation, that iteratively splits and refines the Gaussian mixture models.

The AM provides context dependent semi-tied continuous density HMM using a 6-state topology for each tri-phone. Their emission probabilities are modelled with Gaussian mixtures sharing one common diagonal covariance matrix. A CART ties the HMM states to generalized triphone states.

Based on the broadcast conversational training data, the baseline AM has been estimated passing a number of iterations of re-alignment and intermediate model estimation, whereas LDA and CART are re-estimated twice per iteration.

VTLN Gaussian mixture classifier estimation during training employs solely normalised MFCC.

The iterative training procedure has been enhanced by using Maximimum Likelihood Linear Regression (MLLR) [7] adapted AM during the first Viterbi alignment of acoustic training data within an iteration.

In addition to the speaker independent AM, Speaker Adaptive Training (SAT) [8] has been employed, aiming to model less speaker specific variation in the (SAT) AM. It compensates the loss of speaker specificity of the SAT AM through speaker specific feature space transforms using CMLLR [7]. The transforms are estimated using a compact AM, i.e. a single Gaussian AM, with minimal speaker discriminance. The SAT formalism relies on the concept of acoustic adaptation and is as such applied estimating the feature transforms of corresponding speakers in recognition.

In summary, AM estimation has been carried out for 2 types: a speaker independent AM and a SAT-AM.

Beside the training data of broadcast conversation (ÁGORA) - statistics outlined in Table 1, 2 additional rich context speech corpora were evaluated selectively for training: a read speech corpus (FREESPEECH) and spontaneous utterances of the SpeeCon corpus (SPEECON-S), see Table 2. The FREESPEECH corpus in its entirety displayed a degradation of accuracy, and therefore is not further described.

Table 1. Statistics on acoustic model training data ÀGORA

Transcribed Data [h]	20
# Segments	21420
# Speakers	275
# Running Words	272k

Table 2. Statistics on acoustic model training data SPEECON-S

Transcribed Data [h]	31
# Segments	11190
# Speakers	140
# Running Words	280k

Comparing the ratio of number of running words and total duration in Table 1 and 2 indicate significant differences in speed, although the speaking style for both is considered spontaneous.

The phoneme set contains 39 phonemes + 6 auxiliary units for silence, stationary noise, filled pauses and hesitations, as well as speaker and intermittant noise. Pronunciations were modelled with the UPC rule based phonetizer considering the 4 dialectual regions Eastern, Valencian, Balearic and North-Western Catalan in training and recognition.

4 Language Model and Vocabulary

Language model and vocabulary for recognition are derived from a textual corpus, composed of articles of the online edition of 'El Periodico', a weekly journal published in Catalan and Spanish. It encompasses 10 subsets, each focused on a separate topic with a total size of 43.7 million words, 1.8 million sentences respectively. The 4-gram backing-off language model comprises about 10.1 M multi-grams and achieves minimal perplexity (PPL) with a linear discounting and modified Kneser-Ney smoothing methodology. The estimation of language models is carried out with the SRI LM toolkit [9]. The lexicon contains the 50k most frequent words of the 'El Periodico' corpus. As for AM training, each word received multiple phonetic transcriptions.

5 Recognition and Results

The recognition follows a multi-pass approach, depicted in Figure 1, i.e. a first pass using the speaker independent AM, followed by segmentation and clustering of segments, a second and third pass, both applying the SAT based AM. Whereas the corresponding feature space transforms for a speaker cluster are again estimated using CMLLR. The third pass receives a model parameter adaptation by means of MLLR [10]. Both last passes derive their adaptation transform estimates from unsupervised transcriptions of their previous recognition pass.

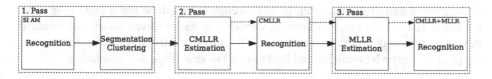

Fig. 1. Multi-pass system architecture for recognition

The overall recognition results in Table 4 and 5 - μ denotes the word-error-rate (WER) across the two sets, μ_s, σ_s the mean and standard deviation of WER across speakers - are fairly high at first glance, but need to be reviewed considering three major aspects: the phenomena of broadcast conversational speech, the amount of available adequate acoustic and language model training data, and the composition of training and testing data.

The development set, although biased due to parameter optimization, poses a larger challenge than the test set. Furthermore, the higher standard deviation across the individual speaker error rates in the development set suggests speakers of particular challenge. A larger perplexity (PPL) and out-of-vocabulary rate (OOV), as indicated in Table 3 may additionally account for the differences.

Table 3. Statistics on development and test set for recognition

	Dev-Set	Test-Set
Duration [h]	0:45	1:15
# Speakers	10	17
# Running Words	8120	14916
μ [s] speaker duration	227	265
σ [s] speaker duration	95	142
OOV [%]	4.2	3.5
PPL	223.7	199.6

Table 4. Recognition results in multi-pass system architecture using ÀGORA Corpus

WER %	Dev-Set			Test-Set		
	μ	μ_s	σ_s	μ	μ_s	σ_s
1. Pass	38.1	37.6	9.8	34.2	33.1	7.6
2. Pass	35.9	35.2	9.7	30.8	29.3	7.4
3. Pass	35.1	34.9	9.5	30.2	28.9	7.3

Table 5. Recognition results in multi-pass system architecture using ÀGORA and SPEECON-S Corpus

WER %	Dev-Set			Test-Set		
	μ	μ_s	σ_s	μ	μ_s	σ_s
1. Pass	34.2	32.2	9.4	28.2	27.5	6.6
2. Pass	33.9	32.0	9.4	26.1	25.5	6.3
3. Pass	33.4	31.5	9.1	25.8	25.2	6.2

Although the SPEECON-S data differ in the level of spontaneity (data collection environment) from those of ÀGORA, the extension of the acoustic training data provides an improvement of relative 17%.

Comparing the results of the speaker independent recognition of the 1. pass with those using the SAT AM in 2nd and 3rd pass in Table 5, there are larger improvements. As both, the 2nd and 3rd pass use speaker adaptation based on previously obtained unsupervised transcriptions, potential improvements tend to be lower due to the overall lower level of accuracy. Moreover, considering the observed mean and standard deviation for speaker durations in Table 3, the estimated transformations for speaker adaptation may be less reliable and lead to non-favourable speaker adaptation.

6   Discussion

In broadcast conversation, speech exhibits various speaking styles with a continuous and frequent change. These can be qualified as planned, extemporaneous or highly spontaneous. Putting the results into context, three major phenomena were assessed: duration reduction, feature distribution reduction, as well as ratios of filled pauses, mispronunciations and word fragments.

In order to qualify the exposed speaking style for the conversational broadcast transcription task, duration and feature space were examined, and compared to those of read speech. The latter was retained from the Catalan FREESPEECH database comprising read-aloud sentences. As auxiliary experiments indicated, the accuracy obtained for this task was above 95% WER.

Duration reduction for both vowels and consonants is a known phenomenon in spontaneous speech [11]. Phoneme durations have been obtained from pruned

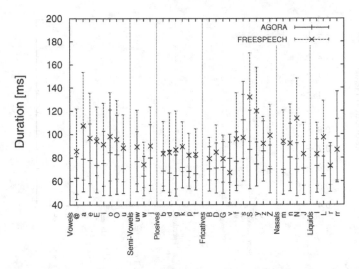

Fig. 2. Mean phoneme durations of broadcast conversational and read speech

forced alignments. Figure 2 depicts the duration of phonemes regarding read speech (FREESPEECH) and spontaneous broadcast conversation (ÁGORA). Speech in conversational broadcast exhibits a significantly lower mean duration for all phonemes and an increased standard deviation compared to read speech. The increased standard deviation suggests a significant higher variability of the exposed speech in broadcast conversation but also an alteration of its style.

The standard deviations indicate a blurred transition between the two. This fact and the noticeable high variation in phoneme duration of broadcast conversation suggests a methodological change in modelling durations. The HMM topology as mentioned above, also referred to as One-Skip HMM, receives a global set of transition probabilities. Noticing variation in speaking style, these parameters should be instantaneously adaptable, specific to phoneme or allophone respectively.

A feature distribution analysis compares feature distributions of each phoneme given spontaneous broadcast conversational and read speech. The phoneme specific feature distributions have been estimated based on labeled feature vectors containing 16 Mel-frequency cepstral coefficients (MFCC), whereas the labels originate from the pruned forced alignments. The ratio of phoneme feature distributions has been defined according to [12] as $||\mu_p(C) - \mu(C)||/||\mu_p(R) - \mu(R)||$, whereas μ_p denotes the center of distribution of phoneme p given broadcast conversational speech (C), and read speech (R) respectively. $\mu(.)$ is the average of the phoneme specific means. The phoneme feature distribution ratios shown in Figure 3 indicate significant differences of MFCC feature distributions for all phonemes in broadcast conversation compared to read speech, in most cases depicting a large reduction. As suggested in [12], the reduction in feature distribution ratio correlates with a loss in accuracy.

At last, the fraction of filled pauses, word fragments and mispronunciations for broadcast conversational speech and read speech was determined from their corresponding transcriptions. Linguistically, the broadcast conversations possess frequent

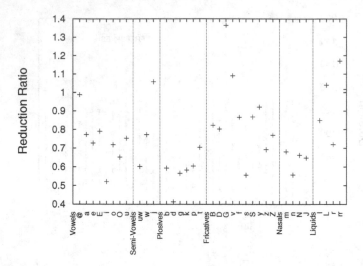

Fig. 3. Phoneme feature distribution ratios between broadcast conversational and read speech

repetition and repairs. Mispronunciations and incompleteness encompass 3.6% of the transcribed spoken events, filled pauses 6.5% - both emphasising the spontaneity of the language. On the other hand, read speech exhibits linguistically neither repetition nor repair. The proportion of mispronunciations and incompleteness is below 0.3%, the one of filled pauses 0.8%. Differences in these ratios emphasises the assessment above.

7 Conclusion

Catalan, as a regional language poses the issue of availability of large amounts of appropriate data. Recent evaluations in broadcast conversational respectively spontaneous speech operate with an amount of AM training data with a factor 4 to 20. Given the high variability in feature space of spontaneous broadcast conversations, larger amounts of acoustic training data are desireable to estimate models and transforms more reliably. As the language model corpus is derived from textual written language, the phenomena addressed above have not been modelled. OOV and PPL still exhibit a lack of appropriate in domain data for both LM and vocabulary.

The results are considered as baseline and encourage for further efforts towards approaches to tackle the problem of acoustic and linguistic data sparseness, discriminativeness of features particular of spontaneous speech.

References

1. Schulz, H., Costa-Jussà, M.R., Fonollosa, J.A.R.: TECNOPARLA - Speech Technologies for Catalan and its Application to Speech-to-Speech Translation. In: Procesamiento del Lenguaje Natural (2008)
2. Herrick, D.: An Acoustic Analysis of Phonological Vowel Reductio. In: SIx Varieties of Catalan. PhD thesis, University of California, Santa Cruz (September 2003)

3. Wheeler, M.W.: The Phonology of Catalan. Oxford University Press, Oxford (2005)
4. Recasens, D.: Place cues for nasal consonants with special reference to Catalan. Journal of the Acoustic Society of America 73, 1346–1353 (1983)
5. N.N.: The RWTH Aachen University Speech Recognition System (November 2008), http://www-i6.informatik.rwth-aachen.de/rwth-asr
6. Lööf, J., Gollan, C., Hahn, S., Heigold, G., Hoffmeister, B., Plahl, C., Rybach, D., Schlüter, R., Ney, H.: The RWTH 2007 TC-STAR Evaluation System for European English and Spanish. In: Interspeech, Antwerp, Belgium, pp. 2145–2148 (August 2007)
7. Gales, M.: Maximum likelihood linear transformations for HMM-based speech recognition. Computer Speech and Language 12(2), 75–98 (1998)
8. Anastasakos, T., McDonough, J., Schwartz, R., Makhoul, J.: A Compact Model for Speaker-Adaptive Training. In: Proc. ICSLP, pp. 1137–1140 (1996)
9. Stolcke, A.: SRILM-an Extensible Language Modeling Toolkit. In: Seventh International Conference on Spoken Language Processing, ISCA (2002)
10. Leggetter, C., Woodland, P.: Maximum likelihood linear regression for speaker adaptation of HMMs. Computer Speech and Language 9, 171–186 (1995)
11. van Son, R.J.J.H., van Santen, J.P.H.: Strong Interaction Between Factors Influencing Consonant Duration. In: EUROSPEECH 1997, pp. 319–322 (1997)
12. Furui, S., Nakamura, M., Ichiba, T., Iwano, K.: Why is the recognition of spontaneous speech so hard? In: Matoušek, V., Mautner, P., Pavelka, T. (eds.) TSD 2005. LNCS (LNAI), vol. 3658, pp. 9–22. Springer, Heidelberg (2005)

An Analysis of the Impact of Ambiguity
on Automatic Humour Recognition*

Antonio Reyes, Davide Buscaldi, and Paolo Rosso

Natural Language Engineering Lab - ELIRF
Departamento de Sistemas Informáticos y Computación
Universidad Politécnica de Valencia, Spain
{areyes,dbuscaldi,prosso}@dsic.upv.es

Abstract. One of the most amazing characteristics that defines the human being is humour. Its analysis implies a set of subjective and fuzzy factors, such
as the linguistic, psychological or sociological variables that produce it. This is
one of the reasons why its automatic processing seems to be not straightforward.
However, recent researches in the Natural Language Processing area have shown
that humour can automatically be generated and recognised with success. On the
basis of those achievements, in this study we present the experiments we have
carried out on a collection of Italian texts in order to investigate how to characterize humour through the study of the ambiguity, especially with respect to morphosyntactic and syntactic ambiguity. The results we have obtained show that it
is possible to differentiate humorous from non humorous data through features
like perplexity or sentence complexity.

1   Introduction

Human behaviour represents a great challenge for all humanistic and scientific disci-
plines. Language, feelings, emotions or beliefs are a small sample within a complex
spiderweb of phenomena that must be studied and analysed in order to be able to understand such behaviour. The task is not easy but a huge effort has been made in several
areas for contributing to the development of knowledge about the different processes,
facts and relations that constitute human beings. On the matter, Natural Language Processing allows us to approach human behaviour from new angles. For instance, NLP has
successfully shown how some human attributes can be simulated by computers. A sample is the language: some time ago the fact of talking with a computer was oneiric.
Today, this dream is reachable, such as the scopes of research. This means that we
may go beyond and be able to identify the mood, the way of thinking or, as in our
case, the humour; task that seems oneiric too. However, recent research works [5], [6],
[10], [11] and [13] have shown that is feasible to generate and recognise humour with
a computer for different purposes such as information retrieval, machine translation,
human-computer interaction, e-commerce, etc.

---

* We would like to thank Fabio Zanzotto for kindly providing the Chaos Parser. The MiDEs
  (CICYT TIN2006-15265-C06) and TeLMoSis (UPV PAID083294) research projects have partially funded this work.

V. Matoušek and P. Mautner (Eds.): TSD 2009, LNAI 5729, pp. 162–169, 2009.

On the basis of the assumptions and results derived from the above research works, we investigated how to identify the features of a set of Italian sentences in order to characterise the humour of this data and to be able to automatically recognise it. In this framework, the rest of the paper is organized as follows. Section 2 introduces some preliminary concepts and describes the state-of-the-art in Computational Humour. Section 3 underlines our initial assumptions on the role of morphosyntactic ambiguity and the aim of our research. Section 4 describes the data sets and all the experiments we carried out. Section 5 presents the discussion of the results and the implications they suggest. Finally, in Section 6 we draw some conclusions and address further work.

2   Computational Humour

One of the characteristics which defines us as human beings and social entities is a very complex concept that we use in our everyday lives and activities: humour. This characteristic, so subjective and fuzzy such as the human behaviour, seems to be hardly translated into a computational perspective. Nevertheless, despite the difficulty the task involves, some recent research works have shown that is feasible to address the problem of automatic humour processing through machine learning techniques [8], [9], [10], [11], [13], [6]. Both in Automatic Humour Recognition (AHR) and Automatic Humour Generation (AHG) tasks, the work has focused on the study of how specific features may be employed to discriminate humorous from non humorous data. The sentence (a) shows, for instance, a structure with some features that can be handled by a computer in order to simulate them and generate a humorous output [3].

(a) What do you use to talk to an elephant? An elly-phone.

For an AHG task this sentence represents a good source of features. Besides its syntactic template is very simple, there is other kind of useful information. Let us see how $elly - phone$ has a phonological similarity with telephone. Moreover, $elly - phone$ is related phonologically and "semantically" to the word which gives it the right meaning: elephant. This kind of structures, named punning riddles, were studied by [3], [4] and [5] for AHG tasks, finding out that features such as the previous ones can be generated by rules in order to automatically produce new humorous sentences. Another example about AHG tasks is the research carried out in [14]. They showed how humour could be generated exploiting incongruity theories. A result of this investigation is the new funny sense for the acronym FBI (Federal Bureau of Investigation) which appears in (b).

(b) Fantastic Bureau of Intimidation.

In AHR the research works described in [9], [10], [11], [8], [13] and [6] have focused on the analysis of another particular kind of structures: one-liners (OLs). According to [8], [9] and [10], such elements are short sentences with comic effects and a very simple syntactic surface structure. For instance, let us consider (c):

(c) In order to get a loan, you must first prove that you don't need it.

This construction seems not to contain much information; nevertheless, the results obtained for automatically discriminating humorous from non humorous sentences have been interesting. In [9] and [10] is described how the authors identified in these sentences features such as alliteration, antonymy and adult slang that, for instance, allow to differentiate between an OL and a proverb. Moreover, the authors have found some semantic "spaces" that are potentially triggers of humour such as human centric vocabulary (pronouns), negative orientation (adjectives) or professional communities (lawyers). Other features they have identified with these structures are: irony, incongruity, idiomatic expressions, common sense knowledge and ambiguity. Their research has established the beginning of new paths to explore. A sample is the work of [13] which claims to recognise humour through the identification of syntactic and semantic ambiguity. The authors have detected some features such as similarity, style or idiomatic expressions which define humorous data. Another example is the work carried out by [6], whose goal was to study whether or not humorous quotes may be separated from non humorous ones through a set of features such as N-grams, length or bag of words. Their results showed how employing features like bag of words or sentence length it is possible to discriminate humorous from non humorous sentences.

3   Perplexity and (Morpho)syntactic Ambiguity

The research work we present in this paper was inspired by assumptions and results of previous investigations in AHR. Our study aims at investigating how valuable information could be extracted from OLs in order to identify features to be used by a classifier for the AHR task. This means, based on the ambiguity phenomenon, to find criteria for classifying new OLs. For instance, [10] investigated the impact of stylistic (alliteration) and content-based (human weakness) features in OLs, nonetheless, according to our data, the sentence (d) below, although presenting alliteration and human weakness features, was not considered as humorous.

(d) Bacco, Tabacco e Venere riducono l'uomo in cenere.[1]

We tried to provide, through automatic measures, information for explaining cases such as (d). One of them is through a measure such as perplexity (Subsect. 4.2) which, according to [7], "given two probabilistic models, the better model is the one that has tighter fit to the test data". The hypothesis is that OLs maximize a dispersion profiling a structural ambiguity, which may be understood as a dispersion in the number of combinations among the words which constitute them. This dispersion functions as a source of ambiguous situations and should appear in a higher degree within OLs. We also considered important the study of the impact that morphosyntactic ambiguity may have on AHR. We think that the number of POS tags of a word may be evidence about the different syntactic functions that this word could play within an OL. Thus, the hypothesis is that morphosyntactic ambiguity could be useful for identifying potentially humorous situations. In Subsect. 4.3 we investigate this issue. Finally, according to [13], it is known that "ambiguity is not only caused by word senses", phonological similarity or by their morphosyntactic functions. Ambiguity also appears codified in the syntactic complexity of a sentence. In Subsect. 4.4 we study, in terms of generating a syntactic tree, how complex and ambiguous the OLs are.

---

[1] Wine, tobacco and women turn men into ash.

4   Experiments

In order to obtain empirical evidence of our assumptions, we ran three kinds of experiments. The first one tried to find out whether a measure such as perplexity (PPL) could be useful for characterizing our humorous data or not. In the second experiment we performed a POS tagging over the corpus for knowing its morphosyntactic information and the ambiguity that exists in the sentences. The last experiment was aimed to investigate how the information extracted from a full syntactic parser could give any hint about the ambiguity at syntactic level.

4.1   Data Set

Our experiments were carried out over the Emoticorpus[2] [6], a corpus which is composed of Italian sentences extracted from the Wikiquote project. This corpus was manually labelled and is made up of 1,966 sentences. Positive examples (OLs) are 471 while negative ones are 1,495.

4.2   Perplexity

We estimated the PPL separating our corpus in three subsets: the training subset made up of 1,475 sentences, the positive test subset made up of 165 OLs, and the negative subset with 326 non humorous sentences. Using the SRILM Toolkit [15], we obtained a language model (LM) composed by 19,923 bigrams employing two smoothing methods: interpolation and Kneser-Ney discount [7]. The two test subsets were compared against this LM. The PPL that we obtained per subset appears in Table 1.

Table 1. Positive and Negative PPL

	Positive Test	Negative Test
PPL	306.29	223.48
OOVs	780	827

As can be seen in this table, the positive PPL is higher than the one obtained for the negative examples, although this subset is bigger. Another interesting aspect we can observe is the number of Out Of the Vocabulary words (OOVs). If we take into account the size of the subsets, we can realize that the LM was obtained on the basis of two thirds of negative samples and, therefore, we would expect that the number of OOVs was smaller in the negative subset. In order to corroborate this behaviour, we performed a second experiment undersampling the data. We selected 600 sentences (300 OLs and 300 non humorous) as training; our test sets were made up of 171 sentences for both positive and negative subsets. A total of 9,297 bigrams integrated the new LM. The PPL obtained is shown in Table 2.

The information in this table corroborate our initial assumptions about the lower predictability of OLs. Also, with a more balanced corpus, the PPL of OLs is still higher.

---

[2] This corpus is available at: http://users.dsic.upv.es/grupos/nle/downloads.html.

Table 2. Balanced Corpus PPL

	Positive Test	Negative Test
PPL	222.21	212.25
OOVs	858	704

Moreover, the fact also that the number of OOVs is higher in the positive subset employing a more balanced LM is a hint of how OLs are intrinsically more difficult to classify.

4.3   Morphosyntactic Ambiguity

In order to investigate whether or not morphosyntactic ambiguity could provide useful information to discriminate humorous sentences (through the number of POS tags), we labelled the positive and negative sets using the TreeTagger software [12]. We employed three different thresholds of probability for getting all the tags that a word could have within the sentence in which appears. These thresholds were selected randomly from a range between 0 and 1. In Table 3 we present the number of tags that could be assigned to each word.

Table 3. Tags assignment per set

	Threshold 0.2		Threshold 0.5		Threshold 0.7	
Tags	Positive	Negative	Positive	Negative	Positive	Negative
2	827	2,092	399	1,006	202	537
3	81	170	9	22	1	4
4	2	5	0	0	0	0

We can see in this table how OLs have a minor range of possibilities to be tagged. However, considering the amount of sentences per set, we may notice that there is a similar distribution in the tags assignment. Therefore, taking into account this difference, we performed another experiment with balanced data. We undersampled our data to 400 sentences extracted randomly for both positive and negative sets. The same thresholds of probability were also used. In Table 4 we show the results.

Table 4. Tags assignment per balanced sets

	Threshold 0.2		Threshold 0.5		Threshold 0.7	
Tags	Positive	Negative	Positive	Negative	Positive	Negative
2	326	302	161	147	97	88
3	32	27	12	7	5	2

It is interesting to notice that, using balanced data, the range of tags that can be assigned to OLs resulted to be wider in comparison to the negative samples. However, this difference is not so relevant from a statistical viewpoint and this issue deserves to be investigated further in the future on a bigger corpus.

4.4 Syntactic Ambiguity

The last experiment is related to the fact of obtaining evidence about the syntactic ambiguity and its impact on the structure of OLs. In order to investigate this issue, we used a measure named Sentence Complexity (SC). This measure, according to [2], "captures aspects like average number of syntactic dependencies". Therefore, we decided to employ the SC measure as a parameter for verifying how complex and ambiguous the syntax of OLs is. The SC was calculated over all our corpus according to the following formula:

$$SC = \frac{\sum V_L, N_L}{\sum Cl} \tag{1}$$

where VL and NL are the number of verbal and nominal links respectively, divided by the number of clauses (Cl) [2]. The results are shown in Figure 1.

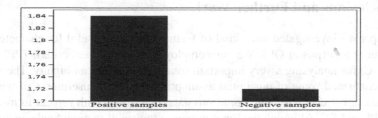

Fig. 1. Sentence Complexity per set

As can be noted, SC for OLs is higher (1.84) than for the non humorous sentences (1.72). This means that OLs have a more complex syntactic structure and, accordingly, it seems they have a major degree of syntactic ambiguity. In the following Sect. we analyse all the results obtained.

5   Discussion

The goal of this research work was to study what kind of information could provide internal features for identifying signs of humour in OLs taking into account different kinds of ambiguities (morphosyntactic and syntactic). The final aim will consist in the generation of a model, integrated by a mixture of linguistic features, capable of automatically recognising humour. On the basis of this objective, we carried out three different experiments whose results suggest interesting findings. The results about PPL establish that, although the amount of OLs is smaller, they have a structure less predictable and more ambiguous than non humorous sentences. This means that, given two different schemas of distribution, OLs are the structures which have a broader range of combinations and, therefore, the ambiguity tends to be higher.

The results of the POS tagging experiment showed how, regardless of the relation 1 to 3 between the positive and negative examples, the words of OLs seem to have higher probability to play different morphosyntactic functions. Considering the results of the balanced experiment reported in Table 4, we may realize that this dispersion in the tags assignment (that we assume as ambiguity) is greater when a tag was assigned to a positive word than when it was assigned to a negative one. On the basis of this behaviour, we may infer that, given an isolated word w_i, the probability to be assigned to various categories can break a logical meaning producing ambiguity and, therefore, humorous effects.

According to [2] about how the information available in the syntactic structure impacts on the ambiguity, we carried out the third experiment trying to verify how the SC measure could contribute to discriminate humorous from non humorous sentences. The results suggest that OLs are syntactically more complex than the negative examples. We infer that this complexity represents a sign about how OLs produce their humorous effect: if SC is greater in humorous sentences, it means that OLs are more ambiguous from a syntactic viewpoint and, therefore, more prone to produce a humorous interpretation.

6   Conclusions and Further Work

In this paper we investigated what kind of features could be useful for characterizing the humour of a corpus of OLs. We have employed three different criteria (PPL, POS tags and SC) for analysing a very important source of humour: ambiguity. The results obtained confirmed some of our initial assumptions about the usefulness of this kind of information for characterizing humorous data, especially, with respect to measures such as PPL and SC. Although the Emoticorpus is too small to draw final conclusions, we believe that the obtained findings are interesting because some of them support the results obtained with the English corpus of OLs employed in the research of [8], [9] and [10]. On the other hand, we must also highlight that these results represent a first approach that attemps to take into account the information that the speakers profile in their non-specialized texts to express different emotions or moods (such as humour) in order to analyse, from another angle, tasks such as information filtering, sentiment analysis or opinion mining. Finally, as further work, we aim at investigating the impact of semantic ambiguity employing the Italian MultiWordNet ontology [1] in order to obtain a set of fine-grained features that may be used not only for having a more robust AHR model but for describing the language in such a way that the scope of results may impact other kinds of NLP tasks beyond AHR.

References

1. Artale, A., Magnini, B., Strapparava, C.: WordNet for Italian and its Use for Lexical Discrimination. In: Proceedings of the 5th Congress of the Italian Association for Artificial Intelligence on Advances in Artificial Intelligence, pp. 346–356 (1997)
2. Basili, R., Zanzotto, F.: Parsing Engineering and Empirical Robustness. Journal of Natural Language Engineering 8(3), 97–120 (2002)
3. Binsted, K.: Machine humour: An implemented model of puns. PhD thesis. University of Edinburgh, Edinburgh, Scotland (1996)

4. Binsted, K., Ritchie, G.: Computational rules for punning riddles. Humor. Walter de Gruyter Co. 10, 25–75 (1997)
5. Binsted, K., Ritchie, G.: Towards a model of story puns. Humor. 14(3), 275–292 (2001)
6. Buscaldi, D., Rosso, P.: Some experiments in humour recognition using the italian wikiquote collection. In: Masulli, F., Mitra, S., Pasi, G. (eds.) WILF 2007. LNCS (LNAI), vol. 4578, pp. 464–468. Springer, Heidelberg (2007)
7. Jurafsky, D., Martin, J.: Speech and Language Processing: An introduction to natural language processing, computational linguistics, and speech recognition. Draft of June 25 (2007)
8. Mihalcea, R.F.: The multidisciplinary facets of research on humour. In: Masulli, F., Mitra, S., Pasi, G. (eds.) WILF 2007. LNCS (LNAI), vol. 4578, pp. 412–421. Springer, Heidelberg (2007)
9. Mihalcea, R., Strapparava, C.: Technologies that make you smile: Adding humour to text-based applications. IEEE Intelligent Systems 21(5), 33–39 (2006)
10. Mihalcea, R., Strapparava, C.: Learning to Laugh (Automatically): Computational Models for Humor Recognition. Journal of Computational Intelligence 22(2), 126–142 (2006)
11. Mihalcea, R.F., Pulman, S.: Characterizing humour: An exploration of features in humorous texts. In: Gelbukh, A. (ed.) CICLing 2007. LNCS, vol. 4394, pp. 337–347. Springer, Heidelberg (2007)
12. Schmid, H.: Improvements in Part-of-Speech Tagging with an Application to German. In: Proceedings of the ACL SIGDAT Workshop (1995)
13. Sjöbergh, J., Araki, K.: Recognizing humor without recognizing meaning. In: Masulli, F., Mitra, S., Pasi, G. (eds.) WILF 2007. LNCS (LNAI), vol. 4578, pp. 469–476. Springer, Heidelberg (2007)
14. Stock, O., Strapparava, C.: Hahacronym: A computational humor system. In: Demo. proc. of the 43rd annual meeting of the Association of Computational Linguistics (ACL 2005), pp. 113–116 (2005)
15. Stolcke, A.: SRILM - An Extensible Language Modeling Toolkit. In: Proc. Intl. Conf. Spoken Language Processing. Denver, Colorado (2002)

Objective vs. Subjective Evaluation of Speakers with and without Complete Dentures

Tino Haderlein[1,2], Tobias Bocklet[1], Andreas Maier[1,2], Elmar Nöth[1],
Christian Knipfer[3], and Florian Stelzle[3]

[1] Universität Erlangen-Nürnberg, Lehrstuhl für Mustererkennung (Informatik 5),
Martensstraße 3, 91058 Erlangen, Germany
Tino.Haderlein@informatik.uni-erlangen.de
http://www5.informatik.uni-erlangen.de
[2] Universität Erlangen-Nürnberg, Abteilung für Phoniatrie und Pädaudiologie,
Bohlenplatz 21, 91054 Erlangen, Germany
[3] Universität Erlangen-Nürnberg, Mund-, Kiefer- und Gesichtschirurgische Klinik,
Glückstraße 11, 91054 Erlangen, Germany

Abstract. For dento-oral rehabilitation of edentulous (toothless) patients, speech intelligibility is an important criterion. 28 persons read a standardized text once with and once without wearing complete dentures. Six experienced raters evaluated the intelligibility subjectively on a 5-point scale and the voice on the 4-point Roughness-Breathiness-Hoarseness (RBH) scales. Objective evaluation was performed by Support Vector Regression (SVR) on the word accuracy (WA) and word recognition rate (WR) of a speech recognition system, and a set of 95 word-based prosodic features. The word accuracy combined with selected prosodic features showed a correlation of up to $r = 0.65$ to the subjective ratings for patients with dentures and $r = 0.72$ for patients without dentures. For the RBH scales, however, the average correlation of the feature subsets to the subjective ratings for both types of recordings was $r < 0.4$.

1 Introduction

Complete loss of teeth can cause a persisting speech disorder by altering dental articulation areas. This reduces the intelligibility of speech severely. Removable complete dentures can partly solve this problem. However, they also disturb speech production as they restrict the flexibility of the tongue, narrow the oral cavity and alter the articulation areas of the palate and teeth.

Objective and independent diagnostic tools for the assessment of speech ability concerning alteration of the dental arch or dento-oral rehabilitation have only been applied for single parameters of speech. They do not evaluate continuous speech but often only single vowels or consonants [1,2,3]. However, this does not reflect real-life communication because no speech but only the voice is examined. Criteria like intelligibility cannot be evaluated in this way. For this study, the test persons read a given standard text which was then analyzed by methods of automatic speech recognition and prosodic analysis.

V. Matoušek and P. Mautner (Eds.): TSD 2009, LNAI 5729, pp. 170–177, 2009.
© Springer-Verlag Berlin Heidelberg 2009

It was the aim of this clinical pilot study to evaluate speech intelligibility of edentulous patients objectively and automatically and to find out whether the impact of complete dentures on speech intelligibility can be evaluated by automatic analysis as part of oro-dental rehabilitation assessment.

In Sect. 2, the speech data used as the test set will be introduced. Section 3 will give some information about the speech recognizer. An overview on the prosodic analysis will be presented in Sect. 4, and Sect. 5 will discuss the results.

2   Test Data and Subjective Evaluation

The study group comprised 28 edentulous, i.e. toothless, patients (13 men, 15 women). Their average age was 64 years; the standard deviation was 10 years. The youngest person was 43, the oldest was 83 years old. They had worn their dentures on average for 59 months (st. dev. 49 months, range: 1 to 240 months) before the recordings. Only patients with removable complete dentures were accepted to participate to avoid the influence of different kinds of dentures. Only patients were chosen who wore them for more than at least one month to ensure patients' habituation to new dentures. All patients were native German speakers using the same local dialect. However, they were asked to speak standard German while being recorded. None of the patients had speech disorders caused by medical problems others than dental or any report of hearing impairment.

Each person read the text "Der Nordwind und die Sonne", a phonetically balanced text with 108 words (71 disjunctive) which is used in German speaking countries in speech therapy. The English version is known as "The North Wind and the Sun". The speech data were sampled with 16 kHz and an amplitude resolution of 16 bit. The patients read the text with their complete dentures inserted at first. The second recording was subsequently performed without dentures.

Six experienced phoniatricians and speech scientists evaluated each speaker's intelligibility in each recording according to a 5-point scale with the labels "very high", "high", "moderate", "low", and "none". Each rater's decision for each patient was converted to an integer number between 1 and 5. The 2·28 recordings were presented to the listeners during one evaluation session in random order.

Since the voice of elderly people is also often hoarse, the RBH scale [4] was applied which is an important rating system for dysphonic speech in German-speaking countries. It allows integer scores between 0 and 3 for the three dimensions "Roughness", "Breathiness", and "Hoarseness". The raters evaluated all recordings also with respect to these criteria. Since the RBH scales evaluate voice quality, it was expected that the dentures merely affect these ratings but rather the intelligibility scores.

3   The Speech Recognition System

The speech recognition system used for the experiments was developed at the Chair of Pattern Recognition in Erlangen [5]. It can handle spontaneous speech with mid-sized vocabularies up to 10,000 words. The system is based on semi-continuous Hidden Markov Models (HMM). It can model phones in a context as large as statistically useful and thus forms the so-called polyphones, a generalization of the well-known bi- or triphones. The HMMs for each polyphone have three to four states; the codebook had 500 classes with full covariance matrices. The short-time analysis applies a Hamming window with a length of 16 ms, the frame rate is 10 ms. The filterbank for the Mel-spectrum consists of 25 triangle filters. For each frame, a 24-dimensional feature vector is computed. It contains short-time energy, 11 Mel-frequency cepstral coefficients, and the first-order derivatives of these 12 static features. The derivatives are approximated by the slope of a linear regression line over 5 consecutive frames (56 ms).

The baseline system for the experiments in this paper was trained with German dialogues from the VERBMOBIL project [6]. The data were recorded with a close-talking microphone at a sampling frequency of 16 kHz and quantized with 16 bit. About 80% of the 578 training speakers (304 male, 274 female) were between 20 and 29 years old, less than 10% were over 40. 11,714 utterances (257,810 words) of the VERBMOBIL-German data (12,030 utterances, 263,633 words, 27.7 hours of speech) were used for training and 48 (1042 words) for the validation set, i.e. the corpus partitions were the same as in [5].

The recognition vocabulary of the recognizer was changed to the 71 words of the standard text. The word accuracy and the word recognition rate were used as basic automatic measures for intelligibility since they had been successful for other voice and speech pathologies [7,8]. They are computed from the comparison between the recognized word sequence and the reference text consisting of the $n_{all} = 108$ words of the read text. With the number of words that were wrongly substituted (n_{sub}), deleted (n_{del}) and inserted (n_{ins}) by the recognizer, the word accuracy in percent is given as

$$WA = [1 - (n_{sub} + n_{del} + n_{ins})/n_{all}] \cdot 100$$

while the word recognition rate omits the wrongly inserted words:

$$WR = [1 - (n_{sub} + n_{del})/n_{all}] \cdot 100$$

Only a unigram language model was used so that the results mainly depend on the acoustic models. A higher-order model would correct too many recognition errors and thus make WA and WR useless as measures for intelligibility.

4   Prosodic Features

In order to find automatically computable counterparts for the subjective rating criteria, also a "prosody module" was used to compute features based upon frequency, duration, and speech energy (intensity) measures. This is state-of-the-art in automatic speech analysis on normal voices [9,10,11].

The input to the prosody module is the speech signal and the output of the word recognition module. In this case the time-alignment of the recognizer and the information about the underlying phoneme classes can be used by the module. For each speech unit which is of interest (here: words), a fixed reference point has to be chosen for the computation of the prosodic features. This point was chosen at the end of a word because the word is a well–defined unit in word recognition, it can be provided by any standard word recognizer, and because this point can be more easily defined than, for example, the middle of the syllable nucleus in word accent position. For each reference point, 95 features are extracted over intervals which contain a word, a word-pause-word interval or the pause between two words. A full description of the features used is beyond the scope of this paper; details and further references are given in [12]. The feature set was also used successfully for other voice and speech pathologies [7,8].

In order to find the best subset of word accuracy, word recognition rate, and the prosodic features to model the subjective ratings, Support Vector Regression (SVR, [13]) was used. The general idea of regression is to use the vectors of a training set to approximate a function which tries to predict the target value of a given vector of the test set. Here, the training set were the automatically computed measures, and the test set consisted of the subjective intelligibility scores or the single dimensions of the RBH scores, respectively. For this study, the sequential minimal optimization algorithm (SMO, [13]) of the Weka toolbox [14] was applied in a 28-fold cross-validation manner due to the 28 available speakers.

5 Results and Discussion

The speech intelligibility of edentulous patients was rated lower by the speech experts on the 5-point scale (Table 1) where lower numbers denote better intelligibility. The average score was 2.19 for patients without teeth and 2.04 for the patients with dentures. The average RBH results, their range and standard deviation for both cases were virtually identical, but a closer analysis revealed that this does not hold for each single patient (Fig. 1). Pearson's correlation r was computed for each rater against the average of the other 5 raters and then averaged (Table 2). For the intelligibility criterion, an average of $r = 0.73$ (edentulous patients) and $r = 0.74$ (with dentures) was reached. The values did not change significantly throughout the study when Spearman's rank-order correlation ρ was computed. For this reason, only r will be given in the following. For the RBH scales, they were in the same range for the R and H scale but only when the patients did not wear their dentures. The reason for this may be the rough 4-point RBH scales that do not allow a better differentiation in rating.

The automatically computed word accuracy and word recognition rate were also lower for the edentulous patients (WA: 55.8%, WR: 63.1%; see Table 1) than for the same persons with complete dentures (WA: 59.4%, WR: 68.2%). The ranges of both measures were shifted by about the same value as the averages. The correlation between subjective evaluation and WA or WR, respectively, was lower than among the rater group (Table 3). For the average rater's intelligibility scores, the WA reached $r = -0.53$ for patients without and $r = -0.60$ for patients with complete denture. The corresponding values for the WR were $r = -0.55$ and $r = -0.46$. The coefficient is negative because high recognition rates came from "good" voices with a low score number and vice versa.

The RBH scores could not achieve satisfying correlation with WA or WR (Table 3). The best correlation was $r = -0.50$ for the WR on patients without teeth. When the patients wore their dentures, the human-machine correlation dropped drastically although there was a high correlation both for the subjective and the objective evaluations when the two recordings of each patient were compared (Table 4). Obviously, WA and WR are not suitable to reflect slight changes in signal quality that the trained listeners can hear.

Table 1. Subjective and objective evaluation results for 28 speakers: intelligibility (int.), roughness (R), breathiness (B), hoarseness (H), all of them averaged across 6 raters, and the word accuracy (WA) and word recognition rate (WR) in percent

	without denture						with denture					
	int.	R	B	H	WA	WR	int.	R	B	H	WA	WR
mean	2.19	0.89	0.31	0.96	55.8	63.1	2.04	0.90	0.32	0.93	59.4	68.2
st. dev.	0.65	0.56	0.29	0.54	12.4	8.8	0.68	0.48	0.28	0.47	15.0	8.6
min.	1.17	0.00	0.00	0.00	13.9	46.3	1.17	0.00	0.00	0.00	5.6	51.9
max.	3.67	2.17	1.17	2.17	75.0	79.6	3.83	2.17	1.00	2.17	85.2	86.1

Table 2. Average inter-rater correlation for intelligibility (int.), roughness (R), breathiness (B), and hoarseness (H)

	without denture				with denture			
	int.	R	B	H	int.	R	B	H
r	0.73	0.73	0.38	0.71	0.74	0.57	0.31	0.56

By using WA, WR, and the prosodic features as input for SVR, higher correlations to the subjective intelligibility score were achieved (Table 5). For edentulous patients $r = 0.72$ was reached when the WA value was combined with the normalized energy computed from the current word and the energy from the two words before the current word and the pause between them. The F_0 value at the end of the last voiced section within a respective word contributes also to these results. For patients with complete denture, $r = 0.65$ was reached. In that case, however, the energy value from the current word was not beneficial. In general, the number of speakers in this study was rather small and the results have to be handled with care. However, the contribution of the mentioned features to the human-machine correlation was evident throughout the experiments.

In order to explain the influence of the speech energy, it would be straightforward to assume that a louder speaker is better intelligible. However, in the prosodic feature set, the energy values are normalized so that a continuously high energy level will have no effect. It is more likely that single phones or phone classes that cannot be uttered properly due to the speech impairment appear in the signal as more noisy and cause local changes in the energy distribution.

The impact of the F_0 value can be explained by the noisy speech that causes octave errors during F_0 detection, i.e. instead of the real fundamental frequency, one of its harmonics one or more octaves higher is found. Again, with more "noisy speech", this may influence the F_0 trajectory and hence the correlation to the subjective results. It is not clear so far, however, why only the end of the voiced sections causes a noticeable effect. There may be a connection to changes in the airstream between the beginning and the end of words or phrases, but this has to be confirmed by more detailed experiments. An aspect that also needs a closer look in the future is that not all phone clases are affected in the same way by missing teeth. Especially the articulation of fricatives, like /s/, is distorted. The word-based analysis will therefore be extended by a phone-based level.

Table 3. Correlation between subjective ratings for intelligibility (int.), roughness (R), breathiness (B), and hoarseness (H) vs. the word accuracy (WA) and word recognition rate (WR) for patients without denture and with full denture, respectively

	without denture				with denture			
	int.	R	B	H	int.	R	B	H
r(WA)	−0.53	−0.42	−0.40	−0.43	−0.60	−0.20	0.04	−0.23
r(WR)	−0.55	−0.49	−0.35	−0.50	−0.46	−0.14	0.01	−0.20

Table 4. Correlation between the ratings for patients without denture and with full denture: intelligibility (int.), roughness (R), breathiness (B), hoarseness (H), word accuracy (WA), and word recognition rate (WR)

	subjective				objective	
	int.	R	B	H	WA	WR
r	0.64	0.80	0.55	0.77	0.80	0.72

Table 5. Correlation of subjective intelligibility scores from 6 raters against a feature subset computed by Support Vector Regression (SVR) from the values of word accuracy (WA), word recognition rate (WR), and 95 prosodic features

	without denture	with denture
r(subset with energy of current word)	0.52	0.72
r(subset without energy of current word)	0.65	0.63

For the RBH scores, the SVR method could not reveal a feature set that showed good results on both types of recordings so far. The average correlations were below $r = 0.40$. Further experiments will be part of future work.

For this study, patients read a standard text, and voice professionals evaluated intelligibility. It is often argued that intelligibility should be evaluated by an "inverse intelligibility test": The patient utters a subset of words and sentences from a carefully built corpus. A naïve listener writes down what he or she heard. The percentage of correctly understood words is a measure for the intelligibility of the patient. However, when automatic speech evaluation is performed for instance with respect to prosodic phenomena, like e.g. word durations or percentage of voiced segments [7], then comparable results for all patients can only be achieved when all the patients read the same defined words or text. This means that an inverse intelligibility test can no longer be performed, and intelligibility has to be rated on a grading scale instead.

The data obtained in this study allow for the following conclusions: There is a significant correlation between subjective rating of intelligibility and automatic evaluation. It also reveals the impact of dentures on intelligibility just as the subjective ratings do. Hence, the method can serve as the basis for more research towards an automatic system that can support oro-dental rehabilitation by objective speech evaluation.

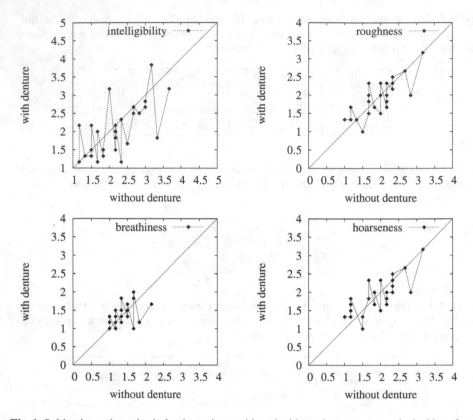

Fig. 1. Subjective rating criteria for the patients with and without dentures, respectively. Note that all measures are ordered independently from each other.

Acknowledgment

This study was partially funded by the German Cancer Aid (Deutsche Krebshilfe) under grant 107873. The responsibility for the contents lies with the authors.

References

1. Runte, C., Tawana, D., Dirksen, D., Runte, B., Lamprecht-Dinnesen, A., Bollmann, F., Seifert, E., Danesh, G.: Spectral analysis of /s/ sound with changing angulation of the maxillary central incisors. Int J Prosthodont 15, 254–258 (2002)
2. Molly, L., Nackaerts, O., Vandewiele, K., Manders, E., van Steenberghe, D., Jacobs, R.: Speech adaptation after treatment of full edentulism through immediate-loaded implant protocols. Clin Oral Implants Res 19, 86–90 (2008)
3. Stojcević, I., Carek, A., Buković, D., Hedjever, M.: Influence of the partial denture on the articulation of dental and postalveolar sounds. Coll Antropol 28, 799–807 (2004)
4. Nawka, T., Anders, L.-C., Wendler, J.: Die auditive Beurteilung heiserer Stimmen nach dem RBH-System. Sprache - Stimme - Gehör 18, 130–133 (1994)

5. Stemmer, G.: Modeling Variability in Speech Recognition. Studien zur Mustererkennung, vol. 19. Logos Verlag, Berlin (2005)
6. Wahlster, W. (ed.): Verbmobil: Foundations of Speech-to-Speech Translation. Springer, Berlin (2000)
7. Haderlein, T.: Automatic Evaluation of Tracheoesophageal Substitute Voices. Studien zur Mustererkennung, vol. 25. Logos Verlag, Berlin (2007)
8. Maier, A.: Speech of Children with Cleft Lip and Palate: Automatic Assessment. Studien zur Mustererkennung, vol. 29. Logos Verlag, Berlin (2009)
9. Nöth, E., Batliner, A., Kießling, A., Kompe, R., Niemann, H.: Verbmobil: The Use of Prosody in the Linguistic Components of a Speech Understanding System. IEEE Trans. on Speech and Audio Processing 8, 519–532 (2000)
10. Chen, K., Hasegawa-Johnson, M., Cohen, A., Borys, S., Kim, S.S., Cole, J., Choi, J.Y.: Prosody dependent speech recognition on radio news corpus of American English. IEEE Trans. Audio, Speech, and Language Processing 14, 232–245 (2006)
11. Shriberg, E., Stolcke, A.: Direct Modeling of Prosody: An Overview of Applications in Automatic Speech Processing. In: Proc. International Conference on Speech Prosody, Nara, Japan, pp. 575–582 (2004)
12. Batliner, A., Buckow, A., Niemann, H., Nöth, E., Warnke, V.: The Prosody Module. In: [6], pp. 106–121
13. Smola, A., Schölkopf, B.: A Tutorial on Support Vector Regression. Statistics and Computing 14, 199–222 (2004)
14. Witten, I., Frank, E., Trigg, L., Hall, M., Holmes, G., Cunningham, S.: Weka: Practical machine learning tools and techniques with java implementations. In: Proc. ICONIP/ANZIIS/ANNES 1999 Future Directions for Intelligent Systems and Information Sciences, pp. 192–196. Morgan Kaufmann, San Francisco (1999)

Automatic Pitch-Synchronous Phonetic Segmentation with Context-Independent HMMs*

Jindřich Matoušek

University of West Bohemia, Faculty of Applied Sciences, Dept. of Cybernetics,
Univerzitní 8, 306 14 Plzeň, Czech Republic
jmatouse@kky.zcu.cz

Abstract. This paper deals with an HMM-based automatic phonetic segmentation (APS) system. In particular, the use of a pitch-synchronous (PS) coding scheme within the context-independent (CI) HMM-based APS system is examined and compared to the "more traditional" pitch-asynchronous (PA) coding schemes for a given Czech male voice. For bootstrap-initialised CI-HMMs, exploited when some (manually) pre-segmented data are available, the proposed PS coding scheme performed best, especially in combination with CART-based refinement of the automatically segmented boundaries. For flat-start-initialised CI-HMMs, an inferior initialisation method used when no pre-segmented data are at disposal, standard PA coding schemes with longer parameterization shifts yielded better results. The results are also compared to the results obtained for APS systems with context-dependent (CD) HMMs. It was shown that, at least for the researched male voice, multiple-mixture CI-HMMs outperform CD-HMMs in the APS task.

1 Introduction

Automatic phonetic segmentation (APS) is a process of detecting boundaries between phones in speech signals. Since manual segmentation is labour-intensive and time-consuming, the automation of the process is very important especially when many speech signals are to be segmented. This is exactly the case of *unit selection*, a very popular and still the most prevalent *text-to-speech* (TTS) synthesis technique. Being a corpus-based concatenative speech synthesis method, the principle of unit selection is to concatenate pre-recorded speech segments (extracted from natural utterances in accordance with the automatically segmented boundaries) carefully selected from a large speech corpus according to phonetic and prosodic criteria imposed by the synthesised utterance. It is evident that automatic phonetic segmentation affects the quality of synthetic speech produced by a unit-selection-based TTS system.

The most successful approaches to the automatic phonetic segmentation are based on *hidden Markov models* (HMMs), a statistical framework adopted from the area of automatic speech recognition. There are many aspects that can affect the performance of the

---

* Support for this work was provided by the Grant Agency of the Czech Republic, project No. GAČR 102/09/0989, and by the Ministry of Education of the Czech Republic, project No. 2C06020.

V. Matoušek and P. Mautner (Eds.): TSD 2009, LNAI 5729, pp. 178–185, 2009.

HMM-based segmentation system such as the context dependency of HMMs, manner of initialisation of HMMs, training strategies, number of Gaussian mixtures used during modelling, speech coding schemes, etc. Various modifications and post-segmentation techniques were also proposed in order to increase the segmentation accuracy of the base APS system, see e.g. [1,2,3,4,5]. In this paper, the refinement based on *classification and regression trees* (CART) [3,5] is also applied when the segmentation accuracy of different coding schemes in a CI-HMM based APS system is evaluated.

In [6], *pitch-synchronous coding scheme* was proposed to enable more precise modelling of spectral properties of speech. It was shown that pitch-synchronous coding scheme outperformed the traditionally utilised pitch-asynchronous coding scheme when single-density context-dependent (CD) HMMs were employed. As some authors advocate for using context-independent (CI) HMMs with multiple Gaussian mixtures [2,5], the impact of different coding schemes on segmentation accuracy of APS system with CI-HMMs is examined in this paper.

The paper is organised as follows. CI-HMM-based APS system is briefly described in Section 2. In Section 3, various speech coding schemes are introduced. Experiments with different coding schemes and the results of the performance evaluation and their discussion are provided in Sections 4 and 5. Finally, conclusions are drawn in Section 6.

2 Automatic Phonetic Segmentation with CI-HMMs

The idea in APS is to apply similar procedures as for speech recognition. However, instead of the recognition, so-called *forced-alignment* is performed to find the best alignment between HMMs and the corresponding speech data, producing a set of boundaries which delimit speech segments belonging to each HMM. Briefly, each phone unit can be modelled by a context-dependent HMM (CD-HMM) or context-independent HMM (CI-HMM). In this paper, CI-HMMs are researched.

Firstly, the model parameters are to be initialised. Two initialisation strategies are usually employed [7]. When some (manually) segmented data are available, so called *bootstrap initialisation* can be applied utilising Baum-Welch algorithm with model boundaries fixed to the manually segmented ones (also called *isolated-unit training*). In this case, each HMM is initialised on its own phone-specific data. When no pre-segmented data are at disposal, so called *flat-start initialisation* is usually performed to set up all HMMs with the same data, typically corresponding to the global mean and variance.

Secondly, parameters of each model are trained on the basis of a collection of all speech data (described by *feature vectors*, often mel-frequency cepstral coefficients, MFCCs) with the corresponding phonetic transcripts. Typically, the *embedded training* strategy is employed in which models associated with the given phonetic transcript are concatenated and parameters of the composite model are simultaneously updated through the Baum-Welch algorithm. Simultaneously, in order to enable more precise modelling, the number of Gaussian mixtures can be incremented.

Finally, the trained HMMs are employed to align a speech signal along the associated phonetic transcript by means of Viterbi decoding. In this way, the best alignment

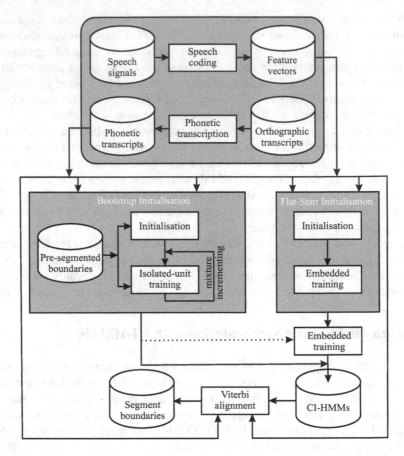

Fig. 1. Schematic view of a base automatic phonetic segmentation system with CI-HMMs, employing bootstrap or flat-start initialisation

between HMMs and the corresponding speech data is found, producing a set of boundaries which delimit speech segments belonging to each HMM. Thus, each phone-like unit is identified in the stream of speech signal and could be used for later purposes (e.g. in unit selection speech synthesis). A simplified scheme of a CI-HMM based APS system is given in Fig. 1.

Optionally, post-processing techniques can be employed to increase the segmentation accuracy by refining the initial segmentations from a base APS system. Statistically motivated approaches like classification and regression trees (CART), neural networks or support vector machines are often used to do the job. In this paper, CART-based refinement was performed similarly as in [3,5]; i.e. firstly, boundary-specific discrepancies between the automatically and manually segmented boundaries were learned by a CART, respecting phonetic type of the boundaries. Secondly, the automatic segmentations were refined by removing the boundary-specific biases according to the trained CART (see [6] for more details).

3   Speech Coding Schemes

The task of speech coding is to produce a sequence of feature vectors from speech signal of each utterance. The feature vectors are then used to train HMMs. It is obvious that the accuracy of boundary detection in an HMM-based APS system is limited by the manner the feature vectors are extracted from speech signals. Traditionally, a *pitch-asynchronous* (PA) coding scheme PA$\{l_u/s_u\}$ is employed for modelling speech. In this scheme, a uniform analysis frame of a given length l_u is defined and slid along the whole speech signal of an utterance with a fixed shift s_u. The length is usually set to comprise frequency characteristics of the speaker ($l_u \approx 2T_0$ where T_0 is a maximum pitch period of the speaker). In automatic speech recognition, the shift is usually set to approx. 8-10 ms which roughly corresponds to T_0. In order to increase the segmentation resolution in APS, smaller shifts (4-6 ms) are also utilised.

In [6], *pitch-synchronous* (PS) coding scheme was proposed. In this scheme, each frame of speech to be extracted for coding is defined both by its position in a speech signal and its length. The positions and lengths of the frames are determined from *pitch-marks*, the locations of principal excitation of vocal tract (corresponding to glottal closure instants) in speech signals. In voiced speech, the position $p_v^{(i)}$ of a frame $f^{(i)}$ ($i = 1, \ldots, N$) corresponds to a pitch-mark and the length $l_v^{(i)}$ is set to $p_v^{(i+1)} - p_v^{(i-1)}$, which roughly corresponds to a double of the local pitch period (T_0). A very robust algorithm of pitch-mark detection was introduced in [8]. In unvoiced speech, no pitch-marks are defined because there is no activity of vocal cords during unvoiced speech regions. Therefore, standard PA coding scheme with a fixed frame length l_u and a fixed frame shift s_u is employed here. Smaller values such as $l_u = 4$–6 ms and $s_u = 2$–4 can be utilised. As a result, a sequence of frames $f^{(i)}$ consisting of the subsequences of

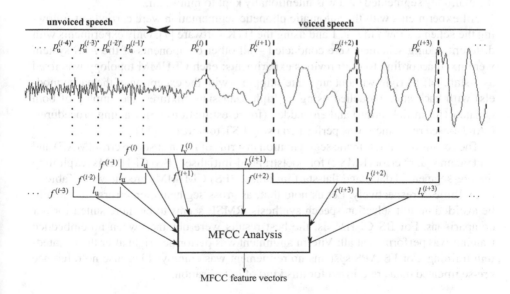

Fig. 2. Illustration of a pitch-synchronous coding scheme

both voiced and unvoiced frames is available for coding. The illustration of the pitch-synchronous coding scheme is given in Fig. 2. More details about the pitch-synchronous coding scheme are given in [6].

4   Experiments and Results

In [6] two Czech phonetically and prosodically rich speech corpora (of a female and male voice) were researched. In this paper, only the male speech corpus is used, as the male voice is currently the main voice of the Czech TTS system ARTIC [9]. Hence, the results presented in this paper will be intentionally more specific for this speech corpus. The utterances included in the corpus were carefully selected, spoken by a professional speaker in an anechoic chamber, recorded at 16-bit precision with 48 kHz sampling frequency (later down-sampled to 16 kHz) and carefully annotated both on the orthographic and phonetic level [10]. Phonetic transcripts for all utterances plus some manual segmentations from a phonetic expert were available. In order to train the APS systems, a feature vector was computed for each frame according to various pitch-asynchronous coding schemes (the length of each frame was set according to our previous experiments to $l_u = 25$ ms and shifts s_u were taken in the range 2-10 ms) and the pitch-synchronous scheme described in Section 3 using 12 MFCCs, log energy and their delta and delta-delta coefficients (39 coefficients for each frame in total). The corpus consists of 12,242 utterances (17.69 hours of speech excluding the pauses, 675,809 phone boundaries in total), 90 of them were segmented manually (11.71 minutes, 7,789 phone boundaries in total). 70 manually segmented utterances were used to initialise APS systems and 20 manually segmented utterances were used for testing. In order to reduce the labour-intensive and time-consuming manual segmentation, the amount of the manually segmented data was intentionally kept to minimum.

All experiments with the automatic phonetic segmentation were carried out following the scheme shown in Fig. 1 and using the HTK software [7]. Only experiments with different coding schemes were conducted – all other components of the APS system were fixed according to our previous experiments: each CI-HMM topology was fixed as 3-state left-to-right without any state skipping (with the exception of the pause models) with each state modelled using multiple Gaussian mixtures, the employ of both isolated (for initialisation) and embedded (for re-estimation) unit training procedures. CART-based refinement was performed using EST tool *wagon* [11].

The results of the automatic segmentation in terms of mean absolute error (MAE) and root mean square error (RMSE) for bootstrapped-initialised (BS) CI-HMMs (exploiting the pre-segmented data) and flat-start initialised (FS) CI-HMMs are shown in Table 1, or in Table 2, respectively. Let us note that, as gross segmentation errors are often to be avoided in unit selection speech synthesis, RMSE seems to be more suited for our comparisons. For BS CI-HMMs, the best results were obtained when no embedded training was performed at all; Viterbi alignment was performed right after the isolated-unit training. For FS APS systems no refinement was employed because no reference pre-segmented data are utilised for this kind of initialisation.

Table 1. Segmentation results for base and CART-refined BS APS systems. The number of mixtures was chosen according to the best segmentation results in terms of RMSE.

Coding scheme	# mixt.	Base MAE [ms]	Base RMSE [ms]	CART MAE [ms]	CART RMSE [ms]
PS	3	6.35	**10.02**	**5.35**	**8.66**
PA{25/10}	10	7.84	11.97	6.74	11.01
PA{25/8}	10	7.38	11.32	6.43	10.63
PA{25/6}	11	6.57	10.69	6.18	10.40
PA{25/4}	8	6.26	10.70	5.83	10.35
PA{25/2}	7	**5.90**	11.99	5.73	11.62

Table 2. Segmentation results for base FS APS systems. The number of mixtures was chosen according to the best segmentation results in terms of RMSE.

Coding scheme	# mixt.	MAE [ms]	RMSE [ms]
PS	2	11.88	18.41
PA{25/10}	1	9.99	17.32
PA{25/8}	1	**9.71**	**17.29**
PA{25/6}	2	10.13	17.71
PA{25/4}	5	10.94	18.35
PA{25/2}	16	13.16	20.84

5   Discussion

Looking at the results in Table 1 and Table 2, we can conclude, at least for the researched male voice:

- Pitch-synchronous coding scheme yields very good results when bootstrapped-initialised CI-HMMs are employed, especially after CART-based refining.
- PA{25/4} is the best one from pitch-asynchronous coding schemes and seems to be a good alternative to the pitch-synchronous coding scheme.
- PA{25/2} yields very good results for BS CI-HMMs in terms of MAE, but is prone to gross segmentation errors (see the RMSE score).
- As expected, BS CI-HMMs are much more superior to FS CI-HMMs. Therefore, one can conclude that it is worth preparing some manual segmentations to bootstrap the segmentation process.
- When no pre-segmented data are available (the case of flat-start initialisation), pitch-asynchronous coding schemes with longer parameterization shifts (8-10 ms) outperform the pitch-synchronous scheme.

Let us recall the segmentation accuracy of context-dependent HMMs for the same male voice, presented in [6]. The best results were achieved for the pitch-synchronous coding scheme, both for BS CD-HMMs (in case of the base system MAE was 8.11 ms, RMSE was 15.19 ms; after the CART-based refinement MAE decreased to 5.36 ms and RMSE to 12.79 ms) and for FS CD-HMMs (MAE was 9.62 ms and RMSE was

20.73 ms). Comparing both CD-HMM based APS and CI-HMM based APS, we can deduce that:

- CI-HMMs clearly outperform CD-HMMs when bootstrap initialisation is employed, both in terms of MAE and RMSE. Pitch-synchronous coding scheme seems to be the best choice in this case.
- As for flat-start initialisation, similar performance for CI-HMMs and CD-HMMs can be reported.

Considering the possibilities of post-segmentation corrections, it is obvious that there is more room for improvement for CD-HMMs. Indeed, analysing the effect of CART-based refinement, the greatest increase in segmentation accuracy was achieved for CD-HMMs (2.75 ms in MAE and 2.40 ms in RMSE; for CI-HMMs it was 1.00 ms in MAE and 1.36 ms in RMSE). As a result, the segmentation accuracy of CART-refined systems is almost the same for CD-HMMs and CI-HMMs in terms of MAE. As for RMSE, CART-refined CI-HMM based APS clearly outperforms CART-refined CD-HMM based APS.

6 Conclusion

In this paper, the use of the pitch-synchronous coding scheme within the context-independent HMM-based APS system was researched. The segmentation accuracy of such a system was compared to the segmentation accuracy of "more traditional" APS systems which utilise different   pitch-asynchronous coding schemes. For bootstrap-initialised CI-HMMs, the proposed pitch-synchronous coding scheme performed best, especially in combination with CART-based refinement of the automatically segmented boundaries. For flat-start-initialised CI-HMMs, an inferior initialisation method used when no pre-segmented data are at disposal, standard pitch-asynchronous coding scheme with longer parameterization shifts yielded better results. The results were also compared to the results obtained for APS systems with context-dependent HMMs (as described in [6]). It can be shown that multiple-mixture CI-HMMs outperform CD-HMMs in the APS task, at least for the particular male voice under research.

The future work will be devoted to the further improvement of the performance of the APS system. Post-segmentation techniques, other than CART-based ones, are planned to be researched. Beside the segmentation accuracy itself, another important aspect of the APS system, the automatic detection of badly segmented boundaries (and especially gross segmentation errors) will be researched. This aspect is very important from the point of view of unit selection speech synthesis, because the badly segmented units (which cause degradation of the resulting speech) can be removed from unit inventories and thus they can be ignored during speech synthesis.

References

1. Matoušek, J., Tihelka, D., Romportl, J.: Automatic Segmentation for Czech Concatenative Speech Synthesis Using Statistical Approach with Boundary-Specific Correction. In: Proceedings of Interspeech, Geneve, Switzerland, pp. 301–304 (2003)

2. Toledano, D., Gómez, L., Grande, L.: Automatic Phonetic Segmentation. IEEE Transactions on Speech and Audio Processing 11(6), 617–625 (2003)
3. Adell, J., Bonafonte, A.: Towards Phone Segmentation for Concatenative Speech Synthesis. In: Proceedings of Speech Synthesis Workshop, Pittsburgh, U.S.A, pp. 139–144 (2004)
4. Lee, K.S.: MLP-Based Phone Boundary Refining for a TTS Database. IEEE Transactions on Audio, Speech and Language Processing 14(3), 981–989 (2006)
5. Park, S.S., Kim, N.S.: On Using Multiple Models for Automatic Speech Segmentation. IEEE Transactions on Audio, Speech and Language Processing 15(8), 2202–2212 (2007)
6. Matoušek, J., Romportl, J.: Automatic Pitch-Synchronous Phonetic Segmentation. In: Proceedings of Interspeech, Brisbane, Australia, pp. 1626–1629 (2008)
7. Young, S., et al.: The HTK Book (for HTK Version 3.4). Cambridge University, Cambridge (2006)
8. Legát, M., Matoušek, J., Tihelka, D.: A Robust Multi-Phase Pitch-Mark Detection Algorithm. In: Proceedings of Interspeech, Antwerp, Belgium, pp. 1641–1644 (2007)
9. Matoušek, J., Tihelka, D., Romportl, J.: Current state of czech text-to-speech system ARTIC. In: Sojka, P., Kopeček, I., Pala, K. (eds.) TSD 2006. LNCS (LNAI), vol. 4188, pp. 439–446. Springer, Heidelberg (2006)
10. Matoušek, J., Tihelka, D., Romportl, J.: Building of a Speech Corpus Optimised for Unit Selection TTS Synthesis. In: Proceedings of International Conference on Language Resources and Evaluation (LREC 2008), Marrakech, Morocco (2008)
11. Taylor, P., Caley, R., Black, A., King, S.: Edinburgh Speech Tools Library: System Documentation (1999),
http://www.cstr.ed.ac.uk/projects/speech_tools/manual-1.2.0/

First Experiments on Text-to-Speech System Personification

Zdeněk Hanzlíček, Jindřich Matoušek, and Daniel Tihelka

University of West Bohemia, Faculty of Applied Sciences, Dept. of Cybernetics,
Univerzitní 8, 306 14 Plzeň, Czech Republic
{zhanzlic,jmatouse,dtihelka}@kky.zcu.cz

Abstract. In the present paper, several experiments on text-to-speech system personification are described. The personification enables TTS system to produce new voices by employing voice conversion methods. The baseline speech synthetizer is a concatenative corpus-based TTS system which utilizes the unit selection method. The voice identity change is performed by the transformation of spectral envelope, spectral detail and pitch. Two different personification approaches are compared in this paper. The former is based on the transformation of the original speech corpus, the latter transforms the output of the synthetizer. Specific advantages and disadvantages of both approaches are discussed and their performance is compared in listening tests.

1 Introduction

Within the concatenative corpus-based speech synthesis framework, a new voice can be obtained by recording a new large speech corpus by the demanded speaker. From that corpus, containing several thousands of utterances, a new unit inventory is created and used within the synthesis process [1]. However, recording of such a great amount of speech data is a difficult task. Usually, a professional speaker is required.

Alternatively, text-to-speech system personification [2] enables this system to produce new voices by employing voice conversion methods. Much fewer speech data are necessary. Our voice conversion system [3] converts spectral envelope and pitch by probabilistic transformation functions; moreover, spectral detail is transformed by employing residual prediction method.

Two different personification approaches are described and compared in this paper. The former is based on the original speech corpus transformation, the latter transforms the output of the synthesizer. Specific advantages and disadvantages of both approaches are discussed and the performance is compared by using preference listening tests.

The paper is organised as follows. In Section 2, the baseline TTS system planned to be personified is described. In Section 3, the voice conversion methods are specified. Section 4 deals with the TTS system personification task. Section 5 describes our first personification experiments. In Section 6, the results are discussed and future work is outlined.

V. Matoušek and P. Mautner (Eds.): TSD 2009, LNAI 5729, pp. 186–193, 2009.

2   Baseline TTS System

The text-to-speech system ARTIC employed in our personification experiments was in detail described in [1]. It has been built on the principles of concatenative speech synthesis. Primarily, it consists of three main modules: acoustic unit inventory, text processing module and speech production module. It is a corpus-based system, i.e. large and carefully prepared speech corpora are used as the ground for the automatic definition of speech synthesis units and the determination of their boundaries as well as for unit selection technique.

Our TTS system was designed for the Czech language, nevertheless many of its parts are language-independent. For our personification experiments, a female speech corpus containing 5,000 sentences (about 13 hours of speech) was employed. The block diagram of our TTS system is shown in Fig. 1.

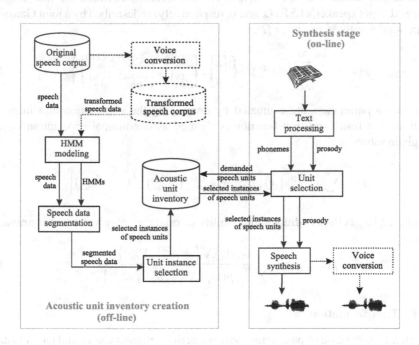

Fig. 1. A scheme of our TTS-system ARTIC including the both personification approaches – see dashed and dotted blocks

3   Voice Conversion System

The voice conversion system utilized for the aforementioned system personification was introduced in [3]. A simplified version of that system is described in this section. For the training of transformation functions, parallel utterances (i.e. pairs of source and target speakers' utterances) are employed. Voiced speech is analysed pitch synchronously; each segment is three pitch periods long and the shift of analysis window is one pitch period. Unvoiced segments are 10 msec long with 5 msec overlap. The spectral envelope of each frame is obtained by using the true envelope estimator [4] and represented by its line spectral frequencies (LSFs). The parameter order is selected individually for each speaker in order that the average envelope approximation error is lower than predefined threshold. Moreover, spectral detail is obtained as a complement of the spectral envelope into the full spectrum. In case of linear prediction analysis, the spectral detail corresponds to the residual signal spectrum. The LSF parameters and the fundamental frequency are transformed by probabilistic transformation functions. The spectral detail is estimated by a residual prediction method.

3.1 Parameter Transformation

Nowadays, the probabilistic (GMM-based) transformation [5] is the most often used transformation function in VC systems. The interrelation between the time-aligned source and target speaker's LSFs (x and y, respectively) is described by a joint Gaussian mixture model with M mixtures Ω_m^{p}

$$p(x, y) = \sum_{m=1}^{M} p(\Omega_m^{\mathrm{p}}) \mathcal{N}\left\{ \begin{bmatrix} x \\ y \end{bmatrix}; \begin{bmatrix} \mu_m^{(x)} \\ \mu_m^{(y)} \end{bmatrix}, \begin{bmatrix} \Sigma_m^{(x)} & \Sigma_m^{(xy)} \\ \Sigma_m^{(yx)} & \Sigma_m^{(y)} \end{bmatrix} \right\}. \tag{1}$$

All unknown parameters are estimated by employing the expectation-maximization algorithm. The transformation function is defined as conditional expectation of target y given source x

$$\tilde{y} = \sum_{m=1}^{M} p(\Omega_m^{\mathrm{p}} \mid x) \left[\mu_m^{(y)} + \Sigma_m^{(yx)} \left(\Sigma_m^{(x)} \right)^{-1} (x - \mu_m^{(x)}) \right], \tag{2}$$

where $p(\Omega_m^{\mathrm{p}} \mid x)$ is the conditional probability of mixture Ω_m^{p} given source parameter vector x

$$p(\Omega_m^{\mathrm{p}} \mid x) = \frac{p(\Omega_m^{\mathrm{p}}) \mathcal{N}\left\{ x; \mu_m^{(x)}, \Sigma_m^{(x)} \right\}}{\sum_{i=1}^{M} p(\Omega_i^{\mathrm{p}}) \mathcal{N}\left\{ x; \mu_i^{(x)}, \Sigma_i^{(x)} \right\}}. \tag{3}$$

3.2 F₀ Transformation

Analogically to the case of parameter conversion, time-aligned source and target instantaneous f$_0$ values $f^{(x)}$ and $f^{(y)}$ are described with a joint GMM

$$p\left(f^{(x)}, f^{(y)}\right) = \sum_{s=1}^{S} p(\Omega_s^{\mathrm{f}}) \mathcal{N}\left\{ \begin{bmatrix} f^{(x)} \\ f^{(y)} \end{bmatrix}; \begin{bmatrix} \mu_s^{(fx)} \\ \mu_s^{(fy)} \end{bmatrix}, \begin{bmatrix} \sigma_s^{(fx)} & \sigma_s^{(fxfy)} \\ \sigma_s^{(fyfx)} & \sigma_s^{(fy)} \end{bmatrix} \right\}. \tag{4}$$

Again, the converted fundamental frequency $\tilde{f}^{(y)}$ is given as the conditional expectation of target $f^{(y)}$ given source $f^{(x)}$

$$\tilde{f}^{(y)} = \sum_{s=1}^{S} p(\Omega_s^{\mathrm{f}} \mid f^{(x)}) \left[\mu_s^{(fy)} + \frac{\sigma_s^{(fyfx)}}{\sigma_s^{(fx)}} \left(f^{(x)} - \mu_s^{(fx)} \right) \right]. \tag{5}$$

3.3 Spectral Details Transformation

Spectral detail is also very important for speaker identity perception. It is a complement of the spectral envelope into the full spectrum and consists of amplitude and phase parts — $A(\omega)$ and $\varphi(\omega)$, which are converted separately. Its transformation usually utilizes the relationship to the shape of spectral envelope, e.g. by employing codebooks [6].

The training stage starts with the clustering of training parameter vectors y into Q classes Ω_q^r; k-means algorithm is employed. Each class Ω_q^r is represented by its centroid $\bar{y}_q$ and covariance matrix S_q. The pertinence of parameter vector y_n to class Ω_q^r is defined as

$$w(\Omega_q^r \mid y_n) = \frac{\left[(y_n - \bar{y}_q)^T S_q^{-1}(y_n - \bar{y}_q)\right]^{-1}}{\sum_{i=1}^{Q}\left[(y_n - \bar{y}_i)^T S_i^{-1}(y_n - \bar{y}_i)\right]^{-1}}. \tag{6}$$

All target speaker's training data are uniquely classified into those classes. For each class Ω_q^r, a set $\mathbb{R}_q$ of pertaining data indexes is established

$$\mathbb{R}_q = \left\{k; \ q = \arg\max_{q=1...Q} w(\Omega_q^r \mid y_k)\right\}. \tag{7}$$

Within each parameter class Ω_q^r, the training data are divided into L_q subclasses $\Omega_{q,\ell}^r$ according to the instantaneous fundamental frequency f_0. Each subclass $\Omega_{q,\ell}^r$ is described by its centroid $\bar{f}_{q,\ell}^{(y)}$. The data belonging into this subclass are defined using a set $\mathbb{R}_{q,\ell}$ of corresponding data indices

$$\mathbb{R}_{q,\ell} = \left\{k; \ k \in \mathbb{R}_q \ \wedge \ \ell = \arg\min_{\ell=1...L_q}\left|f_k^{(y)} - \bar{f}_{q,\ell}\right|\right\}. \tag{8}$$

For each subclass $\Omega_{q,\ell}^r$, a typical spectral detail is determined as follows. Typical amplitude spectrum $\hat{A}_{q,\ell}^{(y)}(\omega)$ is determined as the weighted average over all amplitude spectra $A_n^{(y)}(\omega)$ belonging into that subclass

$$\hat{A}_{q,\ell}^{(y)}(\omega) = \frac{\sum_{n\in\mathbb{R}_{q,\ell}} A_n^{(y)}(\omega)w(\Omega_q^r \mid y_n)}{\sum_{n\in\mathbb{R}_{q,\ell}} w(\Omega_q^r \mid y_n)} \tag{9}$$

and the typical phase spectrum $\hat{\varphi}_{q,\ell}^{(y)}(\omega)$ is selected

$$\hat{\varphi}_{q,\ell}^{(y)}(\omega) = \varphi_{n^*}^{(y)}(\omega) \qquad n^* = \arg\max_{n\in\mathbb{R}_{q,\ell}} w(\Omega_q^r \mid y_n). \tag{10}$$

During the transformation stage, for the transformed parameter vector $\tilde{y}_n$ and fundamental frequency $\tilde{f}_n^{(y)}$, the amplitude spectrum $\tilde{A}_n^{(y)}(\omega)$ is calculated as the weighted average over all classes Ω_q^r. However, for each class Ω_q^r, only one subclass $\Omega_{q,\ell}^r$ is selected in such a way that its centroid $\bar{f}_{q,\ell}^y$ is the nearest to frequency $\tilde{f}_n^{(y)}$

$$\tilde{A}_n^{(y)}(\omega) = \sum_{q=1}^{Q} w(\Omega_q^r \mid \tilde{y}_n)\hat{A}_{q,\ell_q}^{(y)}(\omega) \qquad \ell_q = \arg\min_{\ell=1...L_q}\left|\tilde{f}_n^{(y)} - \bar{f}_{q,\ell}^{(y)}\right|. \tag{11}$$

The phase spectrum $\tilde{\varphi}_n(\omega)$ is selected from the parameter class Ω_q^{r*} with the highest weight $w(\Omega_q^r \mid \tilde{y}_n)$ from that subclass $\Omega_{q,\ell}^{r*}$ having the nearest central frequency $\bar{f}_{q,\ell}^{(y)}$

$$\tilde{\varphi}_n^{(y)}(\omega) = \hat{\varphi}_{q^*,\ell^*}^{(y)}(\omega) \qquad q^* = \arg\max_{q=1\ldots Q} w(\Omega_q^r \mid \tilde{y}_n)$$

$$\ell^* = \arg\min_{\ell=1\ldots L_{q^*}} \left| \tilde{f}_n^{(y)} - \bar{f}_{q^*,\ell}^{(y)} \right|. \tag{12}$$

4 TTS System Personification

4.1 Personification Approaches

In principle, two main approaches to concatenative TTS system personification exists.

1. *Transformation of the original speech corpus* – a new unit inventory is created from the transformed corpus. Thus for each new voice an individual unit inventory is created and ordinarily used for the speech synthesis.
2. *Transformation of TTS system output* – a transformation module is added to the TTS system. The generation of the new voice is performed in two stages: synthesis of the original voice and transformation to the target voice.

Each of these approaches has specific advantages and also disadvantages. The approach based on original corpus transformation can be characterized as follows:

+ The converted corpus can be checked and poorly transformed utterances rejected. Thus the influence of conversion failure can be suppressed.
+ The synthesis process is straightforward and it is not delayed by additional transformation computation.
− The preparation of new voice is time consuming – the whole corpus has to be converted and a new unit inventory built.
− Huge memory requirements for storing several acoustic unit inventories, especially in cases when more different voices should be alternatively synthesized.

Properties of the second approach can be briefly summarized:

+ A new voice can be simply and quickly acquired, only a new set of conversion functions has to be added.
+ Lower memory requirements – only the original unit inventory and conversion functions for other voices have to be stored.
− The resulting system works slower – an extra computation time for transformation is needed.

4.2 Data Origin

In our conversion system, parallel speech data is necessary for the training of conversion function. Within the TTS system personification framework, source speaker's speech data can be obtained in two different ways

- *Natural source speech data* – Source speaker's utterances are selected from the original corpus. The recording of additional utterances by the source speaker is less suitable, especially in cases when a long time has elapsed since original corpus recording, because his/her voice could change since that time.
- *Synthetised source speech data* – Source's speaker utterances are generated by the TTS system. This is necessary in cases when target speaker's utterances are given, but not involved in the source corpus. Moreover, none of both speakers is available for an additional recording.

Considering the training and transformation stage consistency, a natural training data seems to be preferable for the source corpus conversion. However, in the case of TTS system output transformation, a sythetised source training data is more suitable.

5 Experiments

The performance of a conversion system can be evaluated by using so-called performance indices

$$P_{\text{par}} = 1 - \frac{\sum_{i=1}^{N} \mathcal{D}(\tilde{y}_n, y_n)}{\sum_{i-1}^{N} \mathcal{D}(x_n, y_n)} \qquad P_{\text{sp}} = 1 - \frac{\sum_{i=1}^{N} \mathcal{D}\big(\tilde{A}_n^{(y)}(\omega), A_n^{(y)}(\omega)\big)}{\sum_{i-1}^{N} \mathcal{D}\big(A_n^{(x)}(\omega), A_n^{(y)}(\omega)\big)}, \qquad (13)$$

where x_n, y_n and $\tilde{y}_n$ are source, target and transformed parameter vectors, $A_n^{(x)}(\omega)$, $A_n^{(y)}(\omega)$ and $\tilde{A}_n^{(y)}(\omega)$ are corresponding spectral envelopes and $\mathcal{D}$ is usually the Euclidean distance.

The higher are the values of parameter and spectral performance index, the higher is the similarity between transformed and target utterances in comparison with the original similarity between source and target utterances.

In addition to those objective mathematical rates, the speech produced by conversion system or by the personified TTS system can be evaluated in listening tests. For comparison of several system setups a preference test can be employed.

In our experiments, two nonprofessional target speakers recorded 50 quite short sentences (about 6–8 words long), which were selected from the corpus mentioned in Section 2. Thus, parallel training data was available. Within the training stage, 40 utterance pairs were used for the estimation of conversion function parameters.

5.1 The Influence of Data Origin

Regardless of the personification approach, source speaker's training data can either be natural or synthetised by the TTS system (or both together, but that case was not taken into account). Hereinafter, we use notation NTD/STD function for conversion functions trained by using natural/synthetised source training data.

A question arises whether the conversion function trained on natural data could be used for synthetised speech transformation and vice versa. Thus, NTD and STD conversion functions were trained and employed for the transformation of both natural and synthetised speech. An objective comparison of NTD and STD performance, based on

Table 1. The influence of training data origin: natural or synthetised

Training	Testing	Male 1		Male 2	
data	data	P_{par}	P_{sp}	P_{par}	P_{sp}
natural	natural	0.239	0.230	0.354	0.344
synth.	synth.	0.242	0.233	0.331	0.324
synth.	natural	0.194	0.194	0.357	0.347
natural	synth.	0.209	0.200	0.325	0.316

performance indices, is presented in Table 1. The utterances, that were not included in the training set, were used for this assessment.

Moreover, informal listening test was carried out. 10 participants listened to the pairs of utterances transformed by NTD and STD function. The natural and synthetised utterances from both target speakers were evenly occured in the test. In each testing pair, the listeners should select a preferred utterance according to the overall voice quality. The similarity to the real target speaker's voice was not taken into account. The results of this test are presented in Figures 2 and 3.

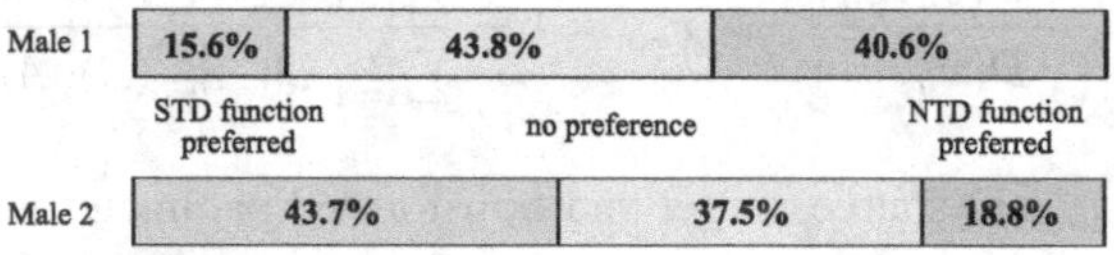

Fig. 2. Preference listening test: Synthetised speech transformation

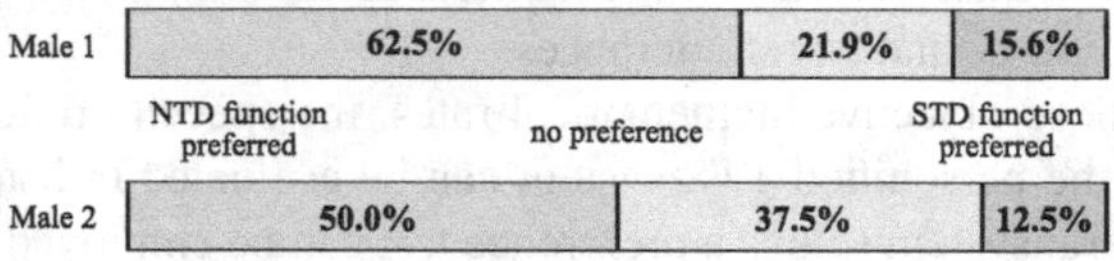

Fig. 3. Preference listening test: Natural speech transformation

The results of the mathematical evaluation and the listening test are consistent. For both speakers, natural speech is better transformed by NTD function. The results for synthetised speech transformation differ for particular speakers. However, the differences between the utterances were mostly insignificant.

5.2 Personification Approaches Comparison

For the comparison of described personification approaches another preference listening test was employed. Again, participants listened to the pairs of utterances produced by the TTS systems personified by the source corpus transformation (approach 1) and synthetiser output transformation (approach 2). The results are presented on Figure 4. For both speakers approach 2 was preferred.

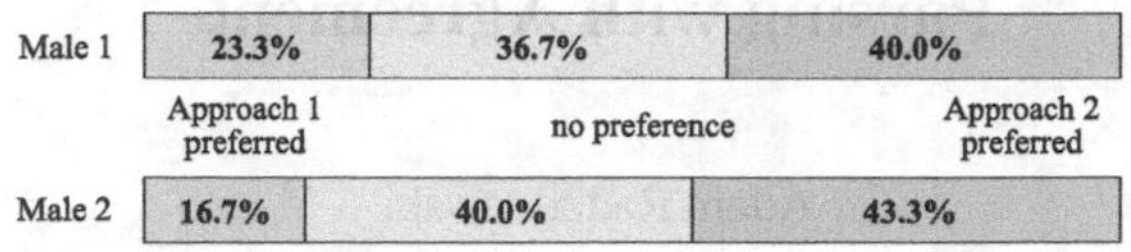

Male 1	23.3%	36.7%	40.0%
	Approach 1 preferred	no preference	Approach 2 preferred
Male 2	16.7%	40.0%	43.3%

Fig. 4. Preference listening test: Personification approaches comparison

6 Conclusion

In this paper, two different approaches to the TTS system personification were compared. The former is based on the original speech corpus transformation and a new unit inventory creation, the latter transforms the output of the original TTS system. In listening tests the corpus transformation approach revealed to be slightly preferred. However, the differences were not too significant. Thus, both approaches are well applicable. Their specific advantages and disadvantages should be considered for concrete applications.

Acknowledegments

Support for this work was provided by the Ministry of Education of the Czech Republic, project No. 2C06020, and by the Grant Agency of the Czech Republic, project No. GAČR 102/09/0989.

References

1. Matoušek, J., Tihelka, D., Romportl, J.: Current state of czech text-to-speech system ARTIC. In: Sojka, P., Kopeček, I., Pala, K. (eds.) TSD 2006. LNCS (LNAI), vol. 4188, pp. 439–446. Springer, Heidelberg (2006)
2. Kain, A., Macon, M.W.: Personalizing a Speech Synthesizer by Voice Adaptation. In: Proceedings of SSW, Blue Mountains, Australia, pp. 225–230 (1998)
3. Hanzlíček, Z., Matoušek, J.: Voice conversion based on probabilistic parameter transformation and extended inter-speaker residual prediction. In: Matoušek, V., Mautner, P. (eds.) TSD 2007. LNCS (LNAI), vol. 4629, pp. 480–487. Springer, Heidelberg (2007)
4. Villavicencio, F., Röbel, A., Rodet, X.: Improving LPC Spectral Envelope Extraction of Voiced Speech by True-Envelope Estimation. In: Proceedings of ICASSP, Toulouse, France, pp. 869–872 (2006)
5. Stylianou, Y., Cappé, O., Moulines, E.: Continuous Probabilistic Transform for Voice Conversion. IEEE Trans. on Speech and Audio Processing 6(2), 131–142 (1998)
6. Kain, A.: High Resolution Voice Transformation. Ph.D. thesis, Oregon Health & Science University, Portland, USA (2001)

Parsing with Agreement

Adam Radziszewski

Institute of Informatics, Wrocław University of Technology,
Wybrzeże Wyspiańskiego 27, Wrocław, Poland
adam.radziszewski@pwr.wroc.pl

Abstract. Shallow parsing has been proposed as a means of arriving at practically useful structures while avoiding the difficulties of full syntactic analysis. According to Abney's principles, it is preferred to leave an ambiguity pending than to make a likely wrong decision. We show that continuous phrase chunking as well as shallow constituency parsing display evident drawbacks when faced with freer word order languages. Those drawbacks may lead to unnecessary data loss as a result of decisions forced by the formalism and therefore diminish practical value of shallow parsers for Slavic languages.

We present an alternate approach to shallow parsing of noun phrases for Slavic languages which follows the original Abney's principles. The proposed approach to parsing is decomposed into several stages, some of which allow for marking discontinuous phrases.

1 Introduction

Shallow parsing is a general term describing a range of tasks and techniques aiming at producing useful yet not complete syntactic analysis of text. The basic assumption is to prefer reliability to completeness, leaving an ambiguity pending wherever the parser has no knowledge necessary to make a decision [1]. Arguably the most popular tasks within the realm of shallow parsing are: *chunking* — marking completely flat phrases without any structuring and *shallow constituency parsing* — constituency parsing with limited depth of the analysis.

Along with a technique based on finite-state transducers, Abney [2] proposed a general philosophy which is now commonly perceived as general principles of shallow parsing. The principles are outlined below.

Easy-first parsing. The easy decisions are made first, leaving more complex problems for next stages.

Containment of ambiguity. Ambiguity is resolved gradually; some cases may remain unsolved.

High precision. As too difficult decisions are not made, it is possible to reach low error rates (at the cost of lower recall).

Growing islands of certainty. The order of resolution reflects the growing complexity of decisions rather than standard bottom-up or top-down manner.

Most of the research in shallow parsing is focused on English. Its characteristics are reflected in the popularity of commonly performed tasks and employed formalisms. We

V. Matoušek and P. Mautner (Eds.): TSD 2009, LNAI 5729, pp. 194–201, 2009.

argue that freer word order and rich inflection typical for Slavic languages bring about a need to reconsider what the basic tasks should assume. We propose an alternative to chunking and shallow constituency parsing of noun phrases which avoids some of the problems posed by discontinuities.

2 Shallow Parsing of Slavic Languages

The relatively free word order of Slavic languages makes the shallow processing harder. For instance, techniques based on identification of n-grams in text are likely to fail as there are numerous permutations allowed. What is more, the rich inflectional morphology implies a multitude of different forms as well as a need for a large tagset to describe the morphological properties of those forms [3].

On the other hand, many dependencies are manifested in morphosyntactic constraints satisfied by the whole phrase, rather than by the linear precedence of its constituents[1]. A typical example is the agreement on number, gender and case — a strong support for the hypothesis that a given set of words belongs to one phrase, even if those words are not adjacent on the surface. Agreement on case (and usually number) is typical for Slavic coordinated phrases.

Slavic languages are typically processed with hand-written grammars. There are several formalisms allowing to mark desired types of word sequences in text and sometimes also select proper morphological tags for them. These formalisms are usually of slightly higher expressive power than regular grammars, allowing for such extensions as agreement on selected features or string concatenation mechanisms for reconstruction of lemmas.

Nenadić and Vitas [5] propose a formalism for capturing Serbo-Croatian noun phrases. The formalism is based on what the authors call *regular morphosyntactic expressions* — regular grammars allowing for referring to particular word forms by specifying some of their morphosyntactic properties. It has also been augmented with means of constraining a constituent to satisfy an agreement on given features [6]. Two grammars for capturing different types of Serbo-Croatian noun chunks have been written manually: chunks satisfying agreement, possibly containing genitive post-modifiers [6], and chunks representing coordinated noun phrases [7]. Later, the same formalism has been used for describing compound verbs in preterite tense [8]. Although an attempt was made to cover discontinuities and permutations of compound verbs, the flexibility is expressed by a small set of explicit grammatical rules. Discontinuity is handled as a rule allowing for particular types of *inserts* between the auxiliary and the main verb. The phrase is marked as a continuous preterite construct (including the intervening inserts) with the verbal parts labelled with morphosyntactic tags, resulting in somewhat cumbersome representation (practically, to abstract from the inserts and phrase ordering one needs yet another grammar). None of these grammars have been evaluated. The

---

[1] The linear precedence is obviously not random and is subjected to many syntactic constraints. This is why we use the term *freer* or *relatively free* word order. Nevertheless, the precedence seems to be governed by a set of constraints and preferences rather than by fixed rewrite rules [4].

only claim about their accuracy is that the grammar for coordinated noun phrases is not complete although it did not mispredict any construct [7].

A similar work is presented in [9] — hand-crafted regular grammars are used to identify noun phrase (NP), prepositional phrase (PP) and verb phrase (VP) chunks in Croatian text. Discontinuity is not supported. NP chunks are sequences of word forms that either (i) agree on number, gender and case (simple NPs) or (ii) sequences of the simple NPs agreed on number and case (appositions) or (iii) any NPs linked with a co-ordinating conjunct, agreed on case only. This task performed on 137-sentence corpus yields the precision of 94.50% and recall of 90.26%. Although these results are promising, the extracted phrases seem surprisingly short (among the given examples only very few exceed four words). This is probably due to the fact that only continuous phrases are included.

Przepiórkowski [10,11] proposes a formalism for shallow constituency parsing inter-mingled with morphosyntactic disambiguation, targeted at Polish language. The formalism is suited for writing rules that can group adjacent words or already marked phrases. Besides grouping, words or *syntactic words* (roughly speaking, multi-word expressions) can be marked as a syntactic or semantic head of a particular group. The matching expressions are conceptually similar to the aforementioned *regular morphosyntactic expressions*, with slightly more expressive power (e.g. we can decide to capture a word whose *all* tags satisfy given specification). The formalism has been applied to specific problems: extraction of valence dictionaries for Polish [12] and sentiment analysis of product reviews [13]. To the best of our knowledge, no general-purpose grammar for shallow parsing of Polish (or Polish NPs) has been written so far.

3   What Do We Lose?

Many syntactic dependencies impose strict constraints on morphosyntactic properties of involved word forms, while limiting their linear order only partially. This results in dependencies between phrases that are non-adjacent on the surface. It is impossible to represent such dependencies as one chunk. Even if this dependency is enclosed within a bigger phrase which we want to mark as a whole (we will call such a case a *local discontinuity*), we are unable to decompose the process into simple steps. We either need a grammar for the whole complex phrase including the dependency in question (which yields in quiet deep analysis) or we evade the dependency and excuse ourselves for not attaching it in terms of the formalism (e.g. it does not belong to the constituent). The problem is that if we decide not to attach one of the parts of the phrase, the ambiguity will appear resolved (some phrase is marked) and the knowledge about the undecided part of the dependency will be lost.

In the rest of this section we discuss some examples of phrases containing disconti-nuities, mostly noun phrases. All examples are from Polish.

3.1   Local Discontinuities

The problem of local discontinuities was discovered during our efforts to manually annotate a reference corpus for chunking of Polish NPs. The assumption was to annotate *simple NP* chunks — limited by the extent of the agreement (following the Croatian chunker [9]), but also *whole NP* chunks — including genitive/dative modifiers as well as PPs. During the chunker performance, *simple NPs* were intended to be identified first and constitute a basis for recognition of *whole NPs*. The examples below[2] show both kinds of chunks.

(1) wbrew [$_{NP}$ niektórym utworom]   u   [$_{NP}$ nas]   [$_M$ publikowanym]
     despite some   compositions at   us   published
     prep   adj-dat-pl-n   n-dat-pl-n   prep pron-dat-pl ppas-dat-pl-n
     'despite some compositions published in our country'

(2) wbrew [$_{WholeNP}$ niektórym utworom u nas publikowanym]

(3) wbrew [$_{NP}$ niektórym utworom publikowanym] u [$_{NP}$ nas]

In the first example the agreed NP *niektórym utworom* is marked. This is where a standard NP chunker would stop: there are no more chunks with a nominal head. Careful analysis gives us a deeper insight: there is agreement between this NP chunk and the modifier *publikowanym* (we will use *M* to mark orphaned NP modifiers). Although these two parts indeed describe one NP, they cannot be joined within one *simple NP* chunk as the agreement would be violated by the intervening PP. Example (2) shows the result of *whole NP* chunking. This task would be easier if we knew that those two phrases belong together. Without this knowledge, attaching the orphaned modifier needs reconsidering the agreement problem. Annotating *simple NPs* is therefore not really useful as we cannot abstract from the problems present at that stage. What is more, if we stop at the first stage (just as the Croatians do), we will suggest that an NP is already recognised and there is no ambiguity left (breaking the *Containment of Ambiguity* principle). There is certain data loss here, which seems even more odd when we consider this very phrase with alternate ordering (which is perfectly grammatical and common) — see example (3).

Some more examples are given below. The phrase in (4) contains a dative modifier — a personal pronoun. Phrases can contain more modifiers, as in (5). It should be stressed that by grouping the agreed parts together we deliberately do not try to attach the insert to the part it modifies. The ambiguity is visible and may be resolved later if needed.

(4) [$_M$ innych   współczesnych] [$_{NP}$ mu]   [$_{NP}$ kierunków]
     other   contemporary   he   trends
     adj-gen-pl-m adj-gen-pl-m   pron-dat-sg n-gen-pl-m
     '(of) other trends contemporary to him'

(5) [$_{NP}$ ruchy] [$_{NP}$ oporu] [$_M$ podległe] [$_{NP}$ rządom]
     movements resistance dependent   governments
     n-nom-pl-m n-gen-sg-m adj-nom-pl-m n-dat-pl-m
     'resistance organisations dependent on governments'

---

[2] The examples in this paper are either taken directly from the *corpus of the frequency dictionary* (*korpus słownika frekwencyjnego*) [14] or cited from Derwojedowa [4].

3.2 Global Discontinuities

Global discontinuities are those that render chunking the whole phrase impossible, i.e. the assumed type of phrase (NP in our considerations) is discontinuous even at the most outer level. They roughly correspond to what the grammarians would call discontinuous constructs.

Such discontinuities within NPs seem rare — at least in written language. Example (6) shows such a construct. The phrase *skrzypcach ulubionych* entwines the main verb. Global discontinuities are quite common in Polish compound verb forms [4]. Example (7), being a common Polish phrase, shows interweaving of the infinitive *zdawać sie* (*to seem*) with the verb form *mogłoby* (*it might*; the conditional particle *by* is detached).

(6) na   [$_{NP}$ skrzypcach] grał   [$_M$ ulubionych]
     on   violin   played   favourite
     prep n-loc-pl-p   v-praet-sg adj-loc-pl-p

     'he played (his) favourite violin'

(7) [$_{V_1}$ zdawać] [$_{V_2}$ by] [$_{V_1}$ się] [$_{V_2}$ mogło]
     seem   might
     v-inf   cond refl   v-praet-sg

     'it might seem'

An interesting problem appears within predicative constructs. Although they do not contain linguistically motivated discontinuity, there are at least two practical reasons for following the agreement between the subject and the predicative adjective: (1) the adjective semantically expresses properties of the subject (just as modifiers do) and (2) doing so is consistent with our approach that agreement is the basis for distinguishing phrases. Consider clause (8). Under these assumptions, the subject (*natura*) is as an NP and the adjective is its orphaned modifier. Some additional motivation for treating them as noun phrases stems from the fact that the order is sometimes altered, resulting in a continuous agreed chunk — cf. (9). It seems difficult to distinguish such cases from regular NPs.

(8) [$_{NP}$ natura] [$_{NP}$ człowieka] jest   [$_M$ społeczna]
     nature   man   is   social
     n-nom-sg-f n-gen-sg-m   v-fin-sg adj-nom-sg-f

     'the nature of man is social'

(9) bywa   [$_{NP}$ to   wszystko   słuszne]
     happens this   all   fair
     v-fin-sg n-nom-sg-n adj-nom-sg-n adj-nom-sg-n

     'all this happens to be fair'

4 Proposed Scheme of Shallow Parsing

The proposed approach is to decompose the process of parsing into several stages of increasing difficulty. The stages are divided into two parts: primary (agreement-based) and secondary (attachment-based). The secondary stages are considered optional and may be adjusted to foreseen applications. The general scheme is presented below. The stages 1–4 are primary.

1. NP and M chunking.
2. Coindexing.
3. Recognition of apposition and coordination.
4. Index unification.
5. Dative/genitive modifier attachment.
6. Recognition of PrepNP and its attachment.

Stages 1 and 3 correspond to a chunking task with several channels (or several chunking tasks, each with one channel). We employ the following channels:

Noun phrases (NP) agreed on number, gender and case. Must contain a nominal head, may contain adverbs and indeclinable forms. Coordination on adjectives seems safe and can be included within a chunk.

Satellites (M) — as above, with no nominal elements. They are potential NP modifiers, although may be something else (e.g. adjectives used nominally).

Appositions (Appo) — adjacent NPs agreed on number and case.

Coordinations (Coord) — NPs or appositions linked with coordinating conjuncts, agreed on case.

Longest NP (LNP) — all the chunks copied from previous channels, preferring longest matches when there are alternatives.

We assume that the input has been morphologically analysed although not necessarily disambiguated. Partial disambiguation will be done along with identification of agreements: when marking an agreement, we constrain the tags of words to those satisfying the agreement (it corresponds to unify action in [10]).

The process begins with chunking of NP and M. It can be driven by a hand-written grammar or using Machine Learning techniques. Next, some NPs and satellites are linked together. The necessary condition for linking is the agreement on number, gender and case. Links can be marked by augmenting chunk representations with indices. The decisions on coindexing can be supported by various sources of information: the distance between chunks, the type of intervening material, the values of morphosyntactic categories of the words involved.

Appositions are chunked in a separate channel. This stage requires some carefully prepared heuristics, especially when dealing with non-disambiguated input. Coordination seems easier as it is limited by the presence of conjuncts.

In the end, LNP channel is populated with all so far recognised phrases. Longer phrases are preferred whenever there is a conflict. The indices of NPs covered by longer phrases are unified and the same is done to all their occurrences. The whole process related to primary stages is presented in Fig. 1.

The primary stages go beyond the standard chunking problem, resulting in a practically useful structure: *simple NP* chunks are recognised along with their orphaned modifiers. What is more, coordination and apposition is explicitly marked in separate channels. Thanks to this information, identifying NP heads and the sets of their modifiers is straightforward. What is more, we achieve partial morphosyntactic disambiguation. If the initially provided morphological information included lemmas, one can

1. [$_M$opracowane i   rozwinięte] przez [$_{NP}$nią] [$_{NP}$metody] i [$_{NP}$techniki] [$_{NP}$badań]
   elaborated and developed by   she   methods and techniques research

```
ppas-nom-sg-n  conj   ppas-nom-sg-n  prep   pron-acc-sg   n-gen-sg-f  conj  n-gen-sg-f   n-gen-pl-n
ppas-acc-sg-n         ppas-acc-sg-n         pron-inst-sg  n-nom-pl-f        n-nom-pl-f
ppas-nom-pl-m         ppas-nom-pl-m                       n-acc-pl-f        n-acc-pl-f
ppas-nom-pl-f         ppas-nom-pl-f                       n-voc-pl-f        n-voc-pl-f
ppas-nom-pl-n         ppas-nom-pl-n
ppas-acc-pl-m         ppas-acc-pl-m
ppas-acc-pl-f         ppas-acc-pl-f
ppas-acc-pl-n         ppas-acc-pl-n
```

2. [$_{M1}$opracowane i   rozwinięte] przez [$_{NP2}$nią] [$_{NP1}$metody] i [$_{NP3}$techniki] [$_{NP4}$badań]

```
ppas-nom-pl-f  conj   ppas-nom-pl-f  prep   pron-acc-sg   n-nom-pl-f  conj  n-gen-sg-f   n-gen-pl-n
ppas-acc-pl-f         ppas-acc-pl-f         pron-inst-sg  n-acc-pl-f        n-nom-pl-f
                                                                            n-acc-pl-f
                                                                            n-voc-pl-f
```

3. [$_{M1}$opracowane i rozwinięte] przez [$_{NP2}$nią] [$_{Coord}$[$_{NP1}$metody] i [$_{NP3}$techniki]][$_{NP4}$badań]

```
ppas-nom-pl-f  conj ppas-nom-pl-f  prep   pron-acc-sg   n-nom-pl-f  conj  n-nom-pl-f   n-gen-pl-n
ppas-acc-pl-f       ppas-acc-pl-f         pron-inst-sg  n-acc-pl-f        n-acc-pl-f
```

4.[$_{MU1}$opracowane i rozwinięte] przez [$_{NP2}$nią]   [$_{NPU1}$metody i   techniki] [$_{NP4}$badań]

```
ppas-nom-pl-f  conj ppas-nom-pl-f  prep   pron-acc-sg   n-nom-pl-f  conj  n-nom-pl-f   n-gen-pl-n
ppas-acc-pl-f       ppas-acc-pl-f         pron-inst-sg  n-acc-pl-f        n-acc-pl-f
```

Fig. 1. Primary stages of parsing an example phrase

transform a sentence into a sequence of noun–set-of-modifiers pairs, which can be very useful for Information Extraction (IE) tasks.

Secondary stages, corresponding to various attachment problems, can be performed disregarding recognised inter-chunk links. This way, the problem is reduced to deciding for each pair of chunks if there is a modifier-head relation between them (and classifying the type of relation). In the case of PP attachment, one needs to recognise PP chunks earlier. Although these problems are stated clearly, the tasks are not trivial and are considered issues for further research (similar problems for English are addressed in [1]). Secondary stages result in an extremely useful structure. The NP chunks can be joined together with all their modifiers, yielding *whole NP* chunks. They may be directly used for IE tasks, particularly for Named Entity Recognition.

5   Conclusions and Further Work

The presented approach allows for performing shallow parsing and partial disambiguation of Slavic NPs. It consists of several stages of increasing difficulty, where the last stages are optional (*easy-first parsing*). The discussed problems with discontinuous phrases are avoided by effectively allowing for discontinuous chunks. Consequently, we are not limiting the recognised phrases to the structures cut by a particular ordering of words (as the word order is subjected to alternations, the cutting point can be somewhat accidental). We therefore avoid unnecessary loss of information. The proposed representation of discontinuities is simple, allowing for performing the attachment-related tasks in the usual way: grouping adjacent chunks together. If we decide not to perform those tasks, the attachment ambiguities remain explicit (according to the *containment of ambiguity* principle).

To evaluate this approach, an implementation is needed. We plan to develop a practical NP chunker for Polish, focusing on the primary stages. The accuracy of disambiguation will also be assessed. Next, we plan a detailed research on attachment-related tasks and finally enhance the chunker with this functionality.

References

1. Abney, S.: Part-of-speech tagging and partial parsing. In: Corpus-Based Methods in Language and Speech, pp. 118–136. Kluwer Academic Publishers, Dordrecht (1997)
2. Abney, S.: Partial parsing via finite-state cascades. In: Natural Language Engineering, pp. 8–15 (1996)
3. Przepiórkowski, A.: Slavic information extraction and partial parsing. In: Proceedings of the Workshop on Balto-Slavonic Natural Language Processing, Prague, Czech Republic, Association for Computational Linguistics, pp. 1–10 (June 2007)
4. Derwojedowa, M.: Porządek linearny składników zdania elementarnego w języku polskim. Dom Wydawniczy Elipsa, Warsaw (2000)
5. Nenadić, G., Vitas, D.: Formal model of noun phrases in serbo-croatian. BULAG (23), 297–311 (1998)
6. Nenadić, G., Vitas, D.: Using local grammars for agreement modeling in highly inflective languages. In: Sojka, P., Kopeček, I., Pala, K. (eds.) Proceedings of TSD 1998, Brno, Czech Republic, pp. 91–96. Springer, Heidelberg (1998)
7. Nenadić, G.: Local grammars and parsing coordination of nouns in serbo-croatian. In: Sojka, P., Kopeček, I., Pala, K., Kopeček, I. (eds.) TSD 2000. LNCS (LNAI), vol. 1902, pp. 57–62. Springer, Heidelberg (2000)
8. Nenadić, G., Vitas, D., Krstev, C.: Local grammars and compound verb lemmatization in serbo-croatian. Current Issues in Formal Slavic Linguistics, 469–477 (1999)
9. Kristina Vučković, M.T., Dovedan, Z.: Rule-based chunker for croatian. In: European Language Resources Association (ELRA) (ed.) Proceedings of the Sixth International Language Resources and Evaluation (LREC 2008), Marrakech, Morocco (May 2008)
10. Przepiórkowski, A.: A preliminary formalism for simultaneous rule-based tagging and partial parsing. In: Data Structures for Linguistic Resources and Applications: Proceedings of the Biennial GLDV Conference 2007, pp. 81–90. Gunter Narr Verlag, Tübingen (2007)
11. Przepiórkowski, A.: Powierzchniowe przetwarzanie języka polskiego. Akademicka Oficyna Wydawnicza EXIT, Warsaw (2008)
12. Przepiórkowski, A.: Towards a partial grammar of Polish for valence extraction. In: Proceedings of Grammar and Corpora 2007, Liblice, Czech Republic (2007)
13. Buczyński, A., Wawer, A.: Shallow parsing in sentiment analysis of product reviews. In: Proceedings of the Partial Parsing workshop at LREC 2008, pp. 14–18 (2008)
14. Ogrodniczuk, M.: Nowa edycja wzbogaconego korpusu słownika frekwencyjnego. Językoznawstwo w Polsce. Stan i perspektywy, 181–190 (2003)

An Adaptive BIC Approach for Robust Speaker Change Detection in Continuous Audio Streams

Janez Žibert[1], Andrej Brodnik[1], and France Mihelič[2]

[1] Primorska Institute of Natural Sciences and Technology, University of Primorska,
Muzejski trg 2, Koper, SI-6000, Slovenia
janez.zibert@upr.si, andrej.brodnik@upr.si
[2] Faculty of Electrical Engineering, University of Ljubljana,
Tržaška 25, Ljubljana, SI-1000, Slovenia
france.mihelic@fe.uni-lj.si

Abstract. In this paper we focus on an audio segmentation. We present a novel method for robust and accurate detection of acoustic change points in continuous audio streams. The presented segmentation procedure was developed as a part of an audio diarization system for broadcast news audio indexing. In the presented approach, we tried to remove a need for using pre-determined decision-thresholds for detecting of segment boundaries, which are usually the case in the standard segmentation procedures. The proposed segmentation aims to estimate decision-thresholds directly from the currently processed audio data and thus reduces a need for additional threshold tuning from development data. It employs change-detection methods from two well-established audio segmentation approaches based on the Bayesian Information Criterion. Combining methods from both approaches enabled us to adaptively tune boundary-detection thresholds from the underlying processing data. All three segmentation procedures are tested and compared on a broadcast news audio database, where our proposed audio segmentation procedure shows its potential.

1 Introduction

In many speech and audio applications, that operate on continuous audio streams, it is necessary to partition and classify different audio sources and acoustic events prior to the application's main processing tasks. For example, audio streams in broadcast news transcription systems should be firstly chopped into smaller acoustic-homogeneous segments, which are then classified as speech and non-speech, since the main task - - speech recognition - is performed only on speech segments [8]. The same pre-processing steps are necessary in audio-diarization and audio-indexing systems, where the obtained speech segments are additionally clustered together by similar acoustic or speaker properties [11], or in speaker detection and tracking systems, that operate in multiple speaker environments, and therefore audio data should be organized in a way to enable speaker recognition on data that belong to one speaker only [7].

The task of partitioning audio streams to homogeneous segments according to similar acoustic properties is known as an audio segmentation task. The aim of the audio segmentation is to detect positions in audio streams, that correspond to change points

V. Matoušek and P. Mautner (Eds.): TSD 2009, LNAI 5729, pp. 202–209, 2009.

between different acoustic events or audio sources. Which change points in an audio stream should be detected, depends on the main application, which is fed by the obtained audio segments, and on the pre-processing steps before audio segmentation, which define the segmentation tasks. Typically, change detection procedures aim to find speech/non-speech segments or speaker change points. In addition, change points between different recording or background conditions can be also considered in a segmentation.

The main development of audio segmentation approaches was achieved along with the development of systems for processing of broadcast news audio data. The task of partitioning the audio streams for speech recognition was first introduced during the NIST Broadcast News Transcription evaluation in years 1997-98 [8]. Audio segmentation and clustering procedures were then fully evaluated in the NIST Rich Transcription speaker-diarization evaluation in years 2003-04 [4,11]. In 2005, the ESTER evaluation was conducted on similar task using French radio BN data [5]. Another evaluation campaign on multilingual BN data was carried on in years 2004-05 during the COST278 action [14]. All these studies have contributed to the development of audio-segmentation approaches as well as they have demonstrated the importance of these tasks in the pre-processing steps of various BN speech processing applications.

In this paper we present a novel audio-segmentation method for robust and accurate detection of acoustic-change points in continuous audio streams. It employs change detection methods from the two well-established audio segmentation approaches based on the Bayesian Information Criterion: a standard approach, introduced in [1], and a *dist-BIC* approach [2]. By combining methods from both approaches we developed a technique, which enabled us to adaptively tune decision-thresholds from the currently processed audio data. These approaches are described in Section 2. All three segmentation procedures were tested and compared on the SiBN audio database. The experiments and the results are discussed in Section 3, while the conclusions are given in the last section.

2 Audio-Segmentation Approaches

2.1 The BIC Segmentation Approaches

The Bayesian Information Criterion (BIC), [10], has become widely used for the audio segmentation and clustering purposes due to its several useful properties, where it was first introduced by Chen and Gopalakrishnan [1].

The BIC is a penalized maximum likelihood model selection criterion. In the general form, it can be written as [10]:

$$BIC(M_i) = L(X; M_i) - \lambda C(M_i), \tag{1}$$

where $L(X; M_i)$ presents a maximized log-likelihood estimation of parameters of the model M_i on a given data set X. $C(M_i)$ accounts for the model complexity and is in the BIC case equal to $\frac{K_i}{2} \log N$, where K_i corresponds to the number of the model parameters and N is the number of data samples in X. The first term accounts for the quality of the match between the model and the data, while the second one is a penalty for the model complexity. The penalty factor λ allows tuning of the balance between both terms. Using this criterion on the same data a model M_i with the highest BIC score is selected out of a set of candidate models.

When the BIC is used as a model selection criterion in the audio segmentation, a problem of change points detection is reformulated as a model selection task between two competing models in the following way. Let us mark a part of an audio stream, where a change point needs to be find, as X. The X defines an analysis window in an audio stream and is usually presented as a sequence of frame-based acoustic feature vectors, i.e., $X = x_1, x_2, \ldots, x_N$. Moreover, it is assumed that each acoustic homogeneous segment of audio data can be modeled by a single Gaussian distribution. Then, for each point $1 < b < N$ within the analysis window X, the BIC difference between two models are produced:

$$\Delta_{BIC}(b) = BIC(M_2) - BIC(M_1)$$
$$= L(X; M_2) - L(X; M_1) - \lambda(C(M_2) - C(M_1)), \qquad (2)$$

where M_1 represents a model, when the data $x_1, \ldots, x_N$ are modeled by a single Gaussian distribution, and M_2 stands for a model, which is composed by two Gaussian distributions: one is estimated from data $x_1, \ldots, x_b$ and the other from data $x_{b+1}, \ldots, x_N$. In the case, when the $\Delta_{BIC}(b) < 0$, the data in the analysis window X is better modeled by a single Gaussian distribution, while the $\Delta_{BIC}(b) > 0$ favors representation of the data by two distributions, which supports the hypothesis that a change in acoustics occurs at the point b.

A detection of multiple segmentation boundaries in an audio stream the BIC measure needs to be applied across the whole range of data points in an audio stream. Since the Δ_{BIC} score is computed for every point within the analysis window X, there exist two general approaches how to construct and move these analysis windows throughout the data stream [11].

The first approach was proposed in [1] and was introduced to the segmentation task together with the Δ_{BIC} measure. This technique works in a sequential manner and tries to find segmentation boundaries within analyzing segments of varying lengths. Segmentation is performed in the following way. At start, initial window X of predetermined length is set at the beginning of a stream and the Δ_{BIC} from (2) is computed for every data point within this window. Candidates for segmentation boundaries are points, where the $\Delta_{BIC} > 0$, and among them a point with the highest Δ_{BIC} score is selected as a change point. In this case, the window X is moved to position of the change point, and the computation of the Δ_{BIC} continues within the new window with the same length. If there are no change points in the initial window X (the $\Delta_{BIC} < 0$ for all the points within the window X), the window X is increased by additional length, and a computation of the Δ_{BIC} is redone on the extended window. These steps are repeated until there are no more data for processing. This approach was proposed together with the Δ_{BIC} measure and it is used as a standard procedure in the most of segmentation systems. We refer it as a *standard BIC approach*.

Another approach of computing segment boundaries from audio data streams is to apply the Δ_{BIC} measure on the fixed-length analysis windows. This approach follows the version of another segmentation procedure, presented in [6,9], where instead of the

BIC measure, the KL2 distance on fixed-length analysis windows was used for change-points detection. The procedure computed the KL2 distance as a symmetric divergence of two Gaussian distributions [10]. The Gaussians were estimated from each half of the fixed-length analysis window, which was centered at the location of the current computation point. Distances were then computed for every point in an audio stream by sliding the analysis window along the data stream. This produced a distance function and the peaks in the distance function were candidates for change points. They were found and marked as change points if their absolute value exceeded a pre-determined threshold chosen on a development data. Additional smoothing of the distance function or elimination of the smaller neighboring peaks within a certain minimum duration was needed to prevent over-generation of change points at true boundaries.

In [2] the symmetric Kullback-Leibler (KL2) distance was exchanged by the Δ_{BIC} measure, defined in (2). This resulted in the BIC-based segmentation with the fixed-length analysis windows. This approach is known as the *distBIC* approach, where in the basic version a generalized log-likelihood ratio was used as a distance measure, but that is actually a simplified version of the Δ_{BIC}, when $\lambda = 0$. Single Gaussians are generally preferred in this approach, but also mixture of Gaussians can be applied [3].

While there have been several solutions proposed to improve the performance of the BIC approaches to be computationally more effective and to increase the accuracy in a detection of short segments, not so many efforts have been made to set proper decision-thresholds for optimal change-points detection. Setting an optimal threshold always presents a trade-off between producing long, pure segments of acoustic homogeneous data, and minimizing the rate of missed change points. A general principle in almost all segmentation procedures is to set thresholds empirically from development data each time, when either the overall audio conditions change and are not known in advance (e.g. different recordings, different BN shows, etc.) or when new (alternative) acoustic features are applied into the segmentation process. In such cases the fixed pre-determined thresholds, estimated from the development data, could not adequately fit to the acoustics in the processing data due to the mismatched conditions between training and real data, which could severely degrade the segmentation performance.

In the next section we propose an approach, that was designed to overcome these problems by introducing adaptively-tuned decision-thresholds into the segmentation process.

2.2 Adaptively-Tuned Thresholds for BIC Segmentation

By combining both BIC segmentation procedures with fixed-and varying-length analysis windows and the post-segmentation refinement procedure we developed a method for estimating thresholds directly from the processed audio data. Accordingly, a need for using a development data for tuning the segmentation parameters was removed.

We joined the *standard BIC* approach with the time-varying analysis windows and the *distBIC* approach, where the Δ_{BIC} measure was computed on the fixed-length analysis windows into two-pass segmentation procedure. In the first pass, the decision thresholds were estimated directly from the processed audio data by using the *distBIC* segmentation procedure, and in the second pass the *standard BIC* approach was applied by using the estimated thresholds from the first pass.

Estimation of the decision threshold in the first pass was performed by using the *distBIC* segmentation procedure, where we computed the distance function for each point in an audio stream by sliding a fixed-length analysis window. Parameters for the window length, a sliding step and the λ in the Δ_{BIC} were set in advance and remained fixed during the computation. All the Δ_{BIC} scores were then collected and among them a global maximum and minimum value was found. The threshold θ for the next segmentation pass was evaluated as:

$$\theta = \max{}_{distBIC}\Delta_{BIC} - \alpha \cdot (\max{}_{distBIC}\Delta_{BIC} - \min{}_{distBIC}\Delta_{BIC}), \quad (3)$$

where $\max{}_{distBIC}\Delta_{BIC}$ presents a maximum and $\min{}_{distBIC}\Delta_{BIC}$ a minimum value among all the Δ_{BIC} scores produced by the sliding fixed-length analysis window on the given audio stream within the *distBIC* approach. The threshold θ was thus determined as a relative shift from the maximum Δ_{BIC} score by a portion of the difference between the maximum and the minimum score, which was controlled by the factor α. The values for α were typically chosen between 0.0 and 0.2.

This threshold was then used in the second pass of the segmentation procedure within the *standard BIC* approach. The decisions about change points in this pass were made in cases when the Δ_{BIC} score exceeded the threshold θ, i.e., the points b, where the $\Delta_{BIC}(b) > \theta$, were chosen as the segment boundaries.

In such a way, we removed a need for tuning the penalty factor λ in the Δ_{BIC} measure by introducing the θ threshold. The θ was evaluated from the Δ_{BIC} scores in the first pass within the *distBIC* approach. The scores distribution was estimated from the current processing of audio data and the threshold θ was evaluated as the relative shift from the maximum Δ_{BIC} score controlled by the parameter α. Consequently, the absolute thresholds, defined by the penalty factor λ in the Δ_{BIC}, were replaced by the relative thresholds, controlled by the α, that is not depend on any other open parameters of the segmentation procedure nor it is influenced by any changes in acoustic environments due to mismatched acoustic conditions between development data and processed audio streams. Consequently, no additional development data were needed for setting the decision thresholds for the BIC segmentation.

3   Evaluation Experiment

We tested all three segmentation approaches: the standard BIC approach (abbreviated as the *standBIC* approach), the *distBIC* approach (abbreviated as the *distBIC* approach), and our version of the standard   BIC segmentation with adaptively-tuned decision-thresholds (abbreviated as the *ATDTstandBIC* approach).

The experiments were performed on the SiBN broadcast news database, which includes 34 hours of BN shows in Slovenian language, [13]. The database was divided to a development and a test set. A development set included 7 hours of BN shows and was used for setting all open parameters of the *standBIC* and the *distBIC* approaches to optimize their segmentation performances. The rest of the SiBN data served for an evaluation of the segmentation approaches.

3.1 Experimental Setup

In all the segmentation procedures, the mel-frequency cepstral coefficients (MFCCs) were used as a basic representation of the audio data. The MFCC features were composed from the first 12 cepstral coefficients (without 0th coefficient) and a short-term energy, estimated directly from the signal.

The probability distributions in the Δ_{BIC} measure, which was used in all the segmentation approaches, were set to be single full-covariance Gaussians. To speed up the segmentation process in all the procedures the estimations of the Gaussians and the computation of BIC measures were performed by implementing several improvements of the base approach [1] as are suggested in [12,11]. All the open parameters of the standBIC and the distBIC approaches were additionally tuned on the development set in a way to optimize the performance of each segmentation procedure.

Segmentation procedures were assessed by using the standard evaluation measures, the recall, the precision and the F-measure, defined as [6]:

$$RCL = \frac{\text{number of correctly found boundaries}}{\text{total number of correct boundaries}} \quad (4)$$

$$PRC = \frac{\text{number of correctly found boundaries}}{\text{total number of found boundaries}}, \quad (5)$$

$$F = \frac{2 \times PRC \times RCL}{PRC + RCL}. \quad (6)$$

The correctly found boundary is defined at the segmentation point, which is spotted in a time tolerance region around a true boundary in the reference segmentation (in our experiment a tolerance region was set to 1.0 s). Note, that the *F-measure* varies from 0 to 1, where higher values indicate better performance.

3.2 Experimental Results

An assessment of the presented segmentation approaches was performed on the SiBN test data, which was composed of 27 hours of BN audio data. In total, it had to be found 6782 segment boundaries in the data. The results are summarized in Table 1.

Table 1. Segmentation results on the SiBN database

segmentation approach / features	RCL (%)	PRC (%)	F-measure (%)
standBIC: MFCC	88.3	74.3	80.7
distBIC: MFCC+ΔMFCC	80.0	79.0	79.5
ATDTstandBIC: MFCC	84.2	80.1	**82.1**

The standBIC and the ATDTstandBIC approaches were performed on the MFCC features, while in the distBIC approach also the ΔMFCC features were included, since the distBIC approach with just the MFCCs did not perform well.

Since the approaches used the same Δ_{BIC} measure for the segmentations, the differences in the results in Table 1 correspond to the settings of the decision-thresholds and the other open parameters in the segmentation procedures. While in the *standBIC* and the *distBIC* approach the penalty factor λ in the Δ_{BIC} measure needed to be set to implicitly define a decision threshold for the segmentation, in the *ATDTstandBIC* approach the λ remained fixed and the open parameter α was set to 0.15 without tuning on the development data.

The *standBIC* approach exposes the biggest difference among the recall and the precision scores among all the tested approaches, even though it performed better than the *distBIC* approach. Overall the highest scores were achieved with the *ATDTstandBIC* approach. This suggests that the performances of the standard segmentation procedures were very sensitive to how well their open parameters were tuned on the training data or how good the development data fit to the real acoustic conditions. On the other hand, the *ATDTstandBIC* approach performed equally well on the development and the test data, which indicates that the BIC segmentation with adaptively-tuned thresholds was more robust and less sensitive to different training and unforeseen acoustic conditions.

We can conclude that the results speak in favor of the proposed BIC segmentation procedure with adaptively-tuned thresholds. Due to the mismatched acoustic conditions between the development and the test set in our experiment the decision thresholds of both standard BIC segmentation procedures were sub-optimally set, which probably caused slightly lower performances of both approaches in comparison to our proposed segmentation procedure, where we did not need any development data for threshold tuning.

4   Conclusion

Our research aimed to explore audio segmentation procedures for speaker change detection in continuous BN audio data streams. We implemented two standard segmentation approaches based on the Bayesian Information Criterion and developed a new approach, that aimed to join two stages of both standard approaches in order to adaptively estimate the decision thresholds from the processing audio data. In such a way we were able to obtain a segmentation procedure, that does not need any additional tuning of the decision thresholds on the development data as is the case in the standard segmentation procedures. The experiment on the BN audio data also showed, that the proposed segmentation approach performed more reliable and stable across different range of acoustic conditions in comparison to the standard BIC and the DISTBIC approaches.

References

1. Chen, S., Gopalakrishnan, P.S.: Speaker, environment and channel change detection and clustering via the Bayesian information criterion. In: Proceedings of the DARPA Speech Recognition Workshop, Lansdowne, USA, pp. 127–132 (1998)
2. Delacourt, P., Wellekens, C.J.: DISTBIC: A speaker-based segmentation for audio data indexing. Speech Communication 32(1-2), 111–126 (2000)
3. Ajmera, J., McCowan, I., Bourlard, H.: Robust speaker change detection. IEEE Signal Processing Letters 11(8) (2004)

4. Fiscus, J.G., Garofolo, J.S., Le, A., Martin, A.F., Pallett, D.S., Przybocki, M.A., Sanders, G.: Results of the Fall 2004 STT and MDE Evaluation. In: Proceedings of the Fall 2004 Rich Transcription Workshop, Palisades, NY, USA (2004)
5. Istrate, D., Scheffer, N., Fredouille, C., Bonastre, J.-F.: Broadcast News Speaker Tracking for ESTER 2005 Campaign. In: Proceedings of Interspeech 2005 - Eurospeech, Lisbon, Portugal, September 2005, pp. 2445–2448 (2005)
6. Kemp, T., Schmidt, M., Westphal, M., Waibel, A.: Strategies for Automatic Segmentation of Audio Data. In: Proc. of the ICASSP, vol. (3), pp. 1423–1426 (2000)
7. Meignier, S., Bonastre, J.-F., Fredouille, C., Merlin, T.: Evolutive HMM for Multi-Speaker Tracking System. In: Proceedings of the IEEE International Conference on Acoustics, Speech and Signal Processing (ICASSP), Istanbul, Turkey (2000)
8. Pallett, D.S., Lamel, L. (eds): Automatic transcription of Broadcast News data. Speech Communication 37(1-2), 1–159 (2002)
9. Siegler, M.A., Jain, U., Raj, B., Stern, R.M.: Automatic Segmentation, Classification and Clustering of Broadcast News. In: Proc. 1997 DARPA Speech Recognition Workshop, Chantilly, VA, February 1997, pp. 97–99 (1997)
10. Theodoridis, S., Koutroumbas, K.: Pattern Recognition, 2nd edn. Academic Press, Elsevier, USA (2003)
11. Tranter, S., Reynolds, D.: An Overview of Automatic Speaker Diarisation Systems. IEEE Transactions on Speech, Audio and Language Processing, Special Issue on Rich Transcription 14(5), 1557–1565 (2006)
12. Tritschler, A., Gopinath, R.: Improved speaker segmentation and segments clustering using the Bayesian information criterion. In: Proceedings of the EUROSPEECH 1999, Budapest, Hungary, September 1999, pp. 679–682 (1999)
13. Žibert, J., Mihelič, F.: Development of Slovenian Broadcast News Speech Database. In: Proceedings of the International Conference on Language Resources and Evaluation (LREC 2004), Lisbon, Portugal, May 2004, pp. 2095–2098 (2004)
14. Žibert, J., et al.: The COST278 Broadcast News Segmentation and Speaker Clustering Evaluation - Overview, Methodology, Systems, Results. In: Proceedings of Interspeech 2005, Lisbon, Portugal, pp. 629–632 (2005)

Fusion of Acoustic and Prosodic Features
for Speaker Clustering

Janez Žibert[1] and France Mihelič[2]

[1] Primorska Institute of Natural Sciences and Technology, University of Primorska,
Muzejski trg 2, Koper, SI-6000, Slovenia
janez.zibert@upr.si
[2] Faculty of Electrical Engineering, University of Ljubljana,
Tržaška 25, Ljubljana, SI-1000, Slovenia
france.mihelic@fe.uni-lj.si

Abstract. This work focus on a speaker clustering methods that are used in speaker diarization systems. The purpose of speaker clustering is to associate together segments that belong to the same speakers. It is usually applied in the last stage of the speaker-diarization process. We concentrate on developing of proper representations of speaker segments for clustering and explore different similarity measures for joining speaker segments together. We realize two different competitive systems. The first is a standard approach using a bottom-up agglomerative clustering principle with the Bayesian Information Criterion (BIC) as a merging criterion. In the next approach a fusion speaker clustering system is developed, where the speaker segments are modeled by acoustic and prosody representations. The idea here is to additionally model the speaker prosody characteristics and add it to basic acoustic information estimated from the speaker segments. We construct 10 basic prosody features derived from the energy of the audio signals, the estimated pitch contours, and the recognized voiced and unvoiced regions in speech. In this way we impose higher-level information in the representations of the speaker segments, which leads to improved clustering of the segments in the case of similar speaker acoustic characteristics or poor acoustic conditions.

1 Introduction

Speaker diarization is the process of partitioning the input audio data into homogeneous segments according to the speaker's identity. The aim of speaker diarization is to improve the readability of an automatic transcription by structuring the audio stream into speaker turns, and in cases when used together with speaker-identification systems, by providing the speaker's true identity. Such information is of interest to several speech- and audio-processing applications. For example, in automatic speech-recognition systems the information can be used for unsupervised speaker adaptation [6], which can significantly improve the performance of speech recognition in large vocabulary, continuous speech-recognition systems [11]. The outputs of a speaker-diarization system can also be used in speaker-identification or speaker-tracking systems, [2].

Most speaker-diarization systems have a similar general architecture. First, the audio data, which are usually derived from continuous audio streams, are segmented into

V. Matoušek and P. Mautner (Eds.): TSD 2009, LNAI 5729, pp. 210–217, 2009.
© Springer-Verlag Berlin Heidelberg 2009

speech and non-speech data. The non-speech segments are discarded and not used in subsequent processing, which is done in a speech-detection module. The speech data are then chopped into homogeneous segments in an audio-segmentation module. The segment boundaries are located by finding the acoustic changes in the signal, and each segment is, as a result, expected to contain speech from only a single speaker. The resulting segments are then clustered so that each cluster corresponds to just a single speaker. This is done in a speaker-clustering module and usually represents the final stage in speaker-diarization systems. The overview of the approaches used in speaker-diarization tasks can be found in [10].

We built a speaker-diarization system that is used for speaker tracking in BN shows [14]. The system was designed in the standard way by including components for speech detection, audio segmentation and speaker clustering. Since we wanted to evaluate and measure the impact of speaker clustering on the overall speaker-diarization performance, we built a system where the components for speech detection and audio segmentation remained fixed during the evaluation process, while different procedures were implemented and tested in the speaker-clustering task. The components for speech detection and audio segmentation were already presented in [14,13]. In this work the speaker clustering module is explored.

The paper is organized as follows: In Section 2, two competitive speaker-clustering approaches are presented. The first is a standard approach using a bottom-up agglomerative clustering principle with the Bayesian information criterion as the similarity measure. In the second system we implemented a novel fusion-based speaker-clustering system, where the speaker segments were modeled by acoustic and prosody representations. The presented clustering procedures were assessed on a Slovenian BN audio database and the evaluation results are presented in Section 3. A discussion of the results and the conclusions are given in Sections 4 and 5.

2 Speaker Clustering

The speaker clustering in speaker-diarization systems aims to associate or cluster together the segments from the same speaker. Ideally, this clustering produces one cluster for each speaker, with all the segments from a given speaker in a single cluster. The dominant approach used in diarization systems is called hierarchical agglomerative clustering [9]; it consists of the following steps [10]:

1. *Initialization:* each segment represents a single cluster;
2. *Similarity measure:* compute the pair-wise distances between each cluster;
3. *Merging step:* merge the closest clusters together and update the distances of the remaining clusters to the new cluster;
4. *Stopping criterion:* iterate step 3 until some stopping criterion is met.

The main issues concerning the above speaker-clustering approach include the choice of a proper similarity measure, the proper representations of the cluster data and usage of a suitable stopping criterion.

2.1 BIC-Based Speaker Clustering

The most common choice for similarity measure in speaker clustering is the Bayesian Information Criterion (BIC), which was first introduced for this task in [1] and then used in several other speaker clustering systems [10,13].

The audio data segments in this case are in general represented by acoustic features consisting of either mel-frequency cepstral coefficients (MFCCs) or perceptual linear-prediction coefficients (PLPCs). The cluster data represented by these features are in the BIC case modeled by single density full-covariance Gaussian distributions. The similarity measure is defined as the Δ_{BIC} measure, that is computed as the difference of the two penalized likelihoods, estimated from the two different clusters, which are compared, [1]. Those clusters that produce the biggest negative difference in terms of Δ_{BIC} among all the pair-wise combinations of clusters are joined together in the merging stage of the clustering. The merging process is stopped when the lowest Δ_{BIC} score from among all the combinations of clusters in the current clustering is higher than a specified threshold, which is set in advance.

We implemented speaker clustering system by using standard MFCC features with the first and the second order derivatives for cluster representations. Gaussians are modeled by single-density full-covariance distributions, while the Δ_{BIC} measure was optimized by varying the penalty factor in the BIC to achieve a maximum performance of the speaker diarization system on a development data.

Note that, when speaker clustering is used as one stage in a speaker-diarization system, several improvements can be made to increase the performance of the speaker diarization, like joint segmentation and clustering [7] and/or cluster re-combination [16]. In our research we focused mainly on an evaluation and a development of the base speaker-clustering approaches, and did not implement any of these methods, even though they could be easily applied in the same manner as they are applied in other systems.

2.2 Fusing of Acoustic and Prosodic Information in Speaker Clustering

The BIC-based speaker-clustering approach perform the clustering by measuring the similarity between the speaker data, based only on the acoustic representations. Since the acoustic representations perform reliably in most speaker-recognition systems, they were an obvious choice in speaker-clustering approaches. Lately, however, several speaker-recognition systems have attempted to include prosodic information as well as acoustics for the representation of the speakers [8]. The fusing of both representations was an attempt to reduce the need for speaker modeling in various acoustic environments and to provide additional information about the speaker's speech characteristics.

We developed an approach to speaker clustering that included both acoustic and prosodic representations of the speakers. The main objectives were to derive the prosodic features from the speaker-cluster data and to integrate them into the basic acoustic representations of the speaker. In order to achieve this, we needed to adopt the presented agglomerative speaker-clustering approach to merge the cluster data by defining a new similarity measure that was able to fuse the similarity scores from both representations.

The development of the *prosodic features* for the speaker clustering was inspired by a derivation of similar features for speaker recognition [8], where they focused on capturing the longer-range stylistic features of a person's speaking behavior. We followed this approach by producing three groups of prosodic features based on the voiced and unvoiced (VU) regions in speech, which were related to pitch, energy and duration measurements in speech signals:

- **Energy features**
  (1) energy mean: the estimated mean of the short-term energy frames in the speech segment;
  (2) energy variance: the estimated variance of the short-term energy frames in the speech segment;
  (3) rising energy frame rate: the number of rising short-term energy frames in the speech segment divided by the total number of energy frames;
  (4) falling energy frame rate: the number of falling short-term energy frames in the speech segment divided by the total number of energy frames.
- **Duration features**
  (5) normalized VU speaking rate: the number of changes of the voiced, unvoiced and silence units in the speech segment divided by the speech-segment duration;
  (6) normalized average VU duration rate: the absolute difference between the average duration of the voiced parts and the average duration of the unvoiced parts, divided by the average duration of all the V, U units in the speech segment;
- **Pitch (f0) features**
  (7) f0 mean: the estimated mean of the f0 frames computed only in the V regions of the speech segment;
  (8) f0 variance: the estimated variance of the f0 frames computed only in the V regions of the speech segment;
  (9) rising f0 frame rate: the number of rising f0 frames in the speech segment divided by the total number of f0 frames;
  (10) falling f0 frame rate: the number of falling f0 frames in the speech segment divided by the total number of f0 frames.

Note that in [8] the prosodic features were extracted from the syllable-based regions of speech, while we decided to use the voiced-unvoiced (VU) regions. Using the VU regions in speaker clustering has several advantages over the syllable-based representation. Both types of sub-word units operate at nearly the same speech-region levels and thus the same techniques for computing prosodic features can be applied, but the VU regions can be detected without the use of large-vocabulary speech-recognition systems and are language independent, which is not the case when the speech units are represented by syllables or words.

All the above features were obtained from the individual speech segments associated within each cluster. The features were designed by following the approach for prosody modeling of speaker data [8] and the development of the prosodic features for word-boundary detection in automatically transcribed speech data [4]. Note that the features in (5) and (6) are the same as those used in speech detection based on phoneme-recognition features [14].

The main reason for integrating the prosodic features into the speaker clustering was to provide information in addition to the basic acoustic features in order to gain some improvement in the speaker clustering in the case of adverse acoustic conditions. Therefore, we developed a new similarity measure for the comparison of the clusters represented by prosodic features and a fusion-based merging criterion to combine the similarity scores of the acoustic and prosodic representations of the clusters.

The prosodic measure was defined on speaker clusters by computing the Mahalanobis distance between the principal components of the speaker segments represented by the prosodic feature vectors. The procedure for obtaining such a measure is fully described in [15].

To make a fusion of the BIC measure and our Mahalanobis-based distance in the merging criterion of a speaker clustering some kind of score normalization needs to be applied to both similarity measures and the appropriate fusion scheme of joining both scores has to be defined. We decided to use the *min-max score normalization* of both similarity measures [5]. We obtained this by computing the minimum and maximum values from among all the pair-wise cluster combinations at the current step of merging. And a controllable fusion of both representations of the speaker clusters was then made by producing a weighted sum of the normalized versions of both similarity measures, where the weights in our experiments were tuned to 0.8 and 0.2 for the acoustic and the prosody scores, respectively.

The speaker clustering was performed by following the same clustering procedure as described in Section 2. The only difference was in the merging step, where a minimum from among all the fusion of scores was used instead of the baseline BIC-derived scores. In this way we were able to include prosodic information in the baseline speaker clustering.

3   Evaluation Experiments

Our experiments were carried out on a BN speech database SiBN, which is composed of 35 hours of BN shows in Slovenian language, [12].

Since we only wanted to assess the performance of the speaker-clustering approaches we used the same speech/non-speech-detection and audio-segmentation procedures in all the evaluation experiments. The speech/non-speech detection used the approach presented in [14], while the audio segmentation used the approach presented in [1].

In the tested speaker-clustering approaches we needed to set different open parameters. The parameters were chosen according to the optimal speaker-diarization performance of the corresponding clustering approaches on the development dataset, which was composed of 7 hours of BN audio data from the SiBN database.

The speaker-clustering approaches were evaluated by measuring the speaker-diarization performance in terms of the diarization error rate (DER), [3].

4   Discussion of the Results

The speaker-diarization results are shown in Figure 1.The DER results, plotted in Figure 1, should be interpreted as follows: the DER results at the evaluation point 0 correspond to the average of the DER across all the evaluated audio files, where the number

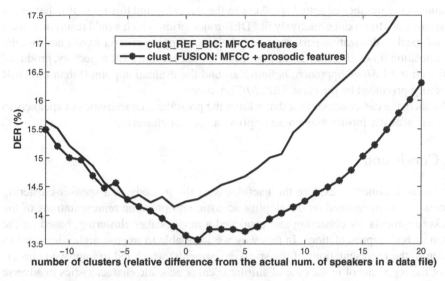

Fig. 1. Speaker-diarization results on the SiBN database, when using different clustering approaches. The lower DER values correspond to a better performance.

of clusters is equal to the actual number of speakers in each file, the DER results at evaluation point +5 correspond to the average of the DER across all the evaluated audio files, where the number of clusters exceeds the actual number of speakers in each file by 5, and analogously, the DER results at evaluation point -5 correspond to the average of the DER across all the evaluated audio files, where the number of clusters is 5 clusters lower than the actual number of speakers in each file, and so on.

The *clust_REF_BIC* and the *clust_FUSION* results in Figure 1 correspond to the baseline BIC approach and the fusion approach, respectively.

The overall performance of the speaker-clustering approaches can be compared by inspecting the difference in the DER results at the points, that correspond to the same number of clusters. The overall performance varies between 13.5% and 16%, measured using the overall DER. The DER trajectories of both approaches achieved their minimum DER values around the evaluation point 0. This means that if the corresponding clustering approaches were to be stopped when the number of clusters is equal to the number of actual speakers in the data, both of the approaches would exhibit their optimum speaker-diarization performance. At that point the best clustering result was achieved with the *clust_FUSION* approach. Also across the whole range of evaluation points the *clust_FUSION* approach perform better than the baseline *clust_REF_BIC* clustering.

Another interesting conclusion can be drawn from observing the flatness of the DER trajectories. Since the proposed evaluation measure aimed to compute the DER values at the relative numbers of clusters in each file, no stopping criteria needed to be applied; however, in practice the proper stopping of the clustering should be ensured. The optimum stopping criteria should end the merging process at the point with the lowest DER, which should coincide with the evaluation point 0, where the number of clusters is equal to the number of actual speakers in the data. Around this point it is better for the approaches to produce relatively flat DER trajectories, which would result in a small loss of speaker-diarization performance, when the stopping criteria would not find the exact position for ending the merging process. In our case, the DER trajectory, produced by the *clust_FUSION* approach, is flatter around the evaluation point 0 than the DER trajectory, produced by the *clust_REF_BIC* approach.

As such, we can conclude that that adding the prosodic characteristics of speakers to the basic acoustic information could improve a speaker clustering.

5   Conclusion

Our research aimed to explore the usefulness of the prosody in a speaker-clustering process. We concentrated on developing acoustic and prosodic representations of the speaker segments for clustering and proposed a new speaker clustering, based on the fusion of both representations. In this way we were able to impose higher-level information in the representations of the speaker segments, which led to improved clustering of the segments of in the case of similar speaker acoustic characteristics in adverse acoustic conditions.

Acknowledgment

This work was supported by Slovenian Research Agency (ARRS), development project M2-0210 (C) entitled "AvID: Audiovisual speaker identification and emotion detection for secure communications."

References

1. Chen, S., Gopalakrishnan, P.S.: Speaker, environment and channel change detection and clustering via the Bayesian information criterion. In: Proceedings of the DARPA Speech Recognition Workshop, Lansdowne, Virginia, USA, pp. 127–132 (1998)
2. Delacourt, P., Bonastre, J., Fredouille, C., Merlin, T., Wellekens, C.: A Speaker Tracking System Based on Speaker Turn Detection for NIST Evaluation. In: Proceedings of International Conference on Acoustics, Speech, and Signal Processing (ICASSP 2000), Istanbul, Turkey (June 2006)
3. Fiscus, J.G., Garofolo, J.S., Le, A., Martin, A.F., Pallett, D.S., Przybocki, M.A., Sanders, G.: Results of the Fall 2004 STT and MDE Evaluation. In: Proceedings of the Fall 2004 Rich Transcription Workshop, Palisades, NY, USA (2004)
4. Gallwitz, F., Niemann, H., Noth, E., Warnke, V.: Integrated recognition of words and prosodic phrase boundaries. Speech Communication 36(1-2), 81–95 (2002)
5. Jain, A., Nandakumar, K., Ross, A.: Score normalization in multimodal biometric systems. Pattern Recognition 38(12), 2270–2285 (2005)
6. Matsoukas, S., Schwartz, R., Jin, H., Nguyen, L.: Practical Implementations of Speaker-Adaptive Training. In: Proceedings of the 1997 DARPA Speech Recognition Workshop, Chantilly VA, USA (February 1997)
7. Meignier, S., Bonastre, J.-F., Fredouille, C., Merlin, T.: Evolutive HMM for Multi-Speaker Tracking System. In: Proceedings of the IEEE International Conference on Acoustics, Speech and Signal Processing (ICASSP), Istanbul, Turkey (2000)

8. Shriberg, E., Ferrer, L., Kajarekar, S., Venkataraman, A., Stolcke, A.: Modeling prosodic feature sequences for speaker recognition. Speech Communication 46(3-4), 455–472 (2005)
9. Theodoridis, S., Koutroumbas, K.: Pattern Recognition, 2nd edn. Academic Press, Elsevier, USA (2003)
10. Tranter, S., Reynolds, D.: An Overview of Automatic Speaker Diarisation Systems. IEEE Transactions on Speech, Audio and Language Processing, Special Issue on Rich Transcription 14(5), 1557–1565 (2006)
11. Woodland, P.C.: The development of the HTK Broadcast News transcription system: An overview. Speech Communication 37(1-2), 47–67 (2002)
12. Žibert, J., Mihelič, F.: Development of Slovenian Broadcast News Speech Database. In: Proceedings of the International Conference on Language Resources and Evaluation (LREC 2004), Lisbon, Portugal, May 2004, pp. 2095–2098 (2004)
13. Žibert, J.: et al.: The COST278 Broadcast News Segmentation and Speaker Clustering Evaluation - Overview, Methodology, Systems, Results. In: Proceedings of Interspeech 2005, Lisbon, Portugal, pp. 629–632 (2005)
14. Žibert, J., Vesnicer, B., Mihelič, F.: Novel Approaches to Speech Detection in the Processing of Continuous Audio Streams. In: Grimm, M., Kroschel, K. (eds.) Robust Speech Recognition and Understanding, pp. 23–48. I-Tech Education and Publishing, Croatia (2007)
15. Žibert, J., Mihelič, F.: Novel approaches to speaker clustering for speaker diarization in audio broadcast news data. In: Mihelič, F., Žibert, J. (eds.) Speech recognition: technologies and applications. Artificial intelligence series, pp. 341–362 (2008) ISBN 978-953-7619-29-9
16. Zhu, X., Barras, C., Meignier, S., Gauvain, J.-L.: Combining Speaker Identification and BIC for Speaker Diarization. In: Proceedings of Interspeech 2005 - Eurospeech, Lisbon, Portugal, pp. 2441–2444 (2005)

Combining Topic Information and Structure Information in a Dynamic Language Model

Pascal Wiggers and Leon Rothkrantz

Man–Machine Interaction Group
Delft University of Technology
Mekelweg 4, 2628 CD Delft, The Netherlands
p.wiggers@tudelft.nl, l.j.m.rothkrantz@tudelft.nl

Abstract. We present a language model implemented with dynamic Bayesian networks that combines topic information and structure information to capture long distance dependencies between the words in a text while maintaining the robustness of standard n-gram models. We show that the model is an extension of sentence level mixture models, thereby providing a Bayesian explanation for these models. We describe a procedure for unsupervised training of the model. Experiments show that it reduces perplexity by 13% compared to an interpolated trigram.

1 Introduction

Much information that is potentially useful for language modelling gets lost in standard n-gram language models. For example, the assumption that the relative frequency of a word combination is the same for all conversations is very useful to ensure that reliable parameter estimates can be found, but is clearly incorrect. A word may be more likely in a particular conversation than on average (in a corpus) and far more likely in that conversation than in other conversations.

This notion can, at least partially, be captured by realising that a discourse is not a random collection of phrases. Following the argument of [1]: if we would pick at random a number of sentences from some text and put them together we would certainly not get anything that looks like a proper paragraph. The difference is, that any discourse will display some coherence. It will contain semantically related words, and particular phrases – typically content words – will reappear in the text. In other words, a discourse has a topic.

In this paper we present a language model implemented with dynamic Bayesian networks that includes topics to model relationships between words that are at an arbitrary distance within a text without sacrificing the strength of n-gram models. In section 2 we briefly describe dynamic Bayesian networks. Our model is introduced in section 3, where we also relate the model to other models found in literature. In section 4 we present a procedure for unsupervised training of the model. Section 5 discusses experimental results.

V. Matoušek and P. Mautner (Eds.): TSD 2009, LNAI 5729, pp. 218–225, 2009.

2   Dynamic Bayesian Networks

Bayesian networks are a method for reasoning with uncertainty based on the formal rules of probability theory [2,3]. Bayesian networks are directed acyclic graphs of which the nodes are random variables and the arcs indicate conditional independence of the variables, i.e. the absence of an arc between two variables signifies that those variables do not directly depend upon each other. Thus a Bayesian network is a factored representation of a joint probability distribution over all variables given by:

$$\prod_V P(V|Parents(V)) \tag{1}$$

Dynamic Bayesian networks (DBNs) model processes that evolve over time [4]. They consist of slices that define the relations between variables at a particular time, implemented as a Bayesian network and a set of arcs that specify how a slice depends on previous time slices. DBNs can be seen as a generalization of both n-gram models and hidden Markov Models. The difference being that HMMs use only one variable to represent the state of the model, whereas a DBN can use any number of variables. Several efficient inference algorithms have been developed for DBNs [5]. Training is usually done with an instance of the Expectation-Maximization (EM) algorithm.

3   The Topic-Based Model

To operationalise the intuitive notion of a topic in a computationally convenient way we consider a conversation a mix of topics, where some topics are more prominent than others. This mix can change over time as topics become more or less prominent. Similar to [6], we define topics as probability distributions over the vocabulary. The idea that a discourse contains a mix of topics can then be implemented as a mixture of topic distributions. In the language of Bayesian networks this can be put as shown in Figure 1, where W_i represents the ith word in a conversation. The words are conditioned on the topic T. When the topic of a discourse is known, assigning a probability to this discourse

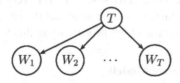

Fig. 1. Conceptual view of a topic model. The word variables W_i are influenced by the topic T.

is straightforward. However, more often than not, the topic of a conversation is not known beforehand. Imagine observing the words in a conversation one by one. Initially, all topics are equally likely or distributed according to some a priori distribution (e.g. people tend to talk about the weather more often than about speech recognition).

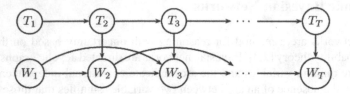

Fig. 2. Topic model with n-gram constraints

Every new word that is observed propagates back through the network as evidence, supporting those topics that predicted it with a high probability and moderating the influence of other topics. As a consequence, the distribution over the topics shifts towards the observed words. This new topic distribution is used in the prediction of subsequent words and if a text is reasonably coherent leads to a better prediction of future words. If the observed words themselves are uncertain, evidence from future words can be used to disambiguate those by also running a backwards pass over the model.

To summarise, the model of Figure 1 can naturally deal with mixtures of topics, can identify the topics of a conversation in the course of that conversation and can adapt when the topic of a conversation changes. One can think of the topic nodes as a means to capture long-distance dependencies: the exact words are not remembered, but the topic mixture provides a summary of the history.

Figure 1 uses a single global topic node that is seen as the common cause of all the word variables. The changes in the topic belief over time can also be modelled explicitly in a DBN as shown in Figure 2. To be completely compatible with the previous model the links between the topic nodes should be deterministic. There are no transitions from a topic to any other topic. Using this representation we can get a better insight in the workings of the model as it is equivalent to a hidden Markov model (HMM) where the topics are the states and the words the observations (ignoring the links between the words for the moment). The total probability of the observation sequence is the sum of the probabilities of the individual topic sequences. As topics have only self-transitions the probability of each topic is the product of the all the topic-dependent word probabilities. Therefore, this model will assign a higher probability to a coherent text than to a text that contains words that have high probabilities in different topics. Note that in general, every word must have a non-zero probability for every topic. Otherwise, a word that has zero probability for a particular topic would set the belief of that particular topic to zero even if all other words have a high probability in this topic.

3.1   The Relation with Mixture Models

Since the topic model defines a distribution over word sequences, it can directly be used as a language model. But taking the trigram as a benchmark it comes of rather poorly, as it is a bag-of-words model that only uses unigram probabilities. Obviously, it can be combined with standard n-gram models to get the best of both worlds. The simplest wat to do so is to condition complete n-grams on the topic. Phrased like that, the model is equivalent to the topic mixture model of [7,8] that combines m topic specific n-gram component models at the sentence level:

$$P(w_{1,T}) = \sum_{k=1}^{m} \lambda_k \left[\prod_{i=1}^{T} P_k(w_i|w_{i-n+1,i-1}) \right]. \tag{2}$$

We can now understand the sentence mixture model from a Bayesian perspective and use this insight to extend the model.

3.2   A More Advanced Model

Making n-grams dependent on the topic may unnecessarily fragment training data. What such models are capturing is not so much the topic of a conversation but rather the distribution of a particular subset of the data, combining information on genre and topics. It seems reasonable to assume that content words are much more topic dependent than function words, whereas function words and local syntactic structure are more dependent on the type of conversation as shown in [9]. Therefore, we only condition the content words on the topic of conversation. This implies that the part-of-speech of a word has to be predicted before the word itself is predicted. The resulting model is shown in Figure 3.

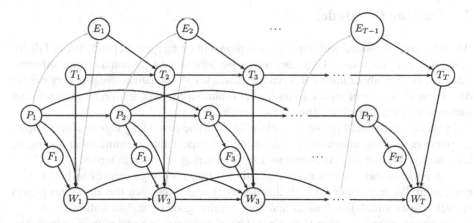

Fig. 3. A topic model that uses part of speech information (P). The F_t variable signals whether the word W_t is a content word in which case it is conditioned on the topic T_t as well. Function words are only conditioned on previous words and part of speech tags. Topic transitions are only allowed at sentence boundaries as indicated by E_t.

In every time slice of this model first the part-of-speech P is predicted based on the parts-of-speech of the previous two words. This has the additional advantage that the model contains an explicit representation of local syntactic structure that is used in the prediction of words as well. The F node is a binary variable that indicates whether the word is a function word or a content word. In case of a content word the word is not only conditioned on its POS-tag and previous two words, but also on the topic. The distribution over the words is found by interpolating the topic and n-gram distributions. In case of a function word the topic distribution is simply a copy of the topic distribution of the previous slice.

Up to now, we have assumed that the transitions between topics are deterministic. And with good reason, as this guarantees that the model prefers coherent texts. Nevertheless, probabilistic transitions between topics are possible, as long as the probability of a self-transition is much larger than that of transitions to other topics there is still a preference for coherent text. The advantage of this approach is that the influence of words on future words reduces as time progresses. In addition, relations between topics can be modelled. A higher belief of a topic then increases the belief of related topics. In theory, probabilistic topic transitions also remove the need for every word to have a non-zero probability in every topic. In practice however, it turns out that it is better to use this constraint. We experimented with several configurations and found that a model that has only self-transitions between topics within a sentence but allows probabilistic transitions to other topics at sentence boundaries (as indicated by variable E) performs best. The probability distribution at topic boundaries typically shows clusters of related topics. Many variations on this model are possible. For example user knowledge can be added to model topic preference. Other modalities, such as computer vision, or background information can influence the topic distribution as well.

4 Training the Model

If a data set is annotated with topic information, obtaining topic-dependent word distributions is straightforward. Unfortunately, most data sets do not contain topic information as it is very labour intensive to create such a set. As argued above it is very hard to decide on a good set of topics. Even if a set is annotated, the question remains whether that annotation is useful for language modelling.

In theory, this should not be a problem, as the topic based language model can learn its parameters from unannotated data using the expectation maximisation algorithm. Essentially, this approach implements a soft-clustering of words in topics.

The major drawback of the EM algorithm is that it only guarantees to find a local maximum. We experimented with this approach and found that the algorithm is very sensitive to its initial distributions and does not find good topic distributions.

There are several solution to this problem. One interesting variation is to introduce the document as a variable in the model to function as a constraint on the topic variable. $P(T|D)$ is initialised in such a way that for every document only a limited number of topics has a high probability. This method is strongly related to probabilistic semantic analysis [10] but adds the time dimension to this model.

Another approach, the one that we took, is to find better initial distributions for the model. We initialised topics with clusters of semantically related documents. To obtain these clusters, we created a vector in lemma space for every document. With every document a weight vector is associated. The length of the vector corresponds to the number of different semantically salient lemma types in the vocabulary that were found by removing all function words and common content words from the vocabulary. We used lemmas rather than words, as inflections are not important for topicality. The entries of the vectors are weights that indicate the relation between the document and the lemmas. We used term frequency-inverse document frequency (TF-IDF) weights as widely used in information retrieval [11]:

$$\text{weight}(i,j) = \begin{cases} (1 + \log(\text{tf}_{ij})) \log(\frac{N}{\text{df}_i}) & \text{tf}_{ij} > 0, \\ 0 & \text{tf}_{ij} = 0, \end{cases} \tag{3}$$

where N is the number of documents and the term frequency tf_{ij} counts the number of times lemma i occurs in document j. High frequency lemmas are thought to be characteristic for the document. Higher counts reflect more saliency of a word for a document but the scale is not linear. Observing a word twice as much does not mean that it is twice as important. Therefore, term frequencies are logarithmically scaled.

This quantity is weighted by the inverse document frequency df_i which gives the number of different documents lemma i occurs in. The idea is that lemmas that occur in many documents are semantically less discriminating. This component is also logarithmically weighted. Note that a word that occurs in all documents will get weight zero.

Together the word vectors span a high-dimensional space in which each dimension corresponds to a lemma. To measure semantic similarity between documents we apply a metric.

To find clusters of related documents we used agglomerative clustering with the cosine as a similarity measure. Agglomerative clustering is a greedy iterative algorithm in which every vector initially has its own single-element cluster. In every iteration, the two most similar clusters are merged. The similarity of two clusters with multiple elements is defined as the distance between the two least similar elements (complete-link clustering).

Every document is annotated with the cluster number of its corresponding vector. This annotated data was used to estimate the initial parameters of the topic model using simple maximum likelihood estimation. The parameters are then interpolated with the global distribution over content words to make sure that all words have a non-zero probability for all topics. One could leave it at this, but then we would not be using the full potential of the topic model. Agglomerative clustering assigns every document to a single cluster but documents may contain several topics. The topic model can represent soft clustering. Therefore, we only selected the most similar half of the documents of every cluster for initialisation of the topic model. The idea is that this will result in relatively coherent topic distributions, while documents that can belong to multiple clusters are not used for initialisation. Next the topic model was retrained on all training data using the EM algorithm. All documents are assigned to all topics weighted by the likelihood of the topics resulting in soft clustering rather than in hard clustering. Recall that our model allows for probabilistic topic transitions at sentence boundaries. This transition distribution, simultaneously trained, enables the model to deal with documents that contain a sequence of topics.

Monitoring the perplexity of a development test set we found that the soft clustering step does result in a better model than using the clusters directly to estimate topic distributions, but the model quickly overtrains. Therefore, we introduced a damping factor [5]:

$$P_k(W|T) = (1 - \delta)\tilde{P}_k(W_t|T) + \delta P_{k-1}(W_k|T), \tag{4}$$

where $\tilde{P}_k(W_t|T)$ is the result of EM iteration k and $P_{k-1}(W_t|T)$ is the topic distribution obtained in the previous iteration. Depending on the weight δ ($0 \leq \delta \leq 1$) distributions are only partially updated in every iteration. If $\delta = 0$ this amounts to normal updating. $\delta = 1$ would result in no updating at all. Because the initial distributions were smoothed by interpolation with the global distributions, the damping factor has additional advantage that it avoids zero probabilities for words that do not occur in the training data in subsequent training iterations.

5   Experiments

We tested the topic model on component f of the Spoken Dutch Corpus (Corpus Gesproken Nederlands or CGN) [12]. This set contains interviews and discussions broadcasted on radio and television. The set contains a total of 790.269 words, 80% of which we use for training, 10% for development testing and tuning and the remaining 10% for evaluation. All words that occur only once in the training set are treated as out-of-vocabulary, resulting in a vocabulary size of 17833 words and 257 POS-tags. The POS-tags include attributes such as number, degree and tense. Table 1 gives the perplexity of the model on the evaluation set. As a baseline the perplexities of standard interpolated bigram and trigram on the same set with the same vocabulary are also shown. In all cases the interpolation weights have been optimised on the same development test set. The topic model clearly outperforms the standard models.

Table 1. Perplexity results

language model	perplexity
interpolated bigram	296.49
interpolated trigram	280.76
topic-based model with 64 topics	242.92

6   Conclusion

We presented a language model that captures contextual coherence. The topic of a conversation is modelled as a mixture of predefined topics. The basic version of the model, that conditions whole n-grams over the words on the topic variable can be interpreted as a sentence level mixture model. Alternatively, it can be seen as a form of probabilistic semantic analysis to which a time dimension is added. The full model includes part-of-speech information and allows for changes of topic. It can be characterized as a generative topic-based language model in which the influence of recent words is stronger than that of words in the more distant past.

References

1. Jurafsky, D., Martin, J.H.: Speech and Language Processing. Prentice-Hall, Englewood Cliffs (2000)
2. Pearl, J.: Probabilistic Reasoning in Intelligent Systems - Networks of Plausible Inference. Morgan Kaufmann Publishers, Inc., San Francisco (1988)

3. Jensen, F.V.: Bayesian Networks and Decision Graphs. Statistics for Engineering and Information Science Series. Springer, Heidelberg (2001)
4. Dean, T., Kanazawa, K.: A model for reasoning about persistence and causation. Computational Intelligence 5(3), 142–150 (1989)
5. Murphy, K.: Dynamic Bayesian Networks: Representation, Inference and Learning. PhD thesis, University of California, Berkeley (2002)
6. Gildea, D., Hofmann, T.: Topic-based language models using EM. In: Proceedings of the 6th European Conference on Speech Communication and Technology, EUROSPEECH (1999)
7. Iyer, R., Ostendorf, M., Rohlicek, J.R.: Language modeling with sentence-level mixtures. In: HLT 1994: Proceedings of the workshop on Human Language Technology, Morristown, NJ, USA, pp. 82–87. Association for Computational Linguistics (1994)
8. Iyer, R., Ostendorf, M.: Modeling long distance dependence in language: Topic mixtures vs. dynamic cache models. In: Proc. ICSLP 1996, Philadelphia, PA, vol. 1, pp. 236–239 (1996)
9. Wiggers, P., Rothkrantz, L.J.M.: Exploring the influence of speaker characteristics on word use in a corpus of spoken language using a data mining approach. In: XII International Conference Speech and Computer, SPECOM 2007 (2007)
10. Hofmann, T.: Probabilistic latent semantic analysis. In: Proc. of Uncertainty in Artificial Intelligence, UAI 1999, Stockholm (1999)
11. Baeza-Yates, R., Ribeiro-Neto, B.: Modern Information Retrieval. Addison-Wesley, Reading (1999)
12. Schuurman, I., Schouppe, M., Hoekstra, H., van der Wouden, T.: CGN, an annotated corpus of spoken Dutch. In: Proceedings of the 4th International Workshop on Linguistically Interpreted Corpora (LINC 2003), Budapest, Hungary, April 14 (2003)

Expanding Topic-Focus Articulation with Boundary and Accent Assignment Rules for Romanian Sentence

Neculai Curteanu[1], Diana Trandabăț[1,2], and Mihai Alex Moruz[1,2]

[1] Institute for Computer Science, Romanian Academy, Iaşi Branch
[2] Faculty of Computer Science, University "Al. I. Cuza" of Iaşi
{curteanu,dtrandabat,mmoruz}@iit.tuiasi.ro

Abstract. The present paper, maintaining the interest for applying Prague School's Topic-Focus Articulation (TFA) algorithm to Romanian, takes the advantage of an experiment of investigating the intonational focus assignment to the Romanian sentence. Using two lines of research in a previous study, it has been showed that TFA behaves better than an inter-clausal selection procedure for *assigning pitch accents* to the Background-Kontrast (Topic-Focus) entities, while the *Inference Boundary* algorithm for computing the Theme-Rheme is more reliable and easier extendable towards *boundary and contour tone assignment* rules, leading to our novel proposal of Sentence Boundary Assignment Rules (SBAR). The main contributions of this paper are: (*a*) The *TFA algorithm* applied for Romanian is extended to inter-clause level, and embedded into a discursive approach for computing the Background-Kontrast entities. (*b*) *Inference Boundary* algorithm, applied to Romanian for clause-level Theme-Rheme span computing, is extended with a set of SBAR rules, which are relying on different *Communicative Dynamism* (CD) *degrees* of the clause constituents. (*c*) On each intonational unit, the extended TFA algorithm is further refined, in order to eliminate ambiguities, with a filter derived from Gussenhoven's SAAR (Sentence Accent Assignment Rule). Remarkably, TFA and CD degrees *proved* again to be resourceful, especially as technical procedures for a new and systematic development of an Information Structure Discourse Theory (ISDT).

1 Introduction

Our investigation is based on fundamental studies and results in the last two decades, which established a reliable association between *intonational phrasing* and meaning of the *information structure* (IS) notions, *viz. Background-Kontrast* entities (alias Topic-Focus in the Prague School's language) and *Theme-Rheme* structures. Papers such as [4], [13 :265-290], [13 :245-264], [12] support and prove, including at experimental and perception level, consistent rules that assign categorical functions of tones and tunes (boundary tones, pitch accents, and contours) to the IS textual entities and spans. The accuracy of the association between IS categories / structures and the corresponding (sequences of) ToBI tags for Romanian is not argued in the present paper. Our main concern is on the *textual* IS-*semantics* side of the language interface within the (Romanian) prosody, with the aim of squeezing all the discursive text meanings which proved to be useful for predicting a human-like prosody.

V. Matoušek and P. Mautner (Eds.): TSD 2009, LNAI 5729, pp. 226–233, 2009.
© Springer-Verlag Berlin Heidelberg 2009

In [7] two lines of investigation were followed for the intonational focus assignment in the Romanian sentence: (**A**) The Prague School's Topic-Focus Articulation (TFA) algorithm is improved at clause level with hints from Van Valin's *linking algorithms(s)* [18] and Gussenhoven's SAAR (Sentence Accent Assignment Rule) [13 :83-100], [16], [17], then extended to inter-clause level. The *information-structural* (IS) textual spans of Theme(s)-Rheme(s) are computed within TFA approach as the lowest-highest *degrees* of *Communicative Dynam*ism (CD) [9], [10], or predicative constituent actual ordering *vs.* the constituent *Systemic Ordering* (SO) [9], [10], [6]. (**B**) The second approach for computing Theme-Rheme and Background-Kontrast (or Topic-Focus in the Prague School's language) is based on IS-semantics discourse theories, actually Asher's SDRT (Segmented Discourse Representation Theory) [2], [15], and Heusinger's attempt to design an Information Structure Discourse Theory (ISDT), which is embedded into SDRT [13 :265-290]. While Background-Kontrast entities are derived from SDRT on the basis of clause-level *Given-New entity* criteria, Theme-Rheme structures are computed with a recursive form of the Leong's *Inference-Boundary* (IB) *algorithm* [14], applied *for the first time* to the Romanian sentence in [7]. The results of the two approaches (A) and (B) are distinct at Background-Kontrast and Theme-Rheme levels, providing different sets of ToBI annotations assigned to as Steedman's rules [13 :245-264], [13 :265-290].

The *main conclusion* in [7] for the two lines of investigation is the following: TFA behaves better than the SDRT (non-balanced) selection procedure for assigning pitch accents to the Background-Kontrast (Topic-Focus) entities, while the IB-algorithm [14] for computing the Theme-Rheme is more reliable and easier extendable towards boundary and contour tone assignement rules. An extension of the IB-algorithm is presented in Section 3 as SBAR (Sentence Boundary Assignment Rules).

The contributions of this paper can be summarized as follows: (*i*) The *TFA algorithm* applied for Romanian is extended to inter-clause level and embedded into a SDRT discursive approach for computing the Background-Kontrast entities. (*ii*) *Inference Boundary* (IB) algorithm [14], applied to Romanian for clause-level Theme-Rheme span computing, is extended with a set of *Sentence Boundary Assignment Rules* (SBAR), which are relying on different CD *degrees* of the clause constituents. (*iii*) On each intonational unit, the extended TFA algorithm is further refined, in order to eliminate ambiguities, with a SAAR-derived filter.

2 Extending Topic-Focus Articulation Algorithm

Prague's School Topic-Focus Articulation (TFA) algorithm [9] attributes Topic-Focus to a sentence, considering its constituent order in communicative dynamism (CD) *vs.* the "standard" systemic ordering (SO), in the framework of *contextual boundness* or *non-boundness*. The TFA algorithm applies on constituency parsed simple (finite) clauses, annotated with part-of-speech information and several semantic features, namely the specificity degrees (*general* – low specificity, contextually non-bound; *specific* – high specificity, contextually non-bound; *indexical* – mid-specificity, contextually-bound) of verbs and temporal / locative complements. The output of the TFA algorithm is determining the appurtenance of an element to a clause Topic or

Focus. In [6], the TFA algorithm was adapted and implemented for Romanian prosodic structures. Examples 1 and 2 present sentences annotated with the TFA algorithm, where T stands for Topic, F for Focus, and T/F for ambiguous cases, and each syntactic constituent is represented within brackets:

Example 1. [Ion]$_T$ [a câştigat]$_F$ [competiţia]$_{T/F}$.
     En: John   won   the competition.
Example 2. [Privit de sus]$_T$, [lichidul]$_T$ [părea negru]$_F$.
     En: Seen from above, the liquid   seemed black.

Applying the TFA algorithm on Romanian raised a number of issues. First, the context considered for the boundness constraints is minimal, addressing only the current sentence (whether the head of a noun group is defined or not, or whether the verb or the complementation are indexical references), although Hajicova *et al.* [9] mentioned that the previous sentence is needed in order to properly analyse a verb context[1]. Another problem is due to the fact that, following the Topic-Focus assignment, some sentences contain no focus at all, or more than one focused constituent. We need a balanced assignment of Topic-Focus entities, observing the TFA criteria, intonational grouping, and sentence level prosody patterns such as SAAR (Sentence Accent Assignment Rule) [3], [13 :83-100], approach which will be discussed in Section 4.

This paper proposes a slightly extended version of the original TFA algorithm in [9], where the syntactic and semantic information required by the original algorithm is completed with IS annotation, obtained by discursive analysis. The starting point of our approach is the fact that the TFA notion of Topic has much in common with the more recently characterized concept of *Background*, while the Focus correspond to the notion of *Kontrast* [13]. Thus, considering the preceding context of a sentence (the sentences already uttered in the discourse, *i.e.* the accessibility domain), and performing co-reference resolution (for instance, using the SDRT algorithm [2]), the contextually-boundness information is obtained using anaphoric relators. Each constituent is thus annotated with *Background* or *Kontrast*, where the *backgrounded* entities are the ones already mentioned in the discourse (represented in many cases by definite noun groups, indexical verbs or complements, personal pronouns, but not only this, as considered by the original TFA algorithm), and *kontrasted* entities are those that *have not* been previously introduced in the discourse (in the original TFA, they correspond to undefined noun groups, specific verbs or complements).

Hajicova *et al.* [9] states that *"In its present form, however, the algorithm has several limitations. It can process only simple sentences."*. Thus, a further step in the development of the TFA algorithm for Romanian was the extension of the TFA algorithm to complex clauses (sentences). A simple and effective method is splitting the complex clause into simple, finite clauses, and then applying the algorithm recursively on each clause. In correlated adjacent clauses, the algorithm is simply applied consecutively for each clause. More attention needs to be assessed in applying the algorithm on subordinate clauses, since the subordinate clause is to be treated as

---

[1] "As for the verb, it is important to have access to the verb of the preceding utterance and to use a systematic semantic classification of the verbs. If the main verb of sentence n has the same meaning as (or a meaning included in) that of sentence n − 1 (in the sense of hyponymy), then it belongs to the topic." (Hajicova et al. [9: pp. 9]).

the corresponding complement of the verb, completing thus the regent sentence. For example, the complex clause in example 3a contains two simple clauses, marked by ^{1}/ and ^{2}/:

Example 3a. [Ion]$_T$ [știa]$_F$ ^{1}/ că [întârziase]$_F$ [mult]$_T$^{2}/.
    En: John   knew   that he was very late.

This example can be transformed into a simple clause, by reducing the completive clause to a direct argument:

Example 3b. [Ion]$_T$ [știa]$_F$ [asta]$_{T/F}$. ^{1}/
    En : John   knew   that.

This compression of the complex clauses into mere complements allows for the appliance of the TFA algorithm on complex sentences. The benefits of this approach become evident if more than one subordinate clause is considered, as in examples 4.

Example 4a. [Ion] [a anunțat] ^{1a}/[ceea ce a descoperit] ^{2}/
    En: John   announced   what   he   has   discovered
[în fața întregii audiențe]. ^{1b}/
in front of the audience.

The complex clause in example 4a is formed by two simple (finite) clauses, but one of them is intercalated in between two others (the clause "*what he has discovered*" is embedded into the second clause "*John announced * in front of the audience*"). If we consider the embedded completive clause a simple complement, the TFA algorithm will yield the result presented in 4b.

Example 4b. [Ion]$_T$ [a anunțat]$_F$ [ceva]$_{T/F}$ [în fața întregii audiențe]$_F$. ^{1}/
    En: John   announced something   in front of the audience.

The examples above show that the Background-Kontrast extended version of the TFA algorithm works on both simple and complex clauses / sentences, providing Background-Kontrast (or Topic-Focus) annotation that can be transformed into pitch accents [17].

3 Towards Sentence Boundary Assignment Rules

Kochanski [12] showed that, besides the pitch accent, the *duration* and the *amplitude* are also important features in prosody. Therefore, we developed an algorithm that delimits intonational groups by assigning boundaries between constituents. The boundaries are classified by their duration in: full boundary break (1%), marking an important pause in spoken language, usually corresponding to an *intonational phrase* [1], half a boundary break (½%), corresponding to an *intermediate phrase* [1], or no boundary at all (0%). The algorithm developed receives as input the set of constituents in the sentence, ordered accordingly to their linear precedence (or real ordering, thus *communicative dynamism*) in the text.

$$C_1\ C_2\ \ldots\ C_{m-1}\ \ C_m\ \ PRED\ \ C_{m+1}\ C_{m+2}\ \ldots\ C_n$$

The *first rule* of assigning intonational boundaries states that the predication is closely linked to the constituent that follows, thus a 0% boundary is present, no matter what the constituent type is (argument, adjunct) or if it respects or *not* the *systemic order* (SO). Then, as moving away from the verb, the boundaries commence to become more obvious, thus the second constituent *after* the verb, and the first constituent *before* the verb, receive half a boundary break (½%), and similarly the third constituent after the verb and the second constituent before the verb receive a full boundary. This rule is illustrated below:

C_1 C_2 ... C_{m-1} 1% C_m ½% PRED 0% C_{m+1} ½% C_{m+2} 1% C_{m+3} ... C_n

A *second rule*, that is more theoretical than practical, due to the language economy, envisages marking boundaries *before* the predication, if more than two constituents are present (*i.e.* for C_1 .. C_{m-1} in the example above), or *after* the predication, if more than three constituents are present (*i.e.* C_{m+3} ... C_n). The rule is drafted below:

- **(i)** if an SO-disordering occurs, the disordered constituent receives an 1% boundary, the rest of the constituents being marked following **(iii)**;
- **(ii)** if the constituents are similar as to their syntactic functions (two direct objects, two indirect objects, two complements, etc.), a ½% boundary break is inserted between the similar constituents;
- **(iii)** all unbounded constituents receive 0% as boundary. Thus, before the verb we will have the sequence C_1 0% C_2 0% .. C_{m-2} 0%, and C_{m+3} 0% ... C_n 0% after the verb.

The boundary degrees are used to build different intonational units: intermediate phrases (*ip*), between ½%, and intonational phrases (*IP*), between 1%, since *IP* are considered to be more general than *ip* [1]. Accordingly, the *IP* boundary breaks are larger than the *ip* ones. Exceptions are infrequent but may happen, *e.g.* [13 :280 (27)].

Several examples of splitting the sentence constituents into intonational units are:

```
Example 5.  [Ion]½%[a câştigat] 0% [competiţia]½%.
            {{[Ion] [a câştigat]  [competiţia]}ip}IP.
      En: John      won          the competition.
Example 6.  [Privit de sus]1%, [lichidul] ½% [părea negru] 0%.
             {[Privit de sus]}IP, {{[lichidul]  [părea negru]}ip}IP.
      En: Seen from above, the liquid    seemed black.
Example 7.
[Ion]½%    [i-a dat]0% [ieri]½%   [Mariei]1%  [o carte]½%    [în parc]1%.
{{[Ion]    [i-a dat]   [ieri]}ip [Mariei]}IP {{[o carte]}ip [în parc]}IP.
En: *John  gave       yesterday  to Mary     a book         in the park.
```

4 Refining Background-Kontrast Assignment Using SAAR

After performing Background-Kontrast (Topic-Focus) assignment on the sentence constituents using the extended TFA presented in section 2, sometimes more than one focus appear, or a sentence may contain no focus at all. We consider that in a sentence may occur several focused constituents, but only if they belong to different intonational groups. In fact, analysing the manually annotated data, we found that each intonational group (*i.e.* intermediate phrase, *ip*) must contain *one* (*and only one*) focus (kontrast), the rest of the elements in the intonational group being topics (backgrounds). In order to assure that each intermediate phrase will receive an unique focus tag, we developed an algorithm that filters if more than one focus appear in an intermediate phrase, or transforms a *t/f* or *t* marking to *f* if no focus is found in the intonational phrase, using a set of rules derived from SAAR (Sentence Accent

Assignment Rule) [13 :83-100, 219-220, 282-283]. The rules presented below apply recursively for all the intonational units in the sentence. In the algorithm presented below *f*, *t* and *t/f* stand for three *sets*, each containing the constituents from the intonational phrase that are marked with the respective tags after applying the extended TFA algorithm.

1. **if f = 0**
/*none of the sentence constituents received focus and the focus set is empty*/
        apply SAAR-derived rules to decide which constituent
turns to f from the t or t/f sets
        The other t/f marked constituents are changed to t.
2. **if f > 1**
/*two or more sentence constituents received focus*/
        Apply SAAR-derived rule to decide which constituent keeps
the focus;
        The other f-marked constituents receive t instead of f.
        The other t/f marked constituents are changed to t.

The **SAAR-derived rules** that we apply are:
(i)   If the verb has adjuncts, then the adjuncts will receive
        focus rather than the arguments.
(ii)   If the verb has arguments, then the adjuncts will receive
        focus rather than the verb.
(iii)   If the verb has more than one adjunct or more than one
        argument, then the rightmost one receives the focus.

The informal meaning of the SAAR-derived rules is that rather the *semantic periphery* is intonationally focused (kontrasted) than its *semantic head*, in the **ascending intonational accentuation** "$<_F$" of *head* $<_F$ *argument* $<_F$ *adjunct* (for finite / non-finite clause) and *head* $<_F$ *modifier* (for nominal, adjectival, verbal groups).

In what follows, several examples of refining the extended version of the TFA algorithm with intonational phrasing sequences are discussed:

Example 8. [Ion]$_T$ [a câştigat]$_F$ [competiţia]$_{T/F}$.
    {{[Ion] [a câştigat] [competiţia]}$_{ip}$}$_{IP}$.
    {{[Ion]$_T$ [a câştigat]$_F$ [competiţia]$_T$}$_{ip}$}$_{IP}$.
  En: John   won   the competition.

In this example, the last constituent changes from T/F to T since the intermediate phrase can have only one focused constituent.

Example 9. [Privit de sus]$_T$, [lichidul]$_T$ [părea negru]$_F$.
    {{[Privit de sus]}$_{IP}$, {{[lichidul] [părea negru]}$_{ip}$}$_{IP}$.
    {{[Privit de sus]$_F$}$_{IP}$, {{[lichidul]$_T$ [părea negru]$_F$}$_{ip}$}$_{IP}$.
  En: Seen from above,   the liquid   seemed black.

In example 9, the first *ip* has no *focus*, hence the constituent "*seen from above*", being the only constituent in the intermediate phrase, receives F instead of T.

Example 10. [Vinul]$_T$ [nu avea]$_T$ [culoarea obişnuită]$_{T/F}$.
    [Vinul]½% [nu avea]0% [culoarea obişnuită]½%.
    {{[Vinul] [nu avea] [culoarea obişnuită]}$_{ip}$}$_{IP}$.
    {{[Vinul]$_T$ [nu avea]$_T$ [culoarea obişnuită]$_F$}$_{ip}$}$_{IP}$.
  En: The wine didn't have the usual color.

In example 10, since there is no focus in the intermediate phrase, the last constituent will receive focus. Although it may seems that the reason for this is that the last constituent is the only constituent to have T/F as compared to the T of other constituents, the real reason of this change is rule (iii) of our SAAR-derived algorithm, *i.e.* the verb in the sentence has two arguments, and the rightmost will receive focus.

Example 11. [Maria]$_T$ [nu a fost]$_T$ [în acea zi]$_T$.
           [Maria]$_{T½\%}$ [nu a fost]$_T$ 0%   [în acea zi]$_T$ ½%.
           {{[Maria]   [nu a fost]   [în acea zi]}$_{ip}$}$_{IP}$.
           {{[Maria]$_T$ [nu a fost]$_T$ [în acea zi]$_F$}$_{ip}$}$_{IP}$.
      En: Mary   hasn't been   that day.

In example 11, the intermediate phrase has no focus, and the verb has an argument (the subject "Mary") and an adjunct (the temporal "that day"). Therefore, the adjunct will receive focus.

Example 12. [Ion]$_T$ [a câştigat]$_F$ [o competiţie]$_F$.
           {{[Ion] [a câştigat]   [o competiţie]}$_{ip}$}$_{IP}$.
           {{[Ion]$_T$ [a câştigat]$_T$ [o competiţie]$_F$}$_{ip}$}$_{IP}$.
      En: John   won   a competition.

In example 12, the intermediate phrase has two focused constituents; therefore the verb will loose its focus in favor of its argument "a competition".

The examples above show that the extended TFA algorithm can be further improved, considering the intonational grouping of the sentence.

5   Discussions

In order to assess the results of the presented algorithms, we compared the output with the manual annotation of the same sentences (a set of spoken Romanian sentences extracted from George Orwell's novel *1984*). The sentences were marked with Theme-Rheme boundaries and with ToBI pitch accents by the Group of Speech Processing within the Institute of Computer Science [1]. The results showed that assigning Background-Kontrast (Topic-Focus) considering the intonational phrasing and the SAAR-derived rules is a very good practice, especially for the cases where the extended TFA algorithm returned no focused constituents. The main differences observed concerned the marking of the last constituent of the sentence: in the TFA approach, it is mostly focused (being the rightmost constituent, is one of the major reasons to focus it); in the gold annotation, the last constituent marks clearly the descending tendency of declarative sentences in Romanian, which are usually marked with a low tone.

We believe that the prediction of prosody from text can be substantially improved by a deeper understanding of the textual discourse theories, with the natural emphasis on the *information structure* (IS) discourse semantics, using also proper *degrees* of *communicative dynamism* for the surface representations of a sentence.

In this paper we presented an extension of the classic TFA algorithm [9] for complex clauses, using also discourse co-referred entities besides syntactical information (nouns determined or not, indexical verbs or complements). We also presented a set of *Sentence Boundary Assignment Rules* (SBAR), relying on the linear precedence of constituents and on different CD degrees of the clause constituents. The extended TFA and the SBAR were linked using SAAR-derived rules in order to refine the Background-Kontrast (Topic-Focus) assignment and to eliminate ambiguities.

Intonational phrasing is a powerful tool in determining the Theme-Rheme structures of a sentence, which, together with Background-Kontrast entities, are inherent parts of a novel *Information Structure Discourse Theory* (ISDT) based on the relations outlined by K. von Heusinger in [11], [13: 265 -290] and developed within the discursive framework of SDRT [2]. The relationship between a systematic theory of intonational phrasing computing and a comprehensive development of ISDT is an essential research topic for prosody prediction and text-to-speech systems. Significantly interesting, TFA and SO-CD degrees proved to be resourceful especially as technical procedures for the development of a new and systematic ISDT.

References

1. Apopei, V.: Analysis of Certain Nonlinear Systems with Applications to Signal Processing. Ph.D. Thesis, Institute of Computer Science, Romanian Academy, Iași Branch (Nov. 2008)
2. Asher, N.: Reference to Abstract Objects in Discourse. Kluwer, Dordrecht (1993)
3. Beaver, D.I., Clark, B.: Sense and Sensitivity. How Focus Determines Meaning. Wiley-Blackwell Publishers, Oxford (2008)
4. Calhoun, S.: The Nature of Theme and Rheme Accents. Centre for Speech Technology Research, Univ. of Edinburgh (2003)
5. Curteanu, N., Trandabăţ, D.: Functional FX bar Projections for Local and Global Text Structures. The Anatomy of Predication. Revue Roumaine de Linguistique, The Publishing House of the Romanian Academy LLI(1-2), 161–194 (2007)
6. Curteanu, N., Trandabăţ, D., Moruz, M.A.: Topic-focus articulation algorithm on the syntax-prosody interface of romanian. In: Matoušek, V., Mautner, P. (eds.) TSD 2007. LNCS (LNAI), vol. 4629, pp. 516–523. Springer, Heidelberg (2007)
7. Curteanu, N., Trandabăţ, D., Moruz, M.A.: Discourse Theories vs. Topic-Focus Articulation Applied to Prosodic Focus Assignment in Romanian. In: Burileanu, C., Teodorescu, H.-N. (eds.) SPED 2009 Conference: Advances in Spoken Language Technology, Constanţa, Romania, The Publishing House of the Romanian Academy, Bucharest, pp. 127–140 (2009)
8. Göbbel, E.: On the Relation between Focus, Prosody, and Word Order in Romanian. In: Quer, J., Schroten, J., Scorretti, M., Sleeman, P., Verheugd, E. (eds.) Romance Languages and Linguistic Theory 2001, pp. 75–92 (2003)
9. Hajičová, E., Skoumalova, H., Sgall, P.: An Automatic Procedure for Topic-Focus Identification. Computational Linguistics 21(1), 81–94 (1995)
10. Hajičová, E., Partee, B.-H., Sgall, P.: Topic-focus Articulation, Tripartite Structures, and Semantic Content. Kluwer Publishers, Dordrecht (1998)
11. von Heusinger, K.: Focus Particles, Sentence Meaning, and Discourse Structure. In: Abraham, W., ter Meulen, A. (eds.) Composing Meaning, pp. 167–193. Benjamins, Amsterdam (2004)
12. Kochanski, G.: Prosody Beyond Fundamental Frequency. In: Sundhoff, S., et al. (eds.) Methods in Empirical Prosody Research. De Gruyter, Berlin (2006)
13. Lee, C., Gordon, M., Büring, D. (eds.): Topic and Focus. Cross-Linguistic Perspective on Meaning and Intonation. Springer, Heidelberg (2007)
14. Leong, P.A.: Delimiting the Theme of the English Clause – An Inference-Boundary Account. SKY Journal of Linguistics (17), 167–187 (2004)
15. Marcu, D.: The Rhetorical Parsing, Summarization, and Generation of Natural Language Texts. Ph.D. Thesis, Univ. of Toronto, Canada, 331 (1997)
16. Selkirk, E., Kratzer, A.: Focuses, Phases and Phrase Stress. Ms., University of Massachusetts at Amherst (2005)
17. Steedman, M.: Information Structure and the Syntax-Phonology Interface. In: Linguistic Inquiry, vol. 34, pp. 649–689. MIT Press, Cambridge (2000)
18. Van Valin Jr., R.D.: Exploring the syntax-semantics interface. Cambridge University Press, Cambridge (2005)

Lexical Affinity Measure between Words

Ivar van Willegen[1], Leon Rothkrantz[1,2], and Pascal Wiggers[1]

[1] Man-Machine Interaction Group
Delft University of Technology
Mekelweg 4, 2628 CD Delft, The Netherlands
l.j.m.rothkrantz@tudelft.nl, p.wiggers@tudelft.nl
[2] SEWACO
The Netherlands Defence Academy,
Het Nieuwe Diep 8, 1781 AC Den Helder, The Netherlands

Abstract. In this paper we research the lexical affinity between words. Our goal is to define a distance measure which corresponds with the semantic affinity between words. This measure is based on WordNet, where two words/concepts can be connected by a string of stepwise synonyms. The number of hops defines the distance between words. In addition we present a natural language processing toolbox, developed in the course of this work, that combines a number of existing tools and adds a number of analysis tools.

1 Introduction

Words in corpora can be assigned to word classes. This plays an important role in text analysis. Another approach of natural language processing is to research the semantic affinity between words. Some words are more similar than others. The similarity measure defines a transitive relation in a word corpus. It is possible to define word graphs of similar words or related concepts. There is evidence that words are stored in our brain as a huge semantic net. Word association games are used by psychologist to reveal the semantic relations between words. In such word association games the most related words pop up first, so probably these words have a stronger link in our brains.

The notion of relating words semantically is implemented by linguists from the University of Princeton in WordNet, a huge corpus of English words with semantic relations such as synonymy between the words.

An alternative way to relate emotional words is to search for common underlying dimensions. A well known solution is the following 2D-space. One dimension of this space corresponds to the valence of the emotional state and the other to the arousal or activation level associated with it. Cowie and colleagues [1] have called this representation the 'activation–evaluation space'. This bipolar affective representation approach is supported in the literature [2], as well being well founded through cognitive appraisal theory. An emotional state is 'valenced', i.e. is perceived to be positive or negative depending on whether the stimulus, event or situation that caused

V. Matoušek and P. Mautner (Eds.): TSD 2009, LNAI 5729, pp. 234–241, 2009.
© Springer-Verlag Berlin Heidelberg 2009

this emotional state to ensue was evaluated (appraised) by the agent of the emotional state as beneficial or detrimental.

The outline of the paper is as follows. In the next section we will discuss related work. Then we present the theory of Kamps, which provides the basic of our model/algorithm in section 3. In section 4 we present our results.

2 Related Work

2.1 WordNet

One of the most significant attempts to realize a large scale lexical knowledge base is WordNet, a thesaurus for the English language based on psycholinguistics principles and developed at the Princeton University by George Miller [3,5]. WordNet organizes lexical information in terms of word meanings, rather than word forms. It has been conceived as a computational resource, improving some of the drawbacks of traditional dictionaries, such as the circularity of the definitions and the ambiguity of sense references. English nouns, verbs, adjectives and adverbs (about 130,000 lemmas for all the parts of speech in version 1.6) are organized into synonym classes (synsets), each representing one underlying lexical concept. Lemmas are organized in synonym classes (about 100,000 synsets). WordNet can be described as a "lexical matrix" with two dimensions: a dimension for lexical relations, that is relations holding among words and thus language-specific, and a dimension for conceptual relations, which hold among senses (in WordNet they are called synsets) and that, at least in part, can be considered independent from a particular language. Word form refers to the physical utterance or inscription; word meaning refers to a lexicalized concept.

The most important lexical relation for WordNet is the similarity of meaning, since the ability to recognize synonymy among words is a prerequisite to build synsets and therefore meaning representation in the lexical matrix. Two expressions are synonymous if substitutivity is valid (in other words if the substitution of one with the other does not change the truth value of a phrase). It is important to note that defining synonyms in terms of substitutivity requires partitioning WordNet into nouns, verbs, adjectives and adverbs. This is consistent with the psycholinguistic evidence that nouns, verbs, adjectives and adverbs are independently organized in the human semantic memory. Obviously, if a word pertains to more than one synset, this gives an indication of its polysemy.

2.2 Whissel's Dictionary of Affect in Language

Whissell's dictionary [8] of affect in language (DAL) is an annotated dictionary of words, which all have been given a value for activation and a value for evaluation. The annotation process for this corpus has been done as follows. Words included in the DAL set were selected in an ecologically valid manner. There were three steps involved in the selection. The Kucera and Francis corpus of 1,000,000 words was sampled from print media in the early 1960's. Words from this corpus with frequencies greater than 10, which also appeared in more than one subsample were included in the DAL list. The word set was then compared to four text samples generated by individuals rather than media. It was also compared to a large sample from juvenile literature. Unique words found in these sources were added to the list. The DAL list which contained approximately 8700 words at the end of step 2 was tested on 16 new, blindly selected, samples. It was also tested on a corpus of 350,000 words of English text collected by Whissell from many sources. The DAL demonstrated a hit rate or matching rate of approximately 90%. The hit rate of 90% means that one would expect nine out of ten words in most English texts to be matched by the DAL. The words of the DAL list were rated along the dimensions of PLEASANTNESS, ACTIVATION and IMAGERY. In each case the scale used was a three-point scale.

2.4 Theory of Kamps

Kamps & Marx [6] proposed a novel way to automatically calculate the affective values for emotional words. The affective dimensions used are based on the factorial analysis of extensive empirical tests that found that the three major factors that play a role in the emotive meaning of a word are evaluation, potency and activity. This research also set the basis for the circumplex of affect by Russell & Lanius [7].

Kamps & Marx try to exploit the lexical relations found in WordNet to measure these factors. As discussed above, the organization of WordNet is not a conventional alphabetical list, but a large interconnected network of words (resembling the organization of human lexical memory). Because of this property, the distance on this graph between words can be seen as the affinity between the words or similarity between word meanings. So by calculating the distance between words, an affinity scale can be set up. This is done as follows.

Two words w_0 and w_n are n-related if there exists an $(n+1)$-long sequence of words $(w_0, w_1, \ldots, w_n)$ such that for each i from 0 to $n-1$ the two words w_i and w_{i+1} are in the same synset.

So for example the adjectives 'good' and 'proper' are 2-related since there exists a 3-long sequence (good, right, proper). Words can of course be related by many different sequences or by none at all. The main interest in their research is on the minimal path length between two words.

Let MPL be a partial function such that $MPL(w_i, w_j) = n$, where n is the smallest number such that w_i and w_j are n-related. According to Kamps & Marx this function is a metric, that is, it gives a non-negative number $MPL(w_i, w_j)$ such that:

i. $MPL(w_i, w_j) = 0$ if and only if $w_i = w_j$,
ii. $MPL(w_i, w_j) = MPL(w_j, w_i)$, and
iii. $MPL(w_i, w_j) + MPL(w_j, w_k) \geq MPL(w_i, w_k)$.

For words that are not connected by any sequence, the result of the MPL function is undefined. The minimal path-length function is a straightforward generalization of the synonymy relation. The synonymy relation connects words with similar meaning, so the minimal distance between words says something on the similarity of their meaning.

However further experimentation quickly revealed that this relation is very weak. It turns out that the similarity of meaning waters down remarkably quick. A striking example of this is that we also find that 'good' and 'bad' themselves are closely related in WordNet; there exists a 5-long sequence (good, sound, heavy, big, bad).

For this reason they set up a different function, which is based on the fact that for every word that is connected to 'good' it is also connected to 'bad'. This function calculates the relative distance from one word to two base words, which are each others antonym (e.g. 'good', 'bad').

Let's define a partial function TRI of w_i, w_j, and w_k (with $w_j \neq w_k$) as:

$$TRI(w_i, w_j, w_k) = \frac{MPL(w_i, w_k) - MPL(w_i, w_j)}{MPL(w_k, w_j)}$$

If any of MPL(w_i, w_j), MPL(w_i, w_k), or MPL(w_k, w_j) is undefined, then TRI(w_i, w_j, w_k) is undefined. By means of this function they set up three different functions to calculate the evaluation, activation and potency, by choosing the words w_j and w_k. For evaluation: Let's define a partial function EVA of w as EVA(w) = TRI (w, good, bad). The function EVA* is defined as follows:

$$EVA^*(\langle w_1, ..., w_n \rangle) = \begin{cases} EVA(w) & if \quad defined \\ 0 & otherwise \end{cases}$$

For activation the function ACT* of a word w is defined as follows:

$$ACT^*(w) = \begin{cases} TRI(w, active, passive) & if \quad defined \\ 0 & otherwise \end{cases}$$

For potency the function POT* of a word w is defined as follows:

$$POT^*(w) = \begin{cases} TRI(w, strong, weak) & if \quad defined \\ 0 & otherwise \end{cases}$$

In the related paper of Kamps & Marx [6], they state clearly that they do not claim that these functions assign a precise measure of for example 'goodness' or 'badness' of individual words, but rather expect that it allows differentiation between words predominantly used for expressing for example positive, negative, or neutral opinions.

3   Model

3.1 Problem Definition

Because Kamps & Marx claim that their functions allow us to differentiate between words that are predominantly used for expressing positive and negative affect we focus on this kind of words. From the DAL corpus lists of words can be obtained that are rated as very positive, negative, active and passive. These lists will be compared to the calculated values of the ACT and EVA functions for these words.

We formulate the following research questions:

- Can the lexical affinity scale as proposed by Kamps & Marx be used to measure the activation and evaluation of words?
- Can the measurement of activation and evaluation be used to classify words as positive, negative, active and passive?

3.2 Algorithm

To answer the research questions of the previous paragraph many MPL calculations need to be done. Because of the size of WordNet [5], version 3.0 has 117659 synsets, the complexity of the MPL function for time and for space is not trivial. To cope with this complexity a number of artificial intelligence search techniques are applied, including depth-first, breadth-first and bidirectional search. Below the various techniques will be explained.

Depth-first search is a very commonly used technique to walk through trees. The search algorithm always expands the deepest node in the current fringe of the search tree. The numbers next to the nodes show the order in which the algorithm checks the nodes. Breadth-first is a strategy in which the root node is expanded first, and then all the successors of the root node are expanded next, and so on. The numbers next to the nodes show the order in which the algorithm checks the nodes again. The idea behind bidirectional search is to run two simultaneous breadth-first searches, one forwards from the initial node and the other backward from the goal node, stopping when the two searches meet in the middle.

As said before, one of the initial problems with the MPL function was that for many words there is no shortest path and so no activation or evaluation can be calculated. A possible solution for this problem is to use the other relations that WordNet has beside to the synset (polysemy/synonymy).

Two possible improvements have been thought up, which differ in the number of extra relations they use. These methods only use the extra relations to expand the number of starting and goal words. This is why the methods are called "shallow cloud method" and "deep cloud method".

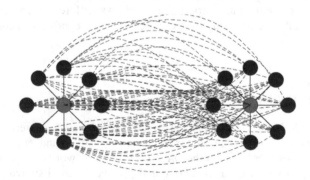

Fig. 1. Cloud method

So graphically represented in Fig. 1, the two light gray nodes (i.e. words) are the starting and goal node. From these nodes, by using the relations as describes above, direct connected nodes are found. The starting node together with its direct connected nodes is called the starting cloud. And the goal node and its direct connected nodes is called the goal cloud. We try to find as many shortest paths from the Ne^N number of paths as possible. We calculate the average path length and use this as the n-relatedness between the clouds. The variance of these paths will help to judge the quality of average calculated relatedness and of course in general the quality of the method.

3.3 Natural Language Processing Toolbox

In order to do these and future experiments, we designed an NLP Affect Toolbox holding a set basic of natural language processing tools and implement various analysis tools. The toolbox also contains a number of corpora, namely "Dictionary of Affect in Language (DAL)" by Whissell [8] and ConceptNet [4]. The natural language processing is mostly done by third party software, namely Proxem Antelope by Proxem [9]. Antelope is an acronym for Advanced Natural Language Object-oriented Processing Environment. With this environment the application is able to do part of speech tagging, chunking, parsing, deep dependency parsing and semantic parsing. Antelope also has an implementation of an object oriented lexicon.

For the cloud distance experiments a large number of calculations needs to be done as fast as possible. The implementation is already using the fastest algorithm (i.e. bidirectional search). To speed up the process even more, a multi threaded version of the cloud distance calculator has been added. Because modern computers have multiple cores in their processors, they can do multiple tasks at the same time. By creating multiple threads, the operating system balances the load evenly over all cores. In this way them process can be (at the current state-of-art) four times faster. The cloud distance experiment uses a semicolon separated value file to queue up all calculations that are needed to be done. The same settings can be applied, as in the normal cloud calculator, plus the number of threads to be used.

4 Experiments

To test our system we selected a number of emotional words used in many psychological studies. We computed the distance between those words using our algorithm. Next we applied Multi Dimensional Scaling technologies to map this data on a 2D space and compared the results with the valence and arousal score. The correlation between these score was 0.67. So less than halve of the variance can be explained by the data. At his moment we are computing the distance between a larger set of words. The problem is that the procedure using the deep cloud method is computationally heavy, and not all the words are available either in the Whissell database or in WordNet, or can be connected in WordNet.

Table 1. Semantic distance between a set of emotional words

	amazed	amused	astonished	bored	concentrated	contempt	curious	disturbed	flabbergasted	hostile	indignant	inspired	irritated	isolated	joyful	melancholy	sad	softened	stimulated
amazed	0	14		14	11	15	14	8	3	18	14	16	14	9	16	12	11	13	12
amused		0	14	11	7	8	3	8	14	14	5	9	7	7	9	9	6	7	11
astonished			0	14	11	15	14	8	3	18	14	16	14	9	16	12	11	13	12
bored				0	10	12	11	8	14	15	11	13	11	8	13	8	7	10	11
concentrated					0	8	7	6	11	12	7	9	6	6	9	6	5	5	9
contempt						0	8	7	15	14	10	8	9	10	4	10	9	11	10
curious							0	8	14	13	4	9	6	6	9	9	6	6	10
disturbed								0	8	11	7	8	6	5	8	5	4	6	4
flabbergasted									0	18	14	16	14	9	16	12	11	13	12
hostile										0	12	15	12	10	15	12	11	12	14
indignant											0	11	8	5	11	8	7	5	10
inspired												0	10	11	9	11	10	11	11
irritated													0	8	10	8	6	8	9
isolated														0	11	5	4	5	8
joyful															0	11	10	12	11
melancholy																0	4	7	8
sad																	0	6	7
softened																		0	9
stimulated																			0

5 Conclusion

In this paper we describe our research on the semantic relation between words. We were able to define a semantic measure between words based on ideas of Kamps and Marx. We implemented our algorithm and tested it on huge datasets. The proposed method is computational heavy. We used special methods to speed up the calculation process.

The first results are reported and are promising but further research is necessary. One of the questions is whether the emotion space is two dimensional as claimed by many psychologists.

References

1. Cowie, R., Douglas-Cowie, E., Tsapatsoulis, N., Votsis, G., Kollias, S., Fellenz, W., et al.: Emotion recognition in human-computer interaction. IEEE Signal Process. Mag. 18, 32–80 (2001)
2. Carver, C.: Affect and the functional bases of behavior: On the dimensional structure of affective experience. Personality and Social Psychology Review 5, 345–356 (2001)

3. Miller, G.: An on-line lexical database. International Journal of Lexicography 13(4), 235–312 (1990)
4. Liu, H., Singh, P.: Commonsense Reasoning in and Over Natural Language. MIT Press, Boston (2004)
5. Fellbaum, C.: WordNet: An Electronic Lexical Database. MIT Press, Massachusetts (1998)
6. Kamps, J., Marx, M.: Words with attitude. CCSOM Working Paper, pp. 01-194 (2001)
7. Russell, J., Lanius, U.: Adaptation Level and the Affective Appraisal of Environments. Journal of Environmental Psychology 4, 119–135 (1984)
8. Whissell, C.: Whissell's Dictionary of Affect in Language - Technical Manual and User's Guide
9. Proxem Antelope: Proxem resources for Natural Language Processing (April 2008), http://www.proxem.com (retrieved 2008)

Multimodal Labeling

Leon Rothkrantz[1,2] and Pascal Wiggers[1]

[1] Man-Machine Interaction Group
Delft University of Technology
Mekelweg 4, 2628 CD Delft, The Netherlands
`l.j.m.rothkrantz@tudelft.nl`, `p.wiggers@tudelft.nl`
[2] SEWACO
The Netherlands Defence Academy,
Het Nieuwe Diep 8, 1781 AC Den Helder, The Netherlands

Abstract. This paper is about automated labeling of emotions. Prototypes have been developed for extraction of features from facial expressions and speech. To train such systems data is needed. In this paper we report about the recordings of semi-spontaneous emotions. Multimodal emotional reactions are evoked in 21 controlled contexts. The purpose of this database is to make it a benchmark for the current and future emotion recognition studies in order to compare the results from different research groups. Validation of the recorded data is done online. Over 60 users scored the apex images (1.272 ratings), audio clips (201 ratings) and video clips (503 ratings) on the valence and arousal scale. Textual validation is done based on Whissell's Dictionary of Affect in Language. A comparison is made between the scores of all four validation methods and the results showed some clusters for distinct emotions.

1 Introduction

Emotions are an integral part of our daily rational decision making and communication. To approach the naturalness of face-to-face interaction machines should be able to emulate the way humans communicate with each other. It is the human face that conveys most of the information about our emotions to the outside world. Considerable research in social psychology has shown that besides speech, non-verbal communicative cues are essential to synchronize the dialogue, to signal certain emotions and intentions and to let the dialogue run smoother and with less interruptions [1]. Non-verbal communication is a process consisting of a range of features including body gesture, posture and touch or paralingual cues, often used together to aid expression. The combination of these features is often a subconscious choice made by native speakers, and interpreted by the listener. Of all different non-verbal communication means, facial expressions are the most important means for interpersonal communication [2]. We learn to recognize faces and facial expressions early in life, long before we learn to communicate verbally. A human face can supply us with important information.

The question of how many emotional states we use in our daily communication has yet to be answered. Very little research has been done to locate and recognize

V. Matoušek and P. Mautner (Eds.): TSD 2009, LNAI 5729, pp. 242–249, 2009.
© Springer-Verlag Berlin Heidelberg 2009

emotions other than the six archetypal emotions as described by Ekman [3]. Therefore most approaches to automatic facial expression analysis attempt to recognize this small set of archetypal emotional expressions. This practice may follow from the work of Darwin [4] and more recently Ekman [6], who has performed extensive studies on human facial expressions. Ekman found evidence to support universality in facial expressions. These "universal (also referred to as 'archetypal' or 'basic') facial expressions" are those representing the six archetypal emotions: Anger, Disgust, Fear, Happiness, Sadness and Surprise. The recognition of these six archetypal emotions has been done separately for every modality. The best results are achieved by looking at facial expressions. Further unimodal research has been done for audio, text and gesture recognition. However, all these approaches use mostly their own, self-created datasets. This means that comparison between the different recognition and classification algorithms is difficult. As a consequence, there have been very little attempts to combine different modalities and approaches in order to fuse these single modalities into one multimodal emotion recognition system.

The main challenge in multimodal emotion recognition is achieving a high recognition rate in various environments under different circumstances. This means that there can be a lot of noise present in a channel, e.g. occlusion of the face by a hand or glasses, the head is rotated in such a way that it is not perfectly visible or multiple persons are speaking at the same time. Using multiple modalities helps in better recognizing the expression as modalities can enhance each other and introduce robustness even if one or more channels are noisy. However, a multimodal dataset containing various environments and different circumstances is yet to be developed.

The lack of a widely used multimodal database with data suitable for emotion recognition for unimodal and multimodal systems made us gather data and develop such a database ourselves. The content from this database should be suitable for multimodal emotion recognition systems as well as unimodal emotion recognition systems, and vocal affect recognition. With this database it should be easier for different research groups to compare their results with other research groups.

The remainder of this paper is structured as follows: section 2 gives a short outline of emotion recognition. Section 3 presents our experimental design and section 4 presents an analysis of the data gathered with this setup. Section 5 describes the data validation through multimodal labeling. The paper ends with a conclusion.

2 Emotion Recognition

Humans use a daunting number of labels to describe emotion [7]. As discussed above most approaches to automatic facial expression recognition attempt to recognize a small set of archetypal emotional facial expressions. Labeling the emotions in discrete categories, such as the six archetypal emotions sometimes is too restricted. One problem with this approach is that the stimuli may contain blended emotions. The choice of words may be too restrictive or culturally dependent too. There is little agreement about a definition of emotion. Many theories of emotion have been proposed. Some of these could not be verified until recently when measurements of physiological signals became available. However, most theories agree that emotions are short-term, whereas moods are long term, and temperaments or personalities are very long-term.

Research from Darwin forward has recognized that emotional states involve dispositions to act in certain ways. The various states that can be expressed are simply rated in terms of the associated activation level. Instead of choosing discrete labels, observers can indicate their impression of each stimulus on several continuous scales. Two common scales are valence and arousal. Valence describes the pleasantness of the stimuli, with positive (or pleasant) on one end, and negative (or unpleasant) on the other. The other dimension is arousal or activation [2, 8]. The vertical axis shows activation level (arousal) and the horizontal axis evaluation (valence). A circumplex can be viewed as implying circular order, such that variables that fall close together are more related than variables that fall further apart on the circle, with opposite variables being negatively related and variables at right angles being unrelated (orthogonal). We can identify the centre as a natural origin or the neutral state. The neutral state is a condition of readiness to respond. Emotional strength can be measured as the distance from the origin to a given point in the activation-evaluation space. An interesting implication is that strong emotions are more sharply distinct from each other than weaker emotions with the same emotional orientation. A related extension is to think of the six archetypal emotions as cardinal points in that space.

Fig. 1. Activation-Evaluation space of the six basic emotions

3   Experimental Design

The database should ideally contain only genuine expressions of emotions. However, as the database should also consist of high-quality video samples (with constant illumination, background, head pose, etc...) to be useful for practical applications, the choice was made to get as close as possible to spontaneous emotions, while keeping at the same time a fully controlled recording environment. To achieve this goal we chose to use pre-defined answers for each situation. Based on the work of [9] 21 emotions were selected. For every emotion a short text (story) describes the context and ends with an emotional reaction of the participants. For each emotion the participant was asked to listen carefully to a short story and to 'immerge' themselves into the situation. Once ready, the participant may read, memorize and pronounce (one at the time) the five proposed utterances, which results in five different reactions to the given situation. The participants are asked to put in as much expressiveness as possible, producing a message that contains only the emotion to be elicited. With this procedure we let our participants take multiple sessions. This way we can fill the database quickly with many different recordings. The goal is to create a balanced database with respect to gender and age. The original texts and recordings are in Dutch, but we also created English versions of the texts. A constant adding of recordings to the database will ensure the expansion of the data corpus and will add more diversity to the recordings. On of the video recordings is shown in the Fig. 2. below.

Fig. 2. An example of the video recordings

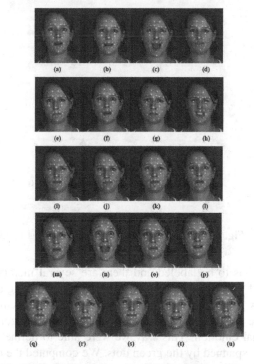

Fig. 3. Examples of recordings of 21 emotions (apex)

4 Data Analysis

Facial expressions are the result of contraction or dilatation of one or more facial muscles. Ekman designed a tool to describe every facial expressions in terms of

Action Units [6]. This labeling should be done by trained human observers and is time consuming. To enable automated labeling of facial expressions we put some green stickers on the faces of respondents on places corresponding to the facial muscles. To extract the position of the green dots automatically we implemented some image processing algorithms. Fig. 4 gives a schematic view of the algorithms used. For details on these algorithms see [10]. As result of the procedure we attach to every frame a vector of the positions of the green dots. Next we implemented an algorithm to track these vectors.

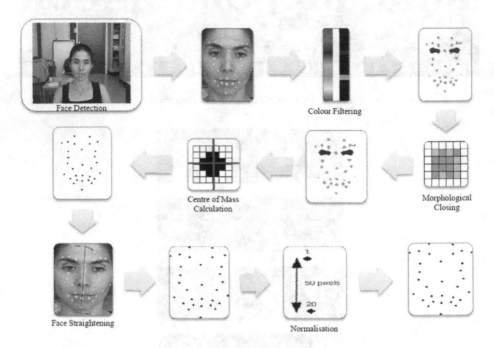

Fig. 4. Model of analysis of video recordings

All recorded data has to be labeled and there are several methods to do so. As a first step we analyzed the trajectories of the vectors corresponding to the positions of the green dots. Every recording was focused on one of the 21 emotions. For every track we localized the apex (see Fig. 3), that is to say the frame corresponding to the expression of maximal emotion. This way we were able to define 21 clusters in the 32-dimensional space spanned by the green dots. We computed the distances between the 21 clusters. Some of the results are displayed in Fig 5. We found that despite the broad clusters there are many individual differences in the ways particular emotions are expressed. For example, in the diagonal of Fig. 5 we can see the distance between the 21 clusters for two respondents named Maaike and Lotte. Secondly, we found that the Euclidean distance we used is not a good measure for emotional similarity in this space. The clusters of related concepts are not close to each other.

Maaike \ Lotte	Admiration	Amusement	Anger	Boredom	Contempt	Desire	Disappointment	Disgust	Dislike	Dissatisfaction	Fascination	Fear	Furious	Happiness	Indignation	Interest	Neutral	Sadness	Satisfaction	Surprise (Pleasant)	Surprise (Unpleasant)
Dislike	0.3686	0.4612	0.0149	0.2207	0.4523	0.2056	0.2245	0.798	0.324	0.3326	0.4236	0.4695	0.6452	0.8638	0.5775	0.0429	0.6108	0.0143	0.2055	0.8194	0.5586
Disgust	1.6881	1.1172	1.3124	1.538	1.8717	1.8338	1.5823	1.1406	0.3861	1.9696	1.7738	1.0627	1.2454	0.5669	0.1476	1.2285	1.1049	2.2597	0.7296	1.2716	1.8112
Disappointment	0.2566	1.1366	1.0443	1.1658	0.8912	0.1166	1.1719	0.7916	1.0833	0.8197	0.9378	0.98	0.9546	0.9602	0.9944	1.1522	1.1293	0.3859	1.1124	0.9711	0.7965
Desire	0.3282	0.6659	0.6686	0.2834	0.6183	3.526	0.0791	1.2133	0.6348	0.0679	0.2492	0.7915	0.633	0.2369	0.8891	0.728	0.6927	0.5964	1.0097	0.5343	0.6756
Contempt	0.5216	0.8643	0.9854	0.791	0.2987	0.3064	0.7884	0.9491	0.9149	0.5098	0.7927	1.1184	0.9001	0.6283	1.1941	0.6166	1.1218	1.1053	0.8578	1.0047	0.3826
Boredom	1.355	0.2603	0.3257	0.2452	0.6174	0.9715	1.1119	0.6679	0.8401	0.9668	0.5343	0.807	0.1092	0.1202	0.9783	0.5897	0.4475	0.349	0.9816	0.1749	1.1916
Anger	1.5121	2.2912	1.9911	2.6162	2.3469	0.9517	2.1459	2.0561	2.5564	2.1031	1.8954	1.7392	2.0073	2.2978	1.4422	2.4101	2.0297	1.9549	2.1302	1.9223	2.0402
Amusement	0.3183	0.0201	0.0969	0.1743	0.2978	0.5376	0.0755	0.0926	0.0643	0.1782	0.2176	0.0297	0.0873	0.3381	0.0956	0.1353	0.0319	0.7941	0.1143	0.0012	0.3891
Admiration	2.5691	2.8545	2.7563	2.2527	1.9061	1.4437	2.6637	2.6481	2.682	2.4749	2.9276	2.7907	2.9669	3.1974	3.0464	2.5363	2.8083	2.0912	2.6967	3.0477	2.3366

Fig. 5. Part of the Euclidean distances between the apex of 21 emotional clusters for two respondents Maaike and Lotte

5 Multimodal Labeling

To validated the recordings, we labeled the data in several ways. Given a trajectory of positions of the green dots we can compute the distances to the barycentre of the 21 clusters. This enables us to label the different tracks. We chose the label corresponding to the minimum cluster distance. In case of blended emotions the trajectory is close to different clusters and it would be better to give a probabilistic label. But more data is necessary to compute the distribution of each cluster.

Next we performed a user experiment to label the multimodal recordings. We used three different modalities: video, audio and text. We split up the validation into three different parts, this to see if there are differences in the validation of the various modalities. We provided raters for each recording only the image showing the apex of the emotion, only the audio clip from the recording or the whole recording as a video clip. We made a website where users could validate sample images, sound clips and videos on two dimensions: valence and arousal. Finally, we used the Whissell database [11] to compute the valence and arousal scores of the 21 emotional labels. The results are displayed in Fig. 6.

Each validation method is represented with its own color. In the ideal case we should see clusters of the same numbers close to each other. But this is not the case for some of the 21 emotions. For example, the number 6, representing the emotion 'contempt' is scattered over three quadrants of the axes. The image validation gives this emotion a positive valence and a negative arousal, the auditory validation a negative valence and a negative arousal, the video validation a positive valence and a

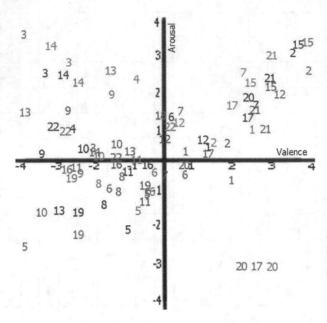

Fig. 6. Comparison of different validation methods – RED: Image validation, GREEN: Audio validation, BLUE: Video validation, and PURPLE: Text validation

positive arousal, and the textual validation gives contempt a negative valence and a negative arousal. This example shows that a lack of information, due to the separation of the unimodal channels can cause the user to misplace the intended emotion. A positive example is, for example, number 8. This number represents the emotion 'disappointment'. It can be clearly seen that all four number eights lay close together. This means that the validation of this emotion does not depend much on multimodal information. Other clear clusters are the numbers 15, 21, 3 and 14, representing happiness, pleasant surprise, anger and furious respectively. These very active emotions were validated very closely by all validation methods, showing that the distinction between active positive and negative emotions is not as subtle as with other emotions. We can also see that the labeling corresponding to the data using video and sound is closest to the Whissellscore. It proves that unimodal labeling is rather difficult. Multimodal data is needed.

6 Conclusion

In this paper we described a setup for the recording of a multimodal emotion database and reported about different labeling methods for such multimodal recordings. To validate our setup, we made video recordings of respondents in 21 different emotional contexts. It proves that unimodal labeling of only video or sound tracks is rather difficult for human observers. Presenting both modalities results in valence and arousal scores close to the the Whissell scores of the 21 emotional labels corresponding to the contexts of the recordings.

Acknowledgements

The authors would like to thank Anna Wojdel and Mathijs van Vulpen for their contributions to this work.

References

1. Boyle, E., Anderson, A.H., Newlands, A.: The effects of visibility on dialogue and performance in a co-operative problem solving task. Language and Speech 37, 1–20 (1994)
2. Russell, J.A., Fernandez-Dols, J.M.: The psychology of Facial Expression 9(3), 185–211 (1990)
3. Ekman, P.: Emotions Revealed. Times Books, New York (2003)
4. Darwin, C.: The Expression of the Emotions in Man and Animals (1872)
5. Martin, O., Kotsia, I., Macq, B., Pitas, I.: The eNTERFACE 2005 Audio-Visual Emotion Database, Atlanta (April 2006)
6. Ekman, P., Friesen, W.V.: The facial action coding system: A technique for measurement of facial movement (1978)
7. Cowie, R., Douglas-Cowie, E.: Emotion Recognition in human-Computer-Interaction. IEEE Signal Processing Magazine (January 2001)
8. Russell, J.A.: A circumplex model of affect. Journal of Personality and Social Psychology 39, 1167–1178 (1980); Vancouver, Canada (2001)
9. Desmet, P.M.A.: Designing Emotions, PhD dissertation, Delft University of Technology, Delft (2002)
10. Wojdel, A.: Knowledge Driven Facial Modelling, PhD dissertation, Delft University of Technology, Delft (2005)
11. Whissell, C.M., Dewson, M.J.: The dictionary of Affect in Language. In: Emotion: Theory, Research and Experience, vol. 18, pp. 113–131. Academic Press, New York (1989)

The ORD Speech Corpus of Russian Everyday Communication "One Speaker's Day": Creation Principles and Annotation

Alexander Asinovsky, Natalia Bogdanova, Marina Rusakova,
Anastassia Ryko, Svetlana Stepanova, and Tatiana Sherstinova

St. Petersburg State University, St. Petersburg,
Universitetsakaya nab. 11, 199034, Russia
{a.s.asinovsky,mvrusakova,sherstinova}@gmail.com,
{nvbogdanova_2005,aryko,stsvet_2002}@mail.ru

Abstract. The main aim of the ORD speech corpus is to fix Russian spontaneous speech in natural communicative situations. The corpus presents the unique linguistic material, allowing to perform fundamental research in many scientific aspects and to solve different practical tasks, especially in speech technologies. The paper concerns methodology and description of the ORD corpus creating and presents the system of annotations.

1 Introduction

Beginning from 1990s national corpora of spontaneous speech are created in many countries of the world. The first audio corpus based on spontaneous oral speech of subjects organized in a demographically balanced sample is a part of the Spoken Component of the British National corpus [12]. So far, there is no really representative corpus of Russian everyday speech, though some databases of oral speech have been elaborated by various scientific groups for at least recent 40 years [1].

Although there is some progress achieved in the exploration of informal speech, this domain certainly calls for further investigation. For example, it is not known how many wordworm (morpheme, sentence) tokens are produced and perceived by a speaker per hour, day, month; how many different language units he (she) processes per various periods of time or what is the average duration of sounding speech within these periods. The lack of linguistic information of this kind is well comprehensible: carrying out investigations needed to achieve such results is extremely labour- and time-consuming, and technical tools for such explorations became available comparatively not long ago. Now both technical means and theoretical background for the exploration of spontaneous oral speech do exist. Nevertheless, these investigations should be treated as highly innovative, because they can be held only by professional teams including many researchers.

During the last decades a great amount of Russian natural oral speech data was collected. Unfortunately, when gathering speech material researchers use different methods of collecting and pursue their individual scientific goals. As a result, the available resources are not uniform, uncoordinated and it is rather difficult to integrate the data

V. Matoušek and P. Mautner (Eds.): TSD 2009, LNAI 5729, pp. 250–257, 2009.

into a single representative corpus open for general use. Therefore, shaping of the full-fledged descriptive database of the Russian oral informal speech is the prerequisite for the progress in those linguistic spheres which are concerned with the native speaker, natural speech and, in a larger scale, communicative behaviour in general.

2 The Main Concepts of the ORD Corpus

2.1 Methodological Background

The main goal of the present investigation is to fix Russian spontaneous speech in natural communicative situations. Firstly, it means that nothing should interfere with the usual habits of speakers' communicative behaviour in particular speech situations during recording. For example, speech communication at breakfast should be realized in those circumstances that are habitual to every individual: in the same setting and with the same interlocutors as usual, with the same level of accompanying noise (with an open window, refrigerator functioning, etc.). Secondly, every subject should realize his speech ability in standard situations, not changing the usual length of utterances, as well as speech topics and repertoire during recording. For example, if a subject got used to read a newspaper, he should not give up this habit and should not organize an unusual for this moment communication (e.g., intentionally talking with other members of the family or inviting unusual guests) in order to enlarge the amount of speech material within the recorded time. At the same time at the first stage of our investigation the participants were asked to make a deliberate choice of a particular day for recordings. For example, it is preferable to make the recording on an ordinary day, rather than on a day when the subject is about to do something unusual (e.g., to go for an excursion, to be absent from home for the whole day because of a seasonal rush at his office, etc.) [2].

2.2 Technical Equipment

Recording is made with the help of a dictaphone. At first the subject makes all the necessary settings and fasten the dictaphone to his clothes (into a pocket or a special bag). Such a mode of recording inevitably causes non-uniformity of the quality of the obtained speech data. In our research we have been using Olympus WS-320M dictaphone, which ensures more that 35 hours of high-quality recordings. A relatively low level of quality in the recording obtained in this way (if compared to studio recordings) is an unavoidable consequence of undertaking a fieldwork aimed at the natural experiment with human speech behaviour.

2.3 Speakers' Selection and Training

At this stage of the research the study is not directly aimed at the description of the Russian language in the whole richness of its manifestations. There is only one form of its functioning that is being studied, viz. the speech of naïve native speakers of Russian who live in a city where the Standard language is dominating, where there is no significant impact on the part of dialects or other languages, and where the population is professionally heterogeneous and unbiased in terms of age and gender distribution.

St. Petersburg is a model city of this type. This is the reason why it was St. Petersburg where the subjects were being selected.

The objective of the present research supposes that subjects should never be chosen among those speakers whose occupation requires any specific level of speech control or self-consciousness in this respect; indeed, it is known that these professional skills might have a strong impact on speech production by individuals.

The preparatory work with the subject consists of two stages: 1) providing the necessary degree of naturalness of the subject's behaviour in communicative situations; 2) providing proper technical quality of the recordings necessary for further speech analysis.

The first of these two stages is indeed very complicated. During a pilot experiment, in which the very members of the research team (i.e., linguists themselves) recorded their days-of-speech, it was detected that despite being highly motivated in terms of striving for adequate results, the subjects were hardly able to "forget" that they were bearing a dictaphone: they could not abandon the idea that their communication with surrounding people, especially with their close relatives, was going to become available to their acquaintances/colleagues.

Finally, it was decided to gather recordings under conditions of full and utter anonymity for the sake of obtaining their maximal naturalness. The following procedure has been elaborated. A non-member of the research team, a psycholinguist by specialization, joins the procedure. He addresses potential subjects, for example, a group of people working in a particular enterprise. When instructing potential subjects he guarantees that under no circumstances he would personally come in touch with the future materials obtained from those subjects.

When the recordings are made, this intermediate collaborator hands them over to the research team. The subjects do not indicate in any form their actual names, but they have to fill in a questionnaire concerning their age, profession, place of birth and other sociological data. As a result the research team obtains recordings of speakers who are absolutely unknown to them; moreover they have never met one another in person. The shortcoming of this approach is that no control on the part of professional linguists is possible at recording stage, hence there is a relatively high ratio of spoilage and technical flaws in recorded files, which make impossible phonetic analysis of correspondent fragments.

The second stage is implemented by way of instructing the subjects-to-be how to use dictaphones for the best results.

2.4   Creating the ORD Corpus

The abbreviation *ORD* stems from Russian *Odin Rechevoj Den'*, literally translated as "one day of speech". A demographically balanced group of subjects was formed for the first series of recordings, including 30 participants (the base speakers or informants) representing various social and age strata in the population of St. Petersburg. After detailed instructions the subjects made recordings of all speech communications in which they took part during one day. Besides, all of them filled in the questionnaires and passed through psychological testing.

The overall length of the recorded material is 240 hours, from which 170 hours contain speech data quite suitable for further analysis. Beside the subjects' speech, that of their 520 interlocutors was also recorded. Interlocutors were people of different ages (from 3 to 68 years), professions and occupations (e.g., salesmen, conductors, managers, lecturers, doctors, librarians, IT-professionals, students, and many others), being in friendly, family, professional or other relations with the subjects. Materials contain diverse genres and styles of speech: professional conversations with colleagues, telephone talks, lectures, practical lessons of foreign languages, communications with friends and relatives during airings, parties, breakfasts, dinners, etc. The topics in these conversations also range from discussions of teeth problems with a dentist to conversations about religion, life and death. Recordings were made at home, while traveling by the public transportation, while walking outside, at the university, in the military college, in coffee bars, in the shops, in the amusement park, etc. The corpus was divided into 2202 communication episodes. 134 episodes are already transcribed in detail [2].

The recordings are processed by professional linguists. The initial stage of this process consists in preliminary description of the material and its orthographic transcription. Generally, when studying oral speech, the burden of deciphering is often put onto the shoulders of the very participants of the dialogues; these are the persons who are often able to interpret the least hearable fragments and to provide the fullest extralinguistic information relevant for the production of the texts in question. In our research, the deciphers could not witness the communicative behaviour of speakers for the sake of speech naturalness. The fact that the recordings are transcribed by non-participants results in partial loss of information. This loss should be viewed as an inevitable toll for the naturalness of the material gathered.

3   System of Annotations

3.1   Annotation Software

Two professional annotation tools – ELAN and Praat – are used to annotate the ORD corpus. The first one is ELAN (EUDICO Linguistic Annotator) developed at the Max Planck Institute for Psycholinguistics (Nijmegen, The Netherlands), with the aim to provide a sound technological basis for annotation and exploitation of multi-media recordings. This professional tool allows to create, edit, visualize and search annotations for video and audio data. Being specifically designed for the analysis of language, ELAN is a convenient tool for multi-level annotations of communication and speech [5]. Figure 1 shows a fragment of multi-level annotation of the ORD corpus for a number of tiers made in ELAN.

ELAN is used for annotation of most tiers (Episodes, Frases, Events, Words, etc.) apart from phonetic ones (Sounds, Syllables, and others presented in phonetic transcription), which are made in Praat.

Praat is a professional phonetic annotator which allows to analyze and manipulate speech. The Praat program was created by Paul Boersma and David Weenink (the Institute of Phonetics Sciences, University of Amsterdam, The Netherlands) [8].

The structures of ELAN and Praat annotations are fully compatible.

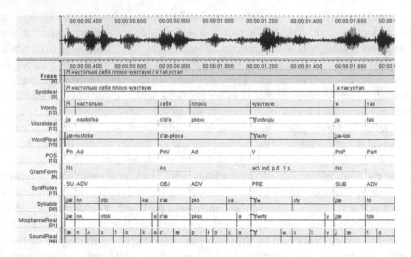

Fig. 1. A fragment of multi-level annotation of the ORD corpus made in ELAN. Tiers: Frase (orthographic transcripts), SynIdeal (syntagmas), Words (orthographic), WordIdeal and WordReal (words in phonetic transcription), POS (part of speech), GramForm (grammatic form), SyntRoles (syntactical role of the word), Syllable (real phonetic transcription), MorphemeReal (morpheme in real phonetic transcription), SoundReal (sounds in real phonetic transcription).

3.2   Annotation Principles

The main principles of multi-level annotation for a spoken corpus were given in [3].

Primary annotation of speech in the ORD corpus is made in ELAN and implies annotation of the followings data types:

Frase - orthographic transcripts of phrases, which are the main units of description. Besides transcripts it contains references to pauses and noisy fragments. Independent type.

Event - nonlanguage audio events (dog barking, squeak of a door, phone ring, radio program, etc.), including as well some voiced events (e.g., cough, yawn, laugh, moan, etc.) not connected with speech. Independent type.

Speaker is a person who pronounced correspondent phrase (Frase) either a base speaker (subject) or one of his/her communicators. This type depends on Frase (Symbolic Association).

Voice is the special characteristic of speech for the given phrase (Frase) or its segment (e.g., hoarse, whisper, scanning, irritated, imitating, ironical, dramatic, etc.). Depends on Frase (Included In).

FraseComment contains all kinds of comments on phrase realisation and researcher's remarks. Depends on Frase (Symbolic Association).

Notes may contain other useful information. Independent type.

Episode (Communication episode) refers to general communicational situation. Independent type.

These main 7 data types are being annotated in ELAN on 7 correspondent tiers: Frase, Events, Speaker, Voice, FraseComments, Notes, Episodes.

Some spelling rules for transcripts (tier Frase):

- Speech is written in standard orthography.
- Speech is divided into conventional sentences [10].
- Transcripts may include the following symbols:

Symbol	Meaning
/ (//)	marks of syntagmatic (phrasal) division
*H	fragment of unintelligible speech
*P	barely intelligible speech
*П	pause
()	short hesitation pause
(...)	long hesitation pause, which may be filled
(m-m), (a-a), (e-e), (a-m)	hesitation pause filled by sounds
!, ?	marks indicating exclamatory and interrogative utterances
ca...	interrupted word
...	unfinished phrases
(?)	questionable or ambiguous transcript
#	change of a speaker in overlapping speech
@	remark inserted by another speaker in overlapping speech fragments

Speaker's code is attached to each phrase (e.g., *M2*). If a person-interlocutor is unknown or unidentified, his/her code is marked by symbol *X* (e.g., *MX1*). In case interlocutors speak simultaneously we use symbol # as a delimiter for individual codes (e.g., *S01#F5*) or symbol @ (e.g., *S07@M2* means that while *S07* is speaking his second male interlocutor pronounces an insert remark).

Figure 2 shows a sample of primary annotation made for three tiers.

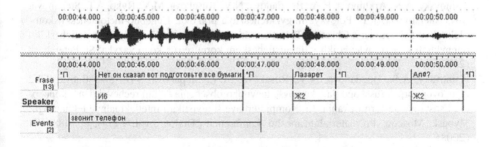

Fig. 2. Sample of primary annotation for three tiers (**Frase**, **Speaker**, **Events**)

The second stage of corpus annotation is made on lexical level and includes tagging for the following main tiers: Words (spelling), POS, GramForm (grammatical form), and SyntRole (syntactic role).

Description of annotations made on phonetic levels in Praat may be found in [9].

4 Conclusion

Though creating of the ORD corpus is still in progress, its diverse speech material and annotations have given birth to a number of linguistic and interdisciplinary researches. Let us mention just some of the papers.

In [4] the most frequent reduced wordforms of spontaneous Russian speech are described. The work [10] discusses general problems of separation of various linguistic units from a real speech stream, whereas the paper [11] describes dynamics of communication episodes in everyday life. Article [6] relates to psycholinguistic studies and concerns the dependence of speech characteristics on speaker's mental state and personality. In [13] one may find sociological description of the speakers recorded for the ORD, and the paper [7] presents study on the everyday rhetoric and its techniques.

The new project which have been started on the material of the ORD corpus is creating an audio dictionary of Russian morphemes.

Acknowledgements

The first recordings and database creating of the ORD corpus were supported by the Russian Foundation for Humanities within the framework of the project *Speech Corpus of Russian Everyday Communication "One Speaker's Day"* (project # 07-04-94515e/Ya). Nowadays creating of the corpus is supported by the program of the Russian Ministry of Education titled *Sound Form of Russian Grammar System in Communicative and Informational Approach* and by the grant of the Russian Foundation for Humanities *Development of an Information System for Monitoring of Russian Spoken Language* (project # 09-04-12115v).

References

1. Asinovsky, A.S., Arkhipova, E.A., Bogdanova, N.V., Rusakova, M.V., Ryko, A.I., Stepanova, S.B., Sherstinova, T.Y.: Polevaya lingvisticheskaya praktika. Uchebno-metodicheskij kompleks slozhnoj struktury. Chast' 1. Teoreticheskie osnovy i metodika sbora lingvisticheskikh dannykh dl'a predstavlenia ikh v lingvisticheskom korpuse russkogo yazyka. St. Petersburg (2007)
2. Asinovsky, A.S., Bogdanova, N.V., Rusakova, M.V., Stepanova, S.B., Sherstinova, T.Y.: Zvukovoj korpus russkogo yazyka povsednevnogo obschenia "Odin rechevoj den": koncepcia i sosytoyanie formirovania. In: Kompjuternaya lingvistika i intellektualnye tekhnologii. Vypusk, Moscow. Po materialam mezhd. konferencii "Dialog", vol. 7 (14), pp. 488–494 (2008)
3. Asinovsky, A.S., Koroleva, I.V., Rusakova, M.V., Ryko, A.I., Philippova, N.S., Stepanova, S.B.: On Integral Multilevel Annotation of a Spoken Russian Corpus. In: Proc. the XIIth International Conference "Speech and Computer" SPECOM 2007, Moscow (2007)
4. Bogdanova, N.V.: Allegrovye formy russkoj rechi: ot proiznositel'noj redukcii k pis'mennoj fiksacii i leksikalizacii v yazyke. Mat-ly XXXVII mezhd. filologicheskoj konferencii. Vypusk 18. "Fonetika". St. Petersburg (2008)
5. ELAN - Linguistic Annotator. Version 3.6,
   http://www.mpi.nl/corpus/manuals/manual-elan.pdf

6. Koroleva, I.V.: Individual'nye sostoyania i svoistva yazykovoj lichnosti: vliyanie na lingvisticheskuju strukturu vyskazyvanij. Mat-ly XXXVII mezhd. filologicheskoj konferencii. Vypusk 21. St. Petersburg. pp. 36–45 (2008)
7. Markasova, E.V.: Ritoricheskaya enantiosemia v korpuse russkogo yazyka povsednevnogo obschenia "Odin rechevoj den". In: Kompjuternaya lingvistika i intellektualnye tekhnologii. Vypusk "Dialog", Moscow, vol. 7(14), pp. 352–356 (2008)
8. Praat: Doing Phonetics by computer, http://www.praat.org
9. Ryko, A.I., Stepanova, S.B.: Mnogourovnevaya lingvisticheskaya razmetka zvukovogo korpusa russkogo yazyka. In: Kompjuternaya lingvistika i intellektualnye tekhnologii. Vypusk. Po materialam mezhd. konferencii "Dialog", Moscow, vol. 7 (14), pp. 460–465 (2008)
10. Ryko, A.I., Stepanova, S.B.: Problemy vychlenenia jedinic analiza spontannogo ustnogo teksta. In: Mat-ly XXXVII mezhd. filologicheskoj konferencii. Vypusk, St. Petersburg, vol. 21, pp. 71–80 (2008)
11. Sherstinova, T.Y.: "Odin rechevoj den" na vremennoj shkale: o perspektivakh issledovania dinamicheskikh processov na materiale zvukovogo korpusa. In: Vestnik Sankt-Peterburgskogo universiteta, Seria 9: Filologia, Vostokovedenie, Zhurnalistika, Chast' 2, St. Petersburg, vol. 4, pp. 227–235 (2008)
12. The British National Corpus http://www.natcorp.ox.ac.uk/
13. Zobnina, E.A.: Social'nye characteristiki govoriaschego: objektivnye dannye i ekspertnaya ocenka rechi (po materialam zvukovogo korpusa "Odin rechevoj den". In: Mat-ly XXXVII mezhd. filologicheskoj konferencii. Vypusk, St. Petersburg, vol. 21, pp. 17–24 (2008)

The Structure of the ORD Speech Corpus of Russian Everyday Communication

Tatiana Sherstinova

St. Petersburg State University,
St. Petersburg, Universitetskaya nab. 11, 199034, Russia
sherstinova@gmail.com

Abstract. The paper presents the structure of the ORD speech corpus of Russian everyday communication, which contains recordings of all spoken episodes recorded during twenty-four hours by a demographically balanced group of people in St. Petersburg. The paper describes the structure of the corpus, consisting of audio files, annotation files and information system and reviews the main communicative episodes presented in the corpus.

1 Introduction: What Is the ORD Corpus?

The abbreviation *ORD* stems from Russian *Odin Rechevoj Den'*, literally translated as "one day of speech". The main aim of creating the ORD corpus is to collect recordings of actual speech which we use in our everyday communication. The ORD creating is an interdisciplinary project, in which specialists in many scientific branches are involved. Primarily, they are linguists – experts in different aspects of Russian language – phoneticians, grammarians, lexicographers, dialectologists. Besides them two psychologists and a sociologist took part in project creation as well as specialists in modern information technologies.

For the first series of recordings a demographically balanced group of 30 subjects representing various social and age strata in the population of St. Petersburg was selected. The subjects spent one day with dictaphones dangling around their necks and recording all their communication. So dictaphones should be turned on in the morning recording breakfast at home with family members, then preparation for going to work, the way to work itself, speaking by cellular telephone, then official and informal conversations at work with colleagues (e.g., about problems with children, world financial crisis, yesterday's football match, etc.), lunch time, shopping, recreation and so on up to the moment when subjects went to bed.

In the result more then 240 hours of recording were obtained, from which 170 hours contain speech data quite suitable for further linguistic analysis and more than 50 hours of recordings are good enough for further phonetic analysis. The corpus was divided into 2202 communication episodes. 134 episodes are already transcribed in detail. At present, orthographic transcription of the corpus numbers more than 50000 wordforms [1].

The corpus presents the unique linguistic material, allowing to perform fundamental research in many aspects including complex behaviour of people in real world. At the

V. Matoušek and P. Mautner (Eds.): TSD 2009, LNAI 5729, pp. 258–265, 2009.
© Springer-Verlag Berlin Heidelberg 2009

same time these utterly natural recordings may be used for practical purposes: for example, for verification of many scientific hypotheses, for adjustment and improvement of speech synthesis and recognition systems, etc.

2   The ORD Corpus Structure

The ORD speech corpus consists of three major components: 1) audio files, 2) correspondent annotation files, and 3) information system.

2.1   Audio Files

The methodology used for collection of speech for the ORD corpus when subjects were asked to keep dictophones on for many hours gave unique material about our everyday speech behaviour (e.g., we have quite rare recordings of people talking just to themselves or informal communication of cadets in barracks of military school). At the same time it inevitably got a great amount of non-acceptable recordings: fragments without speech and fragments with speech in very noisy environment (e.g., background remarks of metro passengers). It was necessary to separate fragments containing speech from that without speech, and to classify speech fragments according to their quality.

Obviously, the archive copy of all recordings is kept in its original form, allowing to reconstitute "speech days" as they were. Every audio file of the other copy of the corpus was carefully listen to and segmented into fragments, which became the main units of audio corpus. It is supposed that each file is not longer than 30 minutes, contains speech recording of the similar quality and refers to the same or adjacent communication episode(s). All fragments without speech longer than several minutes are cut from ORD files, as well as fragments containing just background sounds of working TV or radio. Information on this segmentation is available in an auxiliary database.

The new files have got names, referring to the subject's code and the ordinal number of episode. The phonetic quality of each file was further evaluated and measured in 4-score scale: 1 – the best quality, suitable for precise phonetic analysis, 2 – rather good quality partially suitable for phonetic analysis, 3 – noisy recordings with low quality which is only partially legible (not suitable for phonetic analysis but suitable for other aspects of research), 4 – unintelligible conversations or remarks in extreme noise, which could not be understand without noise reduction. It is planned to annotate first the recordings of the best quality, whereas the most noisy audio files are not to be annotated at all.

2.2   Annotation Files

The ORD corpus is being annotated by means of two professional annotation tools – ELAN [2] and Praat [3]. The main principles of multi-level tagging in the ORD corpus were described in [4]. The annotation formats of ELAN and Praat are fully convertible. ELAN is used for primary and general annotations of the corpus, whereas Praat is used for making real phonetic transcription and other phonetic annotations. The annotations are kept in files of two general types *.eaf (ELAN format) and *.TextGrid (Praat format). Being verified by experts, annotation data are exported into general information system.

2.3   Information System

The information system presents a relational database created on the base of MS Access 2003. All tables of the database are divided into 3 general groups.

Group I – Actual information about speakers, sound files, and communication episodes.

Table 1.1 – *Informants (Speakers/Subjects)*: actual data about all base speakers, presented by the subjects themselves. For example, speaker's code (S01, S02, etc.), his/her nickname; gender; age; place of birth; social group; education; qualification; current occupation; nationality; number and quality of recorded files; total and usable time of recording; comments; etc.

Table 1.2 – *Communicants (Interlocutors)*: some actual data about the main people who communicated with the subjects during their speech day: communicator's code; his/her (nick)name; relation to the subject or his/her social role (e.g., mother, friend, shop assistant, etc.); gender; approximate age; and some other possible information provided by the subjects – interlocutors' place of birth; social group; education; qualification; current occupation; nationality; as well as intelligibility of recorded speech; some comments; etc.

Table 1.3 – *ARCSoundFiles*: information about original (archival) sound files including total duration, that of intelligible speech and illegible or noisy fragments.

Table 1.4 – *ORDSoundFiles*: information about reformatted sound files including reference to correspondent original files, exact position in the original file, total duration, phonetic quality of recording, annotation priority, and the main communication episode, described in three fields: 1) where – 2) doing what – 3) who is (are) the main interlocutor(s).

Table 1.5 – *Episodes*: concise formal description of main voiced communication episodes. Segmentation into episodes was made by expert linguists, its description includes information on interlocutors, time, duration, place, aim and subject of communication, as well as some possible comments. At first segmentation into episodes was made rather arbitrary. Now we try to standardize both segmentation and its description.

Group 2 presents some results of social and psycholinguistic data interpretation. Currently it contains just two tables:

Table 2.1 – *InformantsSocial (Speakers' Social Attributions)* has the same structure as Table 1.2 (*Speakers*), but contains subjective evaluation of speaker's social characteristics by linguists who transcribed their speech. It should be noted that the linguists were not allowed to see actual information from Table 1.1 before filling Table 2.1.

Table 2.2 – *InformantsPsycho (Speakers' Psychological Portraits)*. This table is also filled by linguists who work with recordings and contains speaker's psychological rating in ten-point system for the following aspects: neurotizismus, spontaneous aggression, depression, irritation, sociability, tranquility, responsive aggression, self-consciousness, openness, extroversion/introversion, emotional instability, masculinity/feminism. Besides, here you can find a one-page essay about each base speaker written by researchers in a free style.

Group 3 contains tables for speech transcripts and multi-level annotations. Filling these tables of the database is still in progress.

Table 3.1 – *MiniEpisodes* is used to give reference to smaller real-life episodes within larger episodes described in Table 1.5. For example, Speaker X in Episode N (in the evening – at summer house – with mother) may have the following mini-episodes: 1) searching for matches (2 minutes), 2) trying to set fire to the oven (3 minites), 3) discussing plans for the rest of the evening (5 minutes), etc.

Table 3.2 – *Timeline*: Each (mini-)episode is sequentially subdivided into utterances (remarks/phrases) and pauses. The exact timing of each fragment is given.

Table 3.3 – *Frases* contains orthographic transcript of communication episodes made by linguists using a special system of notation. Reference to the speaking person is given in the "Speaker" field, using the same codes as in Table 1.1 and 1.2. Auxiliary information on starting-ending points of the phrase in milliseconds allows to listen to it in the correspondent database form.

Table 3.4 – *Voice* is linked with the previous table and contains information about possible changes of voice quality in some fragment of the speech (either physiologically or functionally – e.g., hoarsely, smiling, yawning, exciting, imitating, etc.).

Table 3.5 – *Events* describes non-language audio events (squeak of a door, phone ring, etc.).

Table 3.6 – *Notes*: refers to auxiliary information which may be given to some period of time (e.g., "this fragment contains specific youth slang").

Table 3.7 – *Words*: keeps information about each wordform of the corpus (e.g., POS, grammatical form, syntactic role, phonetic transcription, etc.).

Table 3.8 – *PhonWords*: describes phonetic words. Segmentation into phonetic words is being made for sub-corpus of the best quality. The table currently contains orthographic spelling of phonetic words and reference to starting-ending points in correspondent audio files. Phonetic transcription is planned to be added later. It is possible to listen to each segmented phonetic word by means of database utilities.

Table 3.9 – *Morphemes*: contains data on morphemes (e.g., general class, ideal transcription, phonetic transcription, etc.).

Table 3.10 –*Sounds*: contains phonetic information about individual realization of phonemes or indivisible sound groups (ideal and real phonetic transcription, position, etc.).

A number of additional levels of annotation is planned to be included into the database further (e.g., speaker's mental state, emotional connotations, communicational strategies, rhetorical techniques, prosodic models, etc.). Therefore, new tables will appear in the database in the future.

On Fig. 1 you may see examples of orthographic transcripts of phrases in Table Frase for the mini-episode "a story about speaker's driving lesson" (Speaker S05, female, 29 years old, teacher).

Basing on obtained annotations different frequency lists may be built for words and phonetic words for any speaker, episode, group of episodes and corpus in the whole. All occurrences of each word may be found and listen to.

A flexible search system is currently being created for the database of annotations. Further data processing is based on database requests and special applications, which

Annotation	Speak	BeginTin	EndTime
▶ ну рассказать / что я там делала-то на вождении ?	И5	125608	128025
ну что ты доволен?	М13	128315	129103
теорию изучали //	М11	129103	130140
нет / у меня была практика / теория с понедельника начнется //	И5	130420	134200
там меня / дядька посадил / короче говоря / за машину //	И5	135440	138735
и спросил / ну типа умеешь вообще кататься или нет / сидела за рулем / я говорю	И5	139130	145310
ну знаю просто где тормоз / газ / сцепление / всё больше ничего не знаю //	И5	145650	149285
ну ладно / говорит / давай //	И5	149950	151765
отъехали мы / вот / от (...) площади / ну он сам отъехал там на такую дорогу уже / ка	И5	152140	160715
ну вот / он меня посадил за руль //	И5	161720	163505
я вначале тронулась / такая / дын-дын-дын вообще ˚С знаешь //	И5	164310	168375
ну вон он такой / я сказал что сцепление по-другому //	И5	168635	171715
а я думала что ну этот мужик / что он такой как бы сказать / что он псих такой //	И5	171920	175950
о(:) ! я так не хотела к нему идти а такой нормальный // он так ласково разговарива	И5	176015	183425
знаешь по имени сразу / такой / так спросил как по имени //	И5	183765	186325
вот Леночка давай / так отжимай / так тихонечко да-да-да-да / вот так вот / знаешь //	И5	186590	191870
˚С	И5	191870	192270
он ... то есть / чувствуется что он с такими лохушками как я / знаешь работает ˚С	И5	192270	196535
вот //	И5	198220	198730
ну / нормально так //	И5	199220	200290

Запись: [◄] [◄]   168 [►] [►|] [►*] из 521   <   >

Fig. 1. An example of orthographic transcript of phrases in Table *Frase*

are currently being developed. One of such applications is E-Kar utility, which allows complex lexicographic and morphologic data processing.

Speech material of the ORD corpus will be constantly increased. Thus the new recordings are currently being made by new groups of subjects. New functional modules are also being made within the ORD speech corpus – for example, an audio dictionary of Russian morphemes.

3   Summarization of Communicative Episodes

In this section we will briefly describe the content of the ORD corpus from the point of view of episodes' typology. The term episode in the ORD terminology means continuous and preferably long-lasting fragment of one-day-of-speech recordings with the common conditions of communication (time, place, action, interlocutors). Episodes may be further subdivided into mini-episodes which refer to shorter periods of time with the common topic of conversation, simultaneous action, etc.

All 2202 communication episodes detected in the "days of speech" were then divided into 22 general categories. The most frequent types are represented on Fig. 2. Percents shown on this chart refer to total duration of recording made in the given conditions and doesn't reflect the number of utterances (or words) recorded in each situation. It can be seen from Fig. 2, that the largest part of speech records (42% or nearly 93,6 hours) refers to communication related with the main occupation of the subjects (at work or at studies). This category is more than 4 times bigger than the group of the second rank - family conversation at home in the evening (9,92%). The types that follows refer to parties in coffee bars or restaurants (9,80%) and to dialogues on the way to the working place or somewhere else (8,77%). The fifth place is occupied by home conversations in the morning (5,41%). The duration of any other category of episodes does not reach 5% of the averaged speech day.

Such classification of episodes is rather rough. Evidently, some categories should be further reviewed. For example, the huge category "at work" may be divided into business meetings, work with clients, business calls, individual work, and personal contacts, including private and nonbusiness conversations. Moreover, some smaller episodes or situations (e.g., phone calls) may take place within practically any main category.

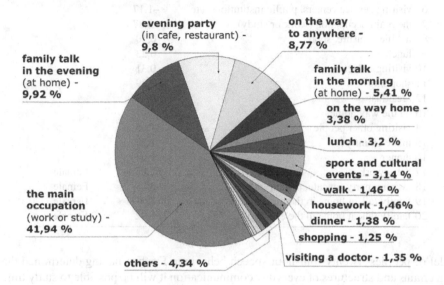

Fig. 2. Summarization of main episodes in the ORD according to the total time of original recordings

3.1 Male and Female Speech Days

Special investigation was made to compare average male and female speech days. In the majority of the categories of episodes the speech day of men does not differ significantly from that of women – the difference is less than 1-2%, such is the case of the dominating category "working/studies" (see Table 1). There are, however, a few categories in which rather substantial difference can be observed. In particular, men spent more than 9% more time than women attending various (sport, cultural, etc.) events; therefore total time of men's being "on a way" is nearly 5% longer. As for women, they spent "this time" at home conversations in the evening (7% more than men on the average), at parties and dinners (both 2% longer than that of men) and in the morning conversations (3% longer). This result is, however, not surprising from both psychological and sociological points of view [5].

When the corpus will be totally annotated and transcribed we can measure more precise the quantity of speech (utterances and words) for each communication episodes and to study its dynamics within days-of-speech. Besides, differentiation of working days and holidays should be also taken into account. Then we can try to built an averaged

Table 1. Comparison of male and female communication episodes in the ORD corpus

General types of episodes	Difference (%)	Male/Female
1 sport and cultural events	−8,69	Male
2 on the way to anywhere	−4,85	Male
3 at home in the daytime	−2,51	Male
4 walk	−1,58	Male
5 corporate party	−1,47	
6 visiting service centers, public institutions, etc.	−1,17	
7 the main occupation (work or study)	−0,87	
8 hobbies, leisure, sport	0,10	
9 lunch	0,23	
10 visiting a doctor	0,31	
11 shopping	0,59	
12 breakfast	0,79	
13 on the way home	0,97	
14 visiting other people	1,06	
15 in the country	1,20	
16 dinner	1,94	
17 housework	2,05	Female
18 evening party (in cafe, restaurant)	2,08	Female
19 family talk at home in the morning	3,01	Female
20 family talk at home in the evening	6,92	Female

model of peoples' twenty-four-hour speech behaviour. Further, having determined the main chains and structures of everyday communication it will be possible to study time series of quantitative variables by means of standard statistical methods and to analyze frequency series (e.g., of lexical, grammatical or semantic units, acoustic phenomena, prosodic contours, etc.) depending on various conditions of communication.

Acknowledgements

The first recordings and database creating of the ORD corpus were supported by the Russian Foundation for Humanities within the framework of the project *Speech Corpus of Russian Everyday Communication "One Speaker's Day"* (project # 07-04-94515e/Ya). Nowadays creating of the corpus is supported by the program of the Russian Ministry of Education titled *Sound Form of Russian Grammar System in Communicative and Informational Approach* and by the grant of the Russian Foundation for Humanities *Development of an Information System for Monitoring of Russian Spoken Language* (project # 09-04-12115v).

References

1. Asinovsky, A.S., Bogdanova, N.V., Rusakova, M.V., Stepanova, S.B., Sherstinova, T.Y.: Zvukovoj korpus russkogo yazyka povsednevnogo obschenia "Odin rechevoj den": koncepcia i sosytojanie formirovania. In: Kompjuternaya lingvistika i intellektualnye tekhnologii. Vypusk. Po materialam mezhd. konferencii "Dialog", Moscow, vol. 7 (14), pp. 488–494 (2008)

The Structure of the ORD Speech Corpus of Russian Everyday Communication   265

2. ELAN - Linguistic Annotator. Version 3.6,
http://www.mpi.nl/corpus/manuals/manual-elan.pdf
3. Praat: Doing Phonetics by computer, http://www.praat.org
4. Ryko, A.I., Stepanova, S.B.: Mnogourovnevaya lingvisticheskaya razmetka zvukovogo korpusa russkogo yazyka. In: Kompjuternaya lingvistika i intellektualnye tekhnologii. Vypusk. Po materialam mezhd. konferencii "Dialog", Moscow, vol. 7 (14), pp. 460–465 (2008)
5. Sherstinova, T.Y.: "Odin rechevoj den" na vremennoj shkale: o perspektivakh issledovania dinamicheskikh processov na materiale zvukovogo korpusa. In: Vestnik Sankt-Peterburgskogo universiteta, Chast' 2, St. Petersburg. Seria 9: Filologia. Vostokovedenie. Zhurnalistika, vol. 4, pp. 227–235 (2008)

Analysis and Assessment of AvID: Multi-Modal Emotional Database

Rok Gajšek[1], Vitomir Štruc[1], Boštjan Vesnicer[1],
Anja Podlesek[2], Luka Komidar[2], and France Mihelič[1]

[1] Faculty of Electrical Engineering, University of Ljubljana,
Tržaška 25, SI-1000 Ljubljana, Slovenia
{rok.gajsek,france.mihelic}@fe.uni-lj.si,
{vitos,bostjanv}@luks.fe.uni-lj.si
http://luks.fe.uni-lj.si/
[2] Faculty of Arts, University of Ljubljana
Aškerčeva 2, SI-1000 Ljubljana, Slovenia
{anja.podlesek,luka.komidar}@ff.uni-lj.si

Abstract. The paper deals with the recording and the evaluation of a multi modal (audio/video) database of spontaneous emotions. Firstly, motivation for this work is given and different recording strategies used are described. Special attention is given to the process of evaluating the emotional database. Different kappa statistics normally used in measuring the agreement between annotators are discussed. Following the problems of standard kappa coefficients, when used in emotional database assessment, a new time-weighted free-marginal kappa is presented. It differs from the other kappa statistics in that it weights each utterance's particular score of agreement based on the duration of the utterance. The new method is evaluated and the superiority over the standard kappa, when dealing with a database of spontaneous emotions, is demonstrated.

1 Introduction

Detecting emotions in speech has become a popular sub-field in speech recognition over recent years. Different tasks such as speech recognition, speaker identification or verification, development of dialog managers, etc., can benefit from added information about the psychological state of the speaker. A telephone based dialog system that would detect if a customer is getting annoyed and dissatisfied with the service could switch to a human operator for further assistance. In telecommunication systems a known emotional state of the person on the other side of the channel would enable a more credible exchange of information. These are just two possible use cases where emotion detection can be used.

A starting point for research in emotion recognition is a quality database. Most of available databases with emotional speech were obtained by recording professional actors as they portray particular emotion ([1,2,5,4,3]). Therefore they do not represent the situations in real life environment very adequately. Moreover to distinguish between normal (relaxed) and non-normal (emotional or aroused) conditions a fair amount of speech in the normal state is needed, which usually is not the case in emotional

V. Matoušek and P. Mautner (Eds.): TSD 2009, LNAI 5729, pp. 266–273, 2009.

databases. Consequently, we decided to record our own audio and video database containing normal speech and spontaneous emotions [11].

In databases with acted emotions labelling of data is simplified since the actors are told which category of emotion to act out whereas with spontaneous emotions there is no a priori knowledge of the emotion expressed in the utterance. Different statistical measures can be used for assessing agreement between labelers, but non of them incorporates the duration of the utterance. We present a modified free-marginal multirater kappa [6] where duration of the utterance is taken into consideration.

The following paper firstly describes the planning and recording and then discusses the assessment of a database of spontaneous emotions. A new time-weighted kappa statistics focused on evaluation of emotional databases is also presented.

2   Recording Strategies

Inducing spontaneous emotions in humans can only be achieved by some sort of deception. One option is to have a hidden camera type of experiment where a participant finds himself in a stressed situation. The advantage of such an approach is that the participant is not aware that he/she is being recorded and thus does not hold back his/her emotions as opposed to being in front of a camera. Drawbacks are that it is technically harder to carry out (audio and video recordings are usually lower quality), some participants do not want to give their consent, actors are needed to play out the scene, etc. For these reasons we decided to take a different approach. A scenario was constructed where participants would be told that the purpose of the experiment is to examine whether different biometric measures could be used in an adaptive test of intelligence. This sessions provided us with recordings of different levels of stress. For recording a particular emotion another scenario consisting of videos and photographs targeting each of the main emotions, was developed.

2.1   Recording Session 1: Adaptive IQ Test

Recording session called adaptive IQ test consisted of four parts. In the first part intended to record a normal state, a number of photographs with neutral content were presented to the participants. They were instructed to describe each photograph in detail, as if he/she would be describing it to a blind person. In this part we supposedly measured the verbal fluency.

In the second part the participants played a game of Tetris but instead of using the keyboard as input they lead the experimenter through the game uttering commands "left", "right", "around" and "down" (actual commands spoken were slovenian translations of these words: "levo", "desno", "okrog" and "dol"). The intent of this part was supposedly to assess the efficiency of his/her verbal instructions given to a teammate in order to achieve a common and specific goal.

Next, the game of Tetris that the participant played in the second part was played back and he/she had to describe what was happening on the screen by using the same four commands.

The final part of the session was the so-called adaptive intelligence test (adaptive meaning that the difficulty of the task will be chosen by the computer according to the correctness of the previous answers, the mental strategies used for solving the problem and the biometric measures). The test consisted of twenty three by three matrices where one element was missing and the participant had to determine which of the six offered answers logically fills the missing spot. The participants had to reason aloud about the principles of the arrangement of the matrix and the logic behind their answer supposedly for the psychologist to assess their mental strategy, but the real goal was to record speech under stress. Further pressure was put on the participants with additional information presented on the screen: current IQ value (constantly dropping regardless of the answers), heart beat (increasing throughout the experiment) and time left to finish the task. Analysis of subjective reports for 15 participants is presented in Fig 1.

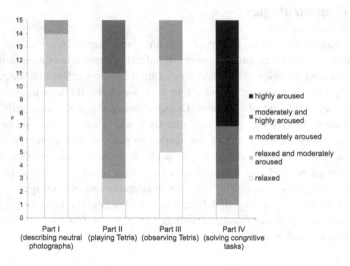

Fig. 1. Levels of arousal during different parts of the experiment based on the participants subjective report

2.2   Recording Session 2: Emotional Videos and Pictures

In the second session the participants watched a short video (approximately 10 minutes) and observed a set of photographs both targeting a particular emotion. After which they presented their thoughts on what they saw, how the videos made them feel, if something from the video or photographs relates to the situations in their life, etc. Emotions covered were happiness, anger, surprise, disgust, fear and sadness.

2.3   Database Description

We recorded 19 participants in the adaptive IQ test session (12 women and 7 men) and 9 participants in the emotional videos session so far. Together that comprises approximately 30 hours of recorded material from which about a third is actual participants speech. The majority of the recordings have already been both transcribed and labelled according to the emotion.

3 Emotion Labelling

With emotions acted out by professional actors there is usually not a big mismatch between the actor's interpretation and the annotator's opinion. This is expected since good acting means exactly that, to adequately represent a particular emotion to the observer. But with spontaneous emotions, that appear at any time during a recording and last a random amount of time, the job of emotion labelling becomes more challenging. If a recording is split into shorter utterances (e.g. sentences), the annotator can focus on one at the time, but this way the information of the context in which the sentence was spoken is lost. Hence in our case the labelers were given a full recording and they were free to put the emotion labels anywhere and of any duration. The utterances defined by transcriptions were extracted from the recording and by comparing different annotator's scores the predominant label was chosen.

3.1 Measuring Annotators Agreement

The problem we are faced with when annotating emotional databases is that there are no references for different types of emotions. Therefore we presume each annotator is partly right. In order to combine their individual labels into a final one we have to examine the agreement between all of them. A predominant way of measuring the agreement between annotators in emotional databases is by using the kappa statistics [10].

3.2 Kappa Coefficients

Different forms of kappa coefficients exist but are all based on the idea by Cohen [8] and shown in Eq. (1).

$$\kappa = \frac{P_o - P_e}{1 - P_e} \qquad (1)$$

P_e is an agreement between the annotators expected by chance and P_o is the actual proportion of agreement between them. Thus the coefficient calculated from Eq. (1) represent the ratio between measured agreement and chance. If the value of κ equals zero this represents a level of agreement equal to chance, values above are better than chance and values below are worst. All various forms of kappa are based on this equation, but the differences between them lie in definitions of P_o and P_e.

The most frequently used kappa coefficient in emotion labelling (as well as other fields) was presented by Fleiss [7], where he generalised Cohen's kappa from two raters to any number of raters. Fleiss defines P_o as

$$P_o = \frac{1}{Nn(n-1)} \left(\left(\sum_{i=1}^{N} \sum_{j=1}^{k} n_{ij}^2 \right) - Nn \right), \qquad (2)$$

where n_{ij} is the number of annotators that assigned to case i the class j, n represents the total number of annotators, N is the number of cases (utterances) and k is the number of classes (emotional states). P_e is defined as:

$$P_e = \sum_{j=1}^{k} \left(\frac{1}{Nn} \sum_{i=1}^{N} n_{ij} \right)^2. \qquad (3)$$

The probability of agreement by chance as defined in Eq. (3) presumes that all the classes (k) are represented equally across all cases. If there is a strong prevalence of one type of class over the other the Fleiss' coefficient drops regardless of the number of labels that are identical between the annotators. As it is demonstrated in [9] there is a quadratic relation between the value of Fleiss' coefficient and the prevalence of one type of class assuming other parameters are held constant (n_{ij}, n, N, k). When recording a database of spontaneous emotions one surely can not expect that all the different emotions or arousal states will be represented equally. Therefore the above calculations of Fleiss' kappa are not appropriate for measuring the agreement between annotators.

In [9] the author proposes a free-marginal multirater kappa as opposed to the standard Fleiss' kappa (which he calls Fleiss' fixed-marginal multirater kappa). The difference is that instead of using Eq. (3) for calculating the probability of agreement by chance he proposes that P_e should be set constant to

$$P_e = 1/k, \tag{4}$$

where k is the number of classes. This way there is no a priori restriction on the distribution of the classes and the above described problem of quadratic effect of prevalence of one class is lost (as it is graphically demonstrated in [9]). The equation of the new free-marginal multirater kappa coefficient thus becomes:

$$\kappa = \frac{(\frac{1}{Nn(n-1)}(\sum_{i=1}^{N}\sum_{j=1}^{k} n_{ij}^2 - Nn)) - \frac{1}{k}}{1 - \frac{1}{k}} \tag{5}$$

Although we avoid the problem of distribution of the classes when using Eq. (5) there is a down side to this approach. The number of possible classes that can be used becomes important since the factor $1/k$ starts decreasing when the number of possible categories increases. Thus the coefficients also increase, regardless of the fact that the data does not change (n_{ij}, n, N).

In order to use the free-marginal multirater kappa in emotional labelling the number of different emotions available to the annotator needs to be carefully selected. Otherwise the values of the coefficient will be always close to 1, not reflecting the actual state. Also if the categories of emotions are equally presented in recordings (as is the case in most acted databases) the author in [9] suggests the use of standard Fleiss' kappa.

3.3   Time-Weighted Free-Marginal Kappa

All kappa statistics described above handle each case equally. In speech databases these means that each utterance being labelled accounts equally to the final matching score regardless of the length. But in emotional databases the duration of the utterance that contains a particular emotion is important. This is the case with spontaneous emotions especially where the normal state usually prevails. Moreover if the annotators agree for example on a sentence lasting 15 seconds and disagree on a 2 second utterance, this means bigger agreement among them as if the case is the other way around. Therefore the duration of the utterance should be taken into consideration when measuring the agreement between annotators. We propose a modified kappa statistics focused specially on speech databases that incorporates the duration of the utterance. The equation for P_e stays the same as for the free-marginal multirater kappa since emotional categories are not represented equally. The change is introduced to the calculation of P_o, where the original equation for one particular case (utterance)

$$P_i = \frac{1}{n(n-1)}\left(\sum_{j=1}^{k} n_{ij}^2 - n\right) \tag{6}$$

is not averaged over all the cases equally. Instead the average is calculated based on the duration of each case (utterance)

$$P_o = \frac{1}{T}\sum_{i=1}^{N} P_i t_i. \tag{7}$$

Equation for P_o thus becomes:

$$P_o = \frac{1}{Tn(n-1)}\sum_{i=1}^{N}\left(\sum_{j=1}^{k} n_{ij}^2 - n\right)t_i, \tag{8}$$

where T is the total time and t_i is a duration of the utterance i.

4   Results

The new time-weighted free-marginal multirater kappa presented in the article was evaluated using a recording session from the AvID database. The new time-weighted free-marginal kappa is calculated in two cases: when emotions are classified in just two groups (normal/aroused) and the other when there are five different emotional categories. The values are compared against the standard Flciss' kappa coefficients.

The quality of the different kappa coefficients cannot be judged by just comparing which score is higher. The superiority of one type needs to be evaluated theoretically and/or supported by some other measures. In order to provide some additional understanding of the annotators agreement, the Table 1 presents the percentage of time that the annotators agreed between each other. It shows that the annotators on average agree on the label around eighty percent of the time which seems like a good result, but the

Table 1. Agreement between five annotators in percentage of time

Annotator	1	2	3	4
5	78.50%	76.97%	84.01%	84.08%
4	79.19%	78.19%	76.53%	
3	72.08%	73.54%		
2	76.57%			

Fleiss' kappa coefficients presented in Table 2. give the impression that the scores are just above (and in some cases even below) chance.

Opposite to the standard Fleiss' Kappa coefficients the new time-weighted free-marginal Kappa values show a stronger agreement between the annotators. This correlates better with the time percentages from Table 1. The slight increase of the new time-weighted free-marginal kappa in the case of five emotional states comes from the above described effect of $1/k$ factor.

Table 2. Kappa statistics between all combinations of two annotators and in the last column between all five

Combinations of annotators	1-2	1-3	1-4	1-5	2-3	2-4	2-5	3-4	3-5	4-5	All 5
Fleiss' Kappa	0.35	0.09	0.34	-0.02	0.21	0.32	-0.04	0.12	0.04	0.05	0.17
Time-weighted free-marginal Kappa (2 cat.)	0.56	0.45	0.59	0.58	0.52	0.57	0.54	0.55	0.68	0.68	0.57
Time-weighted free-marginal Kappa (5 cat.)	**0.72**	**0.66**	**0.75**	**0.74**	**0.70**	**0.73**	**0.71**	**0.72**	**0.80**	**0.80**	**0.72**

5   Conclusion

In the paper we discussed the idea that the utterances should not be treated equally when evaluating agreement in speech databases. Instead the duration of the utterance should have an impact on the final score. Furthermore, Fleiss' kappa normally used for measuring agreement between the annotators, is not appropriate when working with databases of spontaneous emotions since all emotional categories are not represented equally. Therefore, a new time-weighted free-marginal kappa was introduced. Using our own emotional database AvID, discussed in the beginning of the paper, we evaluated the new kappa and compared it against the Fleiss' Kappa. The results show that the new time-weighted free-marginal kappa gives a more realistic measure of agreement between the annotators.

References

1. LDC: SUSAS (Speech Under Simulated and Actual Stress). Language Data Consortium (1999),
   http://www.ldc.upenn.edu/Catalog/
   CatalogEntry.jsp?catalogId=LDC99S78
2. LDC: Emotional Prosody Speech and Transcripts. Language Data Consortium (2002),
   http://www.ldc.upenn.edu/Catalog/
   CatalogEntry.jsp?catalogId=LDC2002S28
3. Martin, O., Kotsia, I., Macq, B., Pitas, I.: The eNTERFACE 2005 Audio-Visual Emotion Database. In: Proc. of the 22nd International Conference on Data Engineering Workshops (ICDEW 2006), p. 8 (2006)

4. Burkhardt, F., Paeschke, A., Rolfes, M., Sendlmeier, W., Weiss, B.: A Database of German Emotional Speech. In: Interspeech 2005, pp. 1517–1520 (2005)
5. Battocchi, A., Pianesi, F.: DAFEX: Un Database Di Espressioni Facciali Dinamiche. In: Proceedings of the SLI-GSCP Workshop "Comunicazione Parlata e Manifestazione delle Emozioni" (2004)
6. Randolph, J.J.: Free-marginal multirater kappa: An alternative to Fleiss' fixed-marginal multirater kappa. In: Joensuu University Learning and Instruction Symposium 2005, Joensuu, Finland (2005)
7. Fleiss, J.L.: Measuring nominal scale agreement among many raters. Psychological Bulletin 76(5), 378–382 (1971)
8. Cohen, J.: A coefficient of agreement for nominal scales. Educational and Psychological Measurement 20(1), 37–46 (1960)
9. Randolph, J.J.: Free-marginal multirater kappa: An alternative to Fleiss' fixed-marginal multirater kappa. In: Joensuu University Learning and Instruction Symposium 2005, Joensuu, Finland (2005)
10. Callejas, Z., López-Cózar, R.: On the Use of Kappa Coefficients to Measure the Reliability of the Annotation of Non-acted Emotions. In: André, E., Dybkjær, L., Minker, W., Neumann, H., Pieraccini, R., Weber, M. (eds.) PIT 2008. LNCS (LNAI), vol. 5078, pp. 221–232. Springer, Heidelberg (2008)
11. Gajšek, R., Štruc, V., Mihelič, F., Podlesek, A., Komidar, L., Sočan, G., Bajec, B.: Multi-Modal Emotional Database: AvID. Informatica (Ljubljana) 33(1), 101–106 (2009)

Refinement Approach for Adaptation Based on Combination of MAP and fMLLR

Zbyněk Zajíc, Lukáš Machlica, and Luděk Müller

University of West Bohemia in Pilsen,
Faculty of Applied Sciences, Department of Cybernetics,
Univerzitní 22, 306 14 Pilsen
{zzajic,machlica,muller}@kky.zcu.cz

Abstract. This paper deals with a combination of basic adaptation techniques of Hidden Markov Model used in the speech recognition. The adaptation methods approach the data only through their statistics, which have to be accumulated before the adaptation process. When performing two adaptations subsequently, the data statistics have to be accumulated twice in each of the adaptation passes. However, when the adaptation methods are chosen with care, the data statistics may be accumulated only once, as proposed in this paper. This significantly reduces the time consumption and avoids the need to store all the adaptation data. Combination of Maximum A-Posteriori Probability and feature Maximum Likelihood Linear Regression adaptation is considered. Motivation for such an approach could be the on-line adaptation, where the time consumption is of big importance.

1 Introduction

Nowadays, systems of speech recognition are based on Hidden Markov Models (HMMs) with output probabilities described mainly by Gaussian Mixture Models (GMMs) [1]. To recognize the speech from a recording one could train a Speaker Dependent (SD) model for each of the speakers present in the recording. However, this is in praxis often intractable because of the need of a large database of utterances coming from one speaker. Instead, so called Speaker Independent (SI) model is trained from large amount of data collected from many speakers, and subsequently, the SI model is adapted to better capture the voice of the talking person. Thus, a SD model is acquired.

More precisely, the adaptation adjusts the SI model so that the probability of the adaptation data would be maximized. Well known adaptation methods are Maximum A-posteriori Probability (MAP) technique (see Section 2.2) and Linear Transformations based on Maximum Likelihood (LTML) (see Section 2.3). All these methods address the data undirectly through their statistics defined in Section 2.1. Because MAP and LTML adaptation work in different way, it would be suitable to combine them in order to gain a superior performance. One of the possibilities would be to accumulate data statistics utilizing the SI model and the adaptation data, run MAP adaptation, accumulate new data statistics based on the MAP adapted model and the same data, and perform one of the LTML adaptations (or vice versa – perform LTML after MAP). Obviously, the main disadvantage consists in the need to store all the adaptation data

V. Matoušek and P. Mautner (Eds.): TSD 2009, LNAI 5729, pp. 274–281, 2009.

and run the system twice. This causes increased time consumption dependent on the amount of processed data. Therefore, in Section 2.4 we proposed an efficient method that avoids the need to accumulate the data statistics twice. In this paper we have chosen out of LTML based adaptations preferably the feature transformations because of the implementation issues (see Section 2.4). In addition, the feature transformations are well suited for on-line adaptation, see [9]. Experimental results discussed in Section 3.2 prove the suitability of the proposed method.

2 Adaptation Techniques

The difference between the adaptation and ordinary training methods stands in the prior knowledge about the distribution of model parameters, usually derived from the SI model [2]. The adaptation adjusts the model in order to maximize the probability of adaptation data. Hence, the new, adapted parameters can be chosen as

$$\lambda^* = \arg\max_{\lambda} p(O|\lambda)p(\lambda), \tag{1}$$

where $p(\lambda)$ stands for the prior information about the distribution of the vector λ containing model parameters, $O = \{o_1, o_2, \ldots, o_T\}$ is the sequence of T feature vectors related to one speaker, λ^* is the best estimation of parameters of the SD model. We will focus on HMMs with output probabilities of states represented by GMMs. GMM of the $j - th$ state is characterized by a set $\lambda_j = \{\omega_{jm}, \mu_{jm}, C_{jm}\}_{m=1}^{M_j}$, where M_j is the number of mixtures, ω_{jm}, μ_{jm} and C_{jm} are weight, mean and variance of the $m - th$ mixture, respectively.

The most know adaptation methods are Maximum A-posteriori Probability (MAP) [4] and Linear Transformations based on the Maximum Likelihood (LTML) [6], the description of these methods will be given in following sections.

2.1 Statistics of Adaptation Data

As will be shown soon, the adaptation techniques do not access the data directly, but only through some statistics, which are accumulated beforehand. Let us define these statistics:

$$\gamma_{jm}(t) = \frac{\omega_{jm} p(o(t)|jm)}{\sum_{m=1}^{M} \omega_{jm} p(o(t)|jm)} \tag{2}$$

stands for the $m - th$ mixtures' posterior of the $j - th$ state of the HMM,

$$c_{jm} = \sum_{t=1}^{T} \gamma_{jm}(t) \tag{3}$$

is the soft count of mixture m,

$$\varepsilon_{jm}(o) = \frac{\sum_{t=1}^{T} \gamma_{jm}(t)o(t)}{\sum_{t=1}^{T} \gamma_{jm}(t)}, \quad \varepsilon_{jm}(oo^{T}) = \frac{\sum_{t=1}^{T} \gamma_{jm}(t)o(t)o(t)^{T}}{\sum_{t=1}^{T} \gamma_{jm}(t)} \tag{4}$$

represent the first and the second moment of features which align to mixture m in the j-th state of the HMM. Note that $\sigma_{jm}^2 = \mathrm{diag}(C_{jm})$ is the diagonal of the covariance matrix C_{jm}.

2.2 Maximum A-posteriori Probability (MAP) Adaptation

MAP is based on the Bayes method for estimation of the acoustic model parameters, with the unit loss function [3]. MAP adapts each of the parameters separately, therefore it is necessary to have for all the parameters enough adaptation data. The result of adaptation is negligible for small amount of data. The parameters are adapted according to formulas

$$\bar{\omega}_{jm} = [\alpha_{jm} c_{jm}/T + (1 - \alpha_{jm})\omega_{jm}]\chi \,, \tag{5}$$

$$\bar{\mu}_{jm} = \alpha_{jm}\varepsilon_{jm}(o) + (1 - \alpha_{jm})\mu_{jm} \,, \tag{6}$$

$$\bar{C}_{jm} = \alpha_{jm}\varepsilon_{jm}(oo^{\mathrm{T}}) + (1 - \alpha_{jm})(\sigma_{jm}^2 + \mu_{jm}\mu_{jm}^{\mathrm{T}}) - \bar{\mu}_{jm}\bar{\mu}_{jm}^{\mathrm{T}} \,, \tag{7}$$

$$\alpha_{jm} = \frac{c_{jm}}{c_{jm} + \tau} \,, \tag{8}$$

where α_{jm} is the adaptation coefficient, which controls the balance between the old and new parameters using empirically determined parameter τ. The parameter τ determines how much the new data have to be "observed" in each mixture till the mixture parameters change (they shift in the direction of new parameters) [4]. χ is a normalization factor, which guarantees that all the new weights of the mixture for one state sum to unity.

2.3 Linear Transformations Based on Maximum Likelihood

These methods are based on linear transformations. The advantage over the MAP technique is that the number of available model parameters is reduced via clustering of similar model components [8]. The transformation is the same for all the parameters from the same cluster $K_n, n = 1, \ldots, N$. Hence, less amount of adaptation data is needed. The first of the methods introduced by Leggeter in [5] is known as Maximum Likelihood Linear Regression (MLLR) and was further investigated by Gales, who introduced feature MLLR (fMLLR). The main difference between these two approaches stands in the area of their interest. MLLR transforms means and covariances of the model, whereas fMLLR transforms directly the acoustic feature vectors. The MLLR method is out of our interest and the adaptation formulas can be found in [5].

Feature Maximum Likelihood Linear Regression (fMLLR)
The method is based on the minimization of the auxiliary function [6]:

$$Q(\lambda, \bar{\lambda}) = const - \frac{1}{2}\sum_{jm}\sum_t \gamma_{jm}(t)(const_{jm} + \log|C_{jm}| + \\ + (\bar{o}(t) - \mu_{jm})^{\mathrm{T}} C_{jm}^{-1}(\bar{o}(t) - \mu_{jm})) \,, \tag{9}$$

where $\bar{o}(t)$ represents the feature vector transformed according to the formula:

$$\bar{o}_t = A_{(n)} o_t + b_{(n)} = W_{(n)}\xi(t) \,, \tag{10}$$

where $W_{(n)} = [A_{(n)}, b_{(n)}]$ stands for the transformation matrix corresponding to the $n - th$ cluster K_n and $\xi(t) = [o_t^T, 1]^T$ represents the extended feature vector. The auxiliary function (9) can be rearranged into the form [7]

$$Q_{W_{(n)}}(\lambda, \bar{\lambda}) = \log|A_{(n)}| - \sum_{i=1}^{I} w_{(n)i}^T k_i - 0.5 w_{(n)i}^T G_{(n)i} w_{(n)i}, \tag{11}$$

where

$$k_{(n)i} = \sum_{m \in K_n} \frac{c_m \mu_{mi} \varepsilon_m(\xi)}{\sigma_{mi}^2}, \quad G_{(n)i} = \sum_{m \in K_n} \frac{c_m \varepsilon_m(\xi\xi^T)}{\sigma_{mi}^2} \tag{12}$$

and

$$\varepsilon_m(\xi) = \left[\varepsilon_m^T(o), 1\right]^T, \quad \varepsilon_m(\xi\xi^T) = \begin{bmatrix} \varepsilon_m(oo^T) & \varepsilon_m(o) \\ \varepsilon_m^T(o) & 1 \end{bmatrix}. \tag{13}$$

To find the solution of equation (11) we have to express $A_{(n)}$ in terms of $W_{(n)}$, e.g. use the equivalency $\log|A_{(n)}| = \log|w_{(n)i}^T v_{(n)i}|$, where $v_{(n)i}$ stands for transpose of the $i - th$ row of cofactors of the matrix $A_{(n)}$ extended with a zero in the last dimension. After the maximization of the auxiliary function (11) we receive

$$\frac{\partial Q(\lambda, \bar{\lambda})}{\partial W_{(n)}} = 0 \Rightarrow w_{(n)i} = G_{(n)i}^{-1} \left(\frac{v_{(n)i}}{\alpha_{(n)}} + k_{(n)i} \right), \tag{14}$$

where $\alpha_{(n)} = w_{(n)i}^T v_{(n)i}$ can be found as the solution of the quadratic function

$$\beta_{(n)} \alpha_{(n)}^2 - \alpha_{(n)} v_{(n)i}^T G_{(n)i}^{-1} k_{(n)i} - v_{(n)i}^T G_{(n)i}^{-1} v_{(n)i} = 0, \tag{15}$$

where

$$\beta_{(n)} = \sum_{m \in K_n} \sum_t \gamma_m(t). \tag{16}$$

Two different solutions $w_{(n)i}^{1,2}$ are obtained, because of the quadratic function (15). The one that maximizes the auxiliary function (11) is chosen. Note that an additional term appears in the log likelihood for fMLLR because of the feature transforms, hence:

$$\log \mathcal{L}\left(o_t | \mu_m, C_m, A_{(n)}, b_{(n)}\right) = \log \mathcal{N}\left(A_{(n)} o_t + b_{(n)}; \mu_m, C_m\right) + 0.5 \log|A_{(n)}|^2. \tag{17}$$

The estimation of $W_{(n)}$ is an iterative procedure. Matrices $A_{(n)}$ and $b_{(n)}$ have to be correctly initialized first, e.g. $A_{(n)}$ can be chosen as a diagonal matrix with ones on the diagonal and $b_{(n)}$ can be initialized as a zero vector. The estimation ends when the change in parameters of transformation matrices is small enough (about 20 iterations are sufficient) [7].

2.4   Combination of MAP and fMLLR

As MAP and fMLLR work in different ways, it would be suitable to combine them. A simple method would be to run MAP and fMLLR subsequently in two passes. There are two possibilities – MAP after fMLLR and fMLLR after MAP. It can be anticipated that the first approach should outperform the second one. The fMLLR method transforms all the mixtures from the same cluster at once, thus also mixtures with insufficient amount of adaptation data. MAP affects each of the mixtures separately, however only mixtures with sufficient amount of adaptation data are improved (shifted towards adaptation data – see equations (5)–(7)). Thus, the second pass (MAP adaptation) can be thought of as a refinement stage of mixtures with sufficient amount of data. In the case when fMLLR succeeds MAP, fMLLR (whole cluster is shifted at once) could disarrange mixtures that were correctly shifted by MAP (each mixture is shifted separately – more precise). This was also proved by the experiments shown in the second part of Table 1, therefore from now on we will focus on fMLLR followed by MAP adaptation. Note that in the case when huge amount of adaptation data would be available, the MAP adaptation would rearrange each of the mixtures, hence the fMLLR would take no effect. However, the first pass of fMLLR could cause more precise estimates of data statistics. On the other hand, having low amount of adaptation data MAP would be negligible and only fMLLR would take effect (see results of experiments depicted in Figure 1). This is a very natural behaviour of a combination of adaptation techniques.

The main disadvantage of such an approach is the need to compute the statistics defined in Section 2.1 twice. The procedure is as follows:

$$\text{SI} \rightarrow stats_1 \text{ for SI} \overset{\text{fMLLR}}{\Longrightarrow} \text{SD}_{\text{fMLLR}} \rightarrow stats_2 \text{ for SD}_{\text{fMLLR}} \overset{\text{MAP}}{\Longrightarrow} \text{SD}_{\text{fMLLR+MAP}} \, . \quad (18)$$

This approach brings an additional improvement of results (see Table 1), but it is not efficient with regard to the time consumption of computing the second statistics, e.g. in on-line adaptation. Hence, it would be suitable to adjust the already accumulated statistics without the need to see the feature vectors once again. Because fMLLR is in use, the adaptation stands for the transformation of feature vectors. Thus, instead of accumulating new statistics of transformed features, it is possible to transform the original statistics in the following way (consider expression (4) and the transformation $\bar{o}_t = A_{(n)} o_t + b_{(n)}$):

$$\bar{\varepsilon}_{jm}(o) = \frac{\sum_{t=1}^{T} \gamma_{jm}(t)(A_{(n)} o(t) + b_{(n)})}{\sum_{t=1}^{T} \gamma_{jm}(t)} = A_{(n)} \varepsilon_{jm} + b_{(n)} \, , \quad (19)$$

$$\bar{\varepsilon}_{jm}(oo^{\text{T}}) = \frac{\sum_{t=1}^{T} \gamma_{jm}(t)(A_{(n)} o(t) + b_{(n)})(A_{(n)} o(t) + b_{(n)})^{\text{T}}}{\sum_{t=1}^{T} \gamma_{jm}(t)} =$$
$$= A_{(n)} \varepsilon_{jm}(oo^{\text{T}}) A_{(n)}^{\text{T}} + 2 A_{(n)} \varepsilon_{jm}(o) b_{(n)}^{\text{T}} + b_{(n)} b_{(n)}^{\text{T}} \, , \quad (20)$$

As can be seen from (19) and (20), the only approximation consists in the use of SI mixtures' posterior defined in (2), which remained unchanged (untransformed). The procedure of the proposed method is as follows:

$$\text{SI} \rightarrow stats_1 \text{ for SI} \overset{\text{fMLLR}}{\Longrightarrow} \text{SD}_{\text{fMLLR}} \rightarrow \text{transform } stats_1 \overset{\text{MAP}}{\Longrightarrow} \text{SD}_{\text{fMLLR+MAP}} \, . \quad (21)$$

Thus, data statistics $stats_1$ are computed according to the SI model. The fMLLR transformation matrices $A_{(n)}$, $b_{(n)}$ defined in (10) are estimated and utilized to transform the statistics $stats_1$ using equations (19), (20). At the end, the transformed statistics are used to refine the model mixtures via MAP adaptation.

3   Experiments

3.1   Test Data

All of the experiments were performed using telephone speech data set. The telephone-based corpus consists of Czech read speech transmitted over a telephone channel. The digitization of an input analog telephone signal was provided by a telephone interface board DIALOGIC D/21D at 8 kHz sample rate and converted to the mu-law 8 bit resolution. The corpus was divided into two parts, the training set and the testing set. The training set consisted of 100 speakers, where each of them read 40 different sentences (length of each sentence was cca 5 sec.). The testing set consisted of 100 speakers not included in the training set, where each of them read the same 20 sentences as the other, further divided into two groups. The first one contained 15 sentences used as adaptation data and the second one contained 5 remaining sentences used for testing of adapted models. The vocabulary in all our test tasks contained 475 different words. Since several words had multiple different phonetic transcriptions the final vocabulary consisted of 528 items. There were no OOV (Out Of Vocabulary) words. The basic speech unit of our system is a triphone. Each individual triphone is represented by a three states HMM; each state is provided by 8 mixtures of multivariate Gaussians. We are considering just diagonal covariance matrices. In all recognition experiments a language model based on zerograms was applied. It means that each word in the vocabulary is equally probable as a next word in the recognized utterance. For that reason the perplexity of the task was 528.

3.2   Results

The results of the experiment are shown in Table 1. The first part of the table contains the Correctness (Corr) and the Accuracy (Acc) of the baseline system (recognition done utilizing only the SI model) and the system adapted only by MAP or fMLLR method. Results of the adaptation process considering two-pass-combination (see (18)) of fMLLR and MAP can be found in the second part of the table. Last part of the table shows results of fMLLR+MAP combination proposed in this paper, where the data statistics were accumulated only once (for SI model) and then transformed using the estimated

Table 1. Correctness (Corr)[%] and Accuracy (Acc)[%] of transcribed words for each type of the adaptation and their combinations

	Corr[%]	Acc[%]
SI model	73.97	66.18
MAP	80.06	74.27
fMLLR	80.27	76.61
two-pass-combination		
MAP+fMLLR	81.91	77.68
fMLLR+MAP	83.05	79.51
one-pass-combination		
fMLLR+MAP	82.69	78.84

fMLLR transformation. At the end, the transformed statistics were used for MAP adaptation (see (21)).

As can be seen from Table 1, both two-pass-combinations MAP+fMLLR and fMLLR+MAP increse the Correctness (Corr) as well as the Accuracy (Acc) of recognized words, but the fMLLR+MAP approach is significantly better (as was supposed in Section 2.4). The result of the one-pass-combination of fMLLR+MAP is very close to the result of the two-pass-combination of fMLLR+MAP, which can be thought of as the upper boundary of the proposed method. We have performed also experiments based on increasing number of adaptation sentences, these can be found in Figure 1.

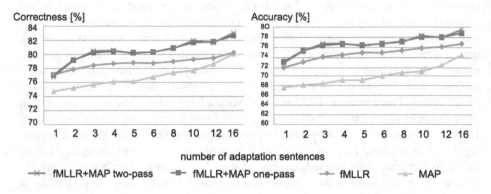

Fig. 1. Correctness (Corr)[%] and Accuracy (Acc)[%] of MAP and fMLLR adapted models and their combinations for increasing number of adaptation sentences. Note that the fMLLR+MAP combination yields the best performance for all different numbers of adaptation sentences and the results of one-pass-combination are very close (in some cases even identical) to the results of the two-pas-combination.

4   Conclusion

In this paper methods for adaptation of an acoustic model and their combinations were presented. More precisely, MAP and fMLLR adaptation methods were described. A simple combination of two different adaptation techniques proved to be convenient, but significantly more time consuming. Hence, we proposed a re-adjustment of primary computed data statistics without the need to see the adaptation data twice. We have demonstrated on experiments that such an one-pass-combination approach of fMLLR and MAP adaptation brings an additional improvement into the speech recognition. We have achieved a 4.57% and 2.23% increase absolutely in the systems' accuracy against the fMLLR and MAP based model, respectively. Let also state that the method proposed in this paper approaches the result of the two-pass-combination of fMLLR followed by MAP, which is however more time consuming.

Acknowledgements

This research was supported by the Grant Agency of the Czech Republic, project No. GAČR 102/08/0707, by the Ministry of Education of the Czech Republic, project No.

MŠMT LC536 and by the Grant Agency of Academy of Sciences of the Czech Republic, project No. 1QS101470516.

References

1. Rabiner, L.R.: A Tutorial on Hidden Markov Models and Selected Applications in Speech Recognition. In: Readings in speech recognition, pp. 267–296 (1990)
2. Psutka, J., Müller, L., Matoušek, J., Radová, V.: Mluvíme s počítačem česky, Academia, Praha (2007) ISBN:80-200-1309-1
3. Gauvain, L., Lee, C.H.: Maximum A-Posteriori Estimation for Multivariate Gaussian Mixture Observations of Markov Chains. IEEE Transactions SAP 2, 291–298 (1994)
4. Alexander, A.: Forensic Automatic Speaker Recognition using Bayesian Interpretation and Statistical Compensation for Mismatched Conditions. Ph.D. thesis in Computer Science and Engineering, pp. 27-29, Indian Institute of Technology, Madras (2005)
5. Leggeter, C.J., Woodland, P.C.: Maximum Likelihood Linear Regression for Speaker Adaption of Continuous Density Hidden Markov Models. Computer Speech and Language 9, 171–185 (1995)
6. Gales, M.J.F.: Maximum Likelihood Linear Transformation for HMM-based Speech Recognition. Tech. Report, CUED/FINFENG/TR291, Cambridge Univ. (1997)
7. Povey, D., Saon, G.: Feature and Model Space Speaker Adaptation with Full Covariance Gaussians. In: Interspeech, paper 2050-Tue2BuP.14 (2006)
8. Gales, M.J.F.: The Generation and use of Regression class Trees for MLLR Adaptation, Cambridge University Engineering Department (1996)
9. Machlica, L., Zajíc, Z., Pražák, A.: Methods of Unsupervised Adaptation in Online Speech Recognition. In: Specom, St. Petersburg (2009)

Towards the Automatic Classification of Reading Disorders in Continuous Text Passages

Andreas Maier[1], Tobias Bocklet[1], Florian Hönig[1],
Stefanie Horndasch[2], and Elmar Nöth[1]

[1] Lehrstuhl für Mustererkennung (Informatik 5),
Friedrich-Alexander Universität Erlangen Nürnberg,
Martensstr.3, 91058 Erlangen, Germany
[2] Kinder- und Jugendabteilung für Psychische Gesundheit,
Universitätsklinikum Erlangen,
Schwabachanlage 6 und 10, 91054 Erlangen, Germany
Andreas.Maier@cs.fau.de

Abstract. In this paper, we present an automatic classification approach to identify reading disorders in children. This identification is based on a standardized test. In the original setup the test is performed by a human supervisor who measures the reading duration and notes down all reading errors of the child at the same time. In this manner we recorded tests of 38 children who were suspected to have reading disorders. The data was confronted to an automatic system which employs speech recognition and prosodic analysis to identify the reading errors. In a subsequent classification experiment — based on the speech recognizer's output, the duration of the test, and prosodic features — 94.7 % of the children could be classified correctly.

1 Introduction

The state-of-the-art approach to examine children for reading disorders is a perceptual evaluation of the children's reading abilities. In all of these reading tests, a list of words or sentences is presented to the child. The child has to read all of the material as fast and as accurate as possible. In order to determine whether the child has a reading disorder two variables are investigated by a human supervisor during the test procedure:

- The duration of the test, i.e. the fluency, and
- The number of reading errors during the reading of the test material, i.e., the accuracy.

Both variables, however, are dependent on the age of the child and related to each other. If a child tries to read very fast, the number of reading errors will increase and vice versa [1]. Furthermore, with increasing age the reading ability of children increases. Hence, appropriate test material has to be chosen according to the age and reading ability of the child. Therefore, reading tests often consist of different sub-tests. While younger children are tested with really existing words and only short sentences, the older children have to be tested with more difficult tasks, such as long complex sentences and pseudo words which may or may not resemble real words. Appropriate subtests are then selected for each tested child. Often this is linked to the child's progress in school.

V. Matoušek and P. Mautner (Eds.): TSD 2009, LNAI 5729, pp. 282–290, 2009.
© Springer-Verlag Berlin Heidelberg 2009

One major drawback of the testing procedure is the intra-rater variablity in the perceptual evaluation procedure. Although the test manual often defines how to differentiate reading errors from normal disfluencies and "allowed" pronunciation alternatives, there is no exact definition of a reading error in terms of its acoustical representation. In order to solve this problem, we propose the use of a speech recognition system to detect the reading errors. This procedure has two major advantages:

- The intra-rater variability of the speech recognizer is zero because it will always produce the same result given the same input.
- The definition of reading errors is standardized by the parameters of the speech recognition system, i.e., the reading ability test can also be performed by lay persons with only little experience in the judgment of readings disorders.

In the literature, different automatic approaches to determine the "reading level" of a child exist. Often the reading level is linked to the perceptual evaluation of expert listeners using five to seven classes. In [2] Black et al. estimate a reading level between 1 and 7 using pronunciation verification methods based on Bayesian Networks. Compared to the human evaluation they achieve correlations between their automatic predictions and the human experts of up to 0.91 on 13 speakers. In [3] the use of finite-state-transducers is proposed to obtain a "reading level" between "A" (best) and "E" (worst). For this five-class problem absolute recognition rates of up to 73.4 % for real words and 62.8 % for pseudo words are reported. In order to remove age-dependent effects from the data, 80 children in the 2nd grade were investigated. Both papers focus on the creation of a "reading tutor" in order to improve children's reading abilities.

In contrast to these studies, we are interested in the diagnosis of reading disorders as they are relevant in a clinical point of view. Currently, we are developing PEAKS (**P**rogram for the **E**valuation of **A**ll **K**inds of **S**peech Disorders [4]) — a client-server-based speech evaluation framework — which was already used to evaluate speech intelligibility in children with cleft lip and palate [5], patients after removal of laryngeal cancer [6], and patients after the removal of oral cancer [7]. PEAKS features interfaces and tools to integrate standardized speech tests easily. After integration of a new test, PEAKS can be used for recording from any PC which is connected to the Internet if Java Runtime Environment version 1.6 or higher is installed. All analyses performed by PEAKS are fully automatic and independent of the supervising person. Hence, it is an ideal framework to integrate an automatic reading disorder classification system.

The paper is organized as follows. First the test material, the recorded speech data and its annotation is described and discussed. Next, the automatic evaluation methods, i.e., the speech recognizer, prosodic features, and the classifiers, are reported. In the results section the classification accuracy is presented in detail. The subsequent section discusses the outcome of the experiments. The paper is concluded by a summary.

2 Speech Data

In order to be able to interpret the results and to compare them to other studies' test material, speech data, and its annotation is described in detail here. Special attention is given to the annotation procedure since the automatic evaluation algorithm aims to be used for clinical diagnosis. Therefore, the annotation should meet clinical standards.

Table 1. Structure of the SLRT test: The table reports all sub-tests of the SLRT with their contents, their number of words, and the school grades in which the respective sub-test is suitable

sub-test	content	# of words	grade
SLRT1	A short list of bisyllabic, single, real words to introduce the test. This part is not analyzed according to the protocol of the test.	8	1–4
SLRT2	A list of mono- and bisyllabic real words	30	1–4
SLRT3	A list of compound words with two to three compounds each	11	3–4
SLRT4	A short story with only mono- and bisyllabic words	30	1–2
SLRT5	A longer story with mainly mono- and bisyllabic words but also a few compound words	57	3–4
SLRT6	A short list of pseudo words with two to three syllables to introduce the pseudo words. This part is not analyzed according to the protocol of the test.	6	3–4
SLRT7	A list of pseudo words with two to three syllables	24	1–4
SLRT8	A list of mono- and bisyllabic pseudo words which resemble real words	30	2–4

2.1 Test Material

The recorded test data is based on a German standardized reading disorder test — the "Salzburger Lese-Rechtschreib-Test" (SLRT, [8]). In total the SLRT consists of eight sub-tests (cf. Table 1). All sub-tests contain 196 words of which 170 are disjoint.

The test is standardized according to the instructions and the evaluation. The test is presented in form of a small book, which is handed to the children to read in. They get the instruction to read the text as fast as possible while doing as little reading mistakes as possible.

In the original setup the supervisor of the test has to measure the time for all sub-tests separately while noting down the reading errors of the child.

We will only report the results obtained for the SLRT4 and SLRT5 sub-tests in the following.

On the one hand, the setup of the perceptual evaluation for all sub-tests is very similar. Therefore, it is not necessary to report the results of all sub-tests. On the other hand, as we also want to investigate prosodic information only continuous texts such as the SLRT4 and SLRT5 sub-tests are suitable. All other sub-tests of the SLRT contain just single words. Hence, prosody was not expected to play a role in these tests.

2.2 Recording Setup

In order to be able to collect the data directly at the PC, the test had to be modified. Instead of a book, the text was presented as a slide on the screen of a PC. The instructions to the child were the same as in the original setup.

All children were recorded with a head-mounted microphone (Plantronics USB 510) at the University Clinic Erlangen. The recordings took place in a separate quiet room without background noises. Hence, appropriate audio quality was achieved in all recordings.

Table 2. 38 Children were recorded with the SLRT: The table shows mean value, standard deviation, minimum, and maximum of the age of the children and the count (#) in the respective group

group	#	mean	std. dev.	min	max
all	38	9.7	0.9	7.8	11.3
girls	12	10.2	0.7	9.0	11.3
boys	26	9.5	0.9	7.8	11.3

Table 3. Overview on the limits of pathology for the SLRT4 and SLRT5 sub-tests

	SLRT 4		SLRT 5	
grade	# of errors	duration [s]	# of errors	duration [s]
1st	4	102	-	-
2nd	3	62	-	-
3rd	-	-	2	64
4th	-	-	2	43

In total 38 children (26 boys and 12 girls) were recorded. The average age of the children was 10.2 ± 0.9 years. A detailed overview regarding the statistics of the children's ages is given in Table 2. All of the children were speculated to have a reading disorder.

2.3 Perceptual Evaluation

For each child the decision whether its reading ability was pathologic or not was determined according to the manual of the SLRT [8]. A child's reading ability is deemed pathologic

- if the duration of the test is longer than an age-dependent standard value or
- if the number of reading errors exceeds an age-dependent standard value.

These limits differ for each sub-test according to the SLRT. Table 3 reports these limits for the sub-tests SLRT4 and SLRT5. In the SLRT4 and the SLRT5 sub-test 30 children were above the time limit.

We assigned each child two different labels: "reading error/normal" and "pathologic/non-pathologic". If only the number of misread words is exceeded, the child is assigned the label "reading error", otherwise "normal". Reading errors are regarded as soon as a single phonemic deviation is found. Errors of the accentuation of the word are also counted as reading errors as described in the manual of the test [8]. In total 18 children exceeded the error limit.

If either of these two boundaries is exceeded by the child, the child is assigned the label "pathologic". 34 of the 38 children were diagnosed to have pathologic reading.

3 Automatic Evaluation System

The automatic evaluation is based on four information sources:

- The total duration of the test
- The reading error and duration limits (cf. Table 3)
- The word accuracy computed by a speech recognition system
- Prosodic information

The test duration can be easily accessed as PEAKS tracks this information automatically during the recording. Prior information about the child — namely the child's age and the respective duration and error limits — can also easily be obtained (cf. Table 3).

3.1 Speech Recognition Engine

For the objective measurement of the reading accuracy, we use an automatic speech recognition system based on Hidden Markov Models (HMM). It is a word recognition system developed at the Chair of Pattern Recognition (Lehrstuhl für Mustererkennung) of the University of Erlangen-Nuremberg. In this study, the latest version as described in detail in [9] and [10] was used.

As features we use 11 Mel-Frequency Cepstrum Coefficients (MFCCs) and the energy of the signal plus their first-order derivatives. The short-time analysis applies a Hamming window with a length of 16 ms, the frame rate is 10 ms. The filter bank for the Mel-spectrum consists of 25 triangular filters. The 12 delta coefficients are computed over a context of 2 time frames to the left and the right side (56 ms in total).

The recognition is performed with semi-continuous HMMs. The codebook contains 500 full covariance Gaussian densities which are shared by all HMM states. The elementary recognition units are polyphones [11], a generalization of triphones. Polyphones use phones in a context as large as possible which can still statistically be modeled well, i.e., the context appears more often than 50 times in the training data. The HMMs for the polyphones have three to four states.

We used a unigram language model to weigh the outcome of each word model. It was trained with the reference of the tests. For our purpose it was necessary to emphasize the acoustic features in the decoding process. In [12] a comparison between unigram and zerogram language models was conducted. It was shown that intelligibility can be predicted using word recognition accuracies computed using either zero- or unigram language models. The unigram, however, is computationally more efficient because it can be used to reduce the search space. The use of higher n-gram models was not beneficial.

The result of the recognition is a word lattice. In order to get an estimate of the quality of the recognition, the word accuracy (WA) is computed. Based on the number of correctly recognized words C and the number of words R in the reference, the WA is further dependent on the number or wrongly inserted words I:

$$\mathrm{WA} = \frac{C - I}{R} \cdot 100\,\%$$

Hence, the WA can take values between minus infinity and 100 %.

The speech recognition system had been trained with acoustic information from 23 male and 30 female children from a local school who were between 10 and 14 years old (6.9 hours of speech). To make the recognizer more robust, we added data from 85 male and 47 female adult speakers from all over Germany (2.3 hours of spontaneous speech from the VERBMOBIL project, [13]). The data were recorded with a close-talk microphone with 16 kHz sampling frequency and 16 bit resolution. The adult speakers were from all over Germany and thus covered most dialect regions. However, they were asked to speak standard German. The adults' data were adapted by vocal tract length normalization as proposed in [14]. During training an evaluation set was used that only contained children's speech. MLLR adaptation (cf. [15,16]) with the patients' test data led to further improvement of the speech recognition system.

3.2 Prosodic Features

The prosody module used in these experiments was originally developed within the VERBMOBIL project [18], mainly to speed up the linguistic analysis [19,20]. It assigns a vector of prosodic features to each word in a word hypothesis graph which is then used to classify a word w.r.t., e.g. carrying the phrasal accent and being the last word in a phrase. For this paper, the prosody module takes the text reference and the audio signal as input and returns 37 prosodic features for each word and then calculates the mean, the maximum, the minimum, and the variance of these features for each speaker, i.e. the prosody of the whole speech of a speaker is characterized by a 148-dimensional vector. These features differ in the manner in which the information is combined (cf. Fig. 1):

1. onset
2. onset position
3. offset
4. offset position
5. maximum
6. position of maximum

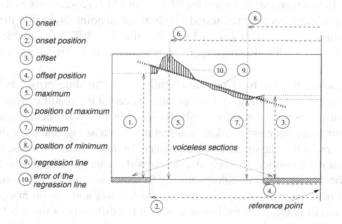

Fig. 1. Computation of prosodic features within one word (after [17])

7. minimum
8. position of minimum
9. regression line
10. mean square error of the regression line

These features are computed for the fundamental frequency (F_0) and the energy (absolute and normalized). Additional features are obtained from the duration and the length of pauses before and after the respective word. Furthermore jitter, shimmer and the length of voiced (V) and unvoiced (UV) segments are calculated as prosodic features.

3.3   Classification System

Classification was performed in a leave-one-speaker-out (LOO) manner since there was only little training and test data available. We chose two popular measures in order to report the classification accuracy.

– **RR:** The total recognition rate determined as the fraction of correctly identified speakers c divided by the number of speakers n:

$$RR = \frac{c}{n} \cdot 100\%$$ (1)

The RR reports the overall performance of the classifier including the class distribution of the data.
– **ROC** denotes the area under the Receiver-Operating-Characteristic (ROC) curve [21]. A random classifier yields an area of 0.5 while the perfect classifier would yield an area of 1.0.

As classification system we decided for Ada-Boost [22] in combination with an LDA-Classifier as simple classifier as it was already successfully applied in [23].

4   Results and Discussion

In the following evaluation we regard the SLRT4 and SLRT5 sub-tests as a single classification experiment because the tested children are disjoint. Note that the SLRT4 is suitable for children in school grades 1 and 2 while the SLRT5 is suitable for grades 3 and 4 (cf. Table 1). All following experiments were conducted in a leave-one-speaker-out manner.

Table 4 shows the results of the classification task "reading error" and "pathologic". Only 63.2 % of the children who actually exceeded the reading error limit could actually classified as such. Therefore, only the duration, the word accuracy, and the age-dependent limits are necessary. Additional features, such as age and prosodic information, even decrease the classification performance. In this case the prosodic information even confuses the classifier so much, that it learns the opposite of the actual classification task. The classification rate drops to 47.4 % which is actually worse than random guessing. Hence, one can conclude that prosodic features and age do not contain help in the detection of reading errors. Please note that the difficulty of the sub-tests SLRT4 and SLRT5 are already adjusted to the school grade of the children (and therewith also to the age).

Table 4. Overview on the classification results for the two tasks "reading error" and "pathologic". RR is the absolute recognition rate and ROC the area under the ROC curve.

	"reading error"		"pathologic"	
feature set	RR [%]	ROC	RR [%]	ROC
duration and accuracy	60.5	0.61	78.9	0.58
+ age-dependent limits	**63.2**	**0.63**	81.6	0.84
+ age	55.3	0.59	89.5	0.67
+ prosodic information	47.4	0.52	**94.7**	**0.96**

However, prosodic information plays an important, yet rarely investigated role for the detection of reading pathologies. For the classification task "pathologic" the classification performance is maximal at 94.7 % if prosodic features are employed in addition to the other features. This observation is important because current state-of-the-art tests for reading pathologies do not take any prosodic analyses into account.

In future work we want to investigate the other sub-tests of the SLRT and automate them. In this manner we will create a reliable and automatic test for reading pathologies. This will help in clinical daily routine-use as automatic methods can save time and money.

5 Summary

In this paper we presented an automatic approach for the classification of reading disorders based on automatic speech recognition. The evaluation is performed on a standardized German reading capability test that contains pseudo words. To our knowledge such a system has not been published before. The system is web-based and can be accessed from any PC which is connected to the Internet.

Using a database with 38 children classification rates of up to 94.7 % (RR) could be achieved. The system is suitable for the automatic classification of reading disorders.

References

1. Dennis, I., Evans, J.S.B.T.: The speed-error trade-off problem in psychometric testing. British Journal of Psychology 87, 105–129 (1996)
2. Black, M., Tepperman, J., Lee, S., Narayanan, S.: Estimation of children's reading ability by fusion of automatic pronunciation verification and fluency detection. In: Interspeech 2008 – Proc. Int. Conf. on Spoken Language Processing, 11th International Conference on Spoken Language Processing, Brisbane, Australia, Proceedings, September 25-28, pp. 2779–2782 (2008)
3. Duchateau, J., Cleuren, L., Hamme, H.V., Ghesquiere, P.: Automatic assessment of children's reading level. In: Interspeech 2007 – Proc. Int. Conf. on Spoken Language Processing, 10th European Conference on Spoken Language Processing, Antwerp, Belgium, August 27-31, pp. 1210–1213 (2007)
4. Maier, A., Haderlein, T., Eysholdt, U., Rosanowski, F., Batliner, A., Schuster, M., Nöth, E.: PEAKS – A System for the Automatic Evaluation of Voice and Speech Disorders. Speech Communication 51(5), 425–437 (2009)
5. Maier, A., Nöth, E., Batliner, A., Nkenke, E., Schuster, M.: Fully Automatic Assessment of Speech of Children with Cleft Lip and Palate. Informatica 30(4), 477–482 (2006)

6. Schuster, M., Haderlein, T., Nöth, E., Lohscheller, J., Eysholdt, U., Rosanowski, F.: Intelligibility of laryngectomees' substitute speech: automatic speech recognition and subjective rating. Eur. Arch. Otorhinolaryngol. 263(2), 188–193 (2006)
7. Windrich, M., Maier, A., Kohler, R., Nöth, E., Nkenke, E., Eysholdt, U., Schuster, M.: Automatic Quantification of Speech Intelligibility of Adults with Oral Squamous Cell Carcinoma. Folia Phoniatr Logop 60, 151–156 (2008)
8. Landerl, K., Wimmer, H., Moser, E.: Salzburger Lese- und Rechtschreibtest. Verfahren zur Differentialdiagnose von Störungen des Lesens und des Schreibens für die 1. bis 4. Schulstufe, Huber, Bern (1997)
9. Gallwitz, F.: Integrated Stochastic Models for Spontaneous Speech Recognition. Studien zur Mustererkennung, vol. 6. Logos Verlag, Berlin (2002)
10. Stemmer, G.: Modeling Variability in Speech Recognition. Studien zur Mustererkennung, vol. 19. Logos Verlag, Berlin (2005)
11. Schukat-Talamazzini, E., Niemann, H., Eckert, W., Kuhn, T., Rieck, S.: Automatic Speech Recognition without Phonemes. In: Proc. European Conf. on Speech Communication and Technology (Eurospeech), Berlin, Germany, vol. 1, pp. 129–132 (1993)
12. Riedhammer, K., Stemmer, G., Haderlein, T., Schuster, M., Rosanowski, F., Nöth, E., Maier, A.: Towards Robust Automatic Evaluation of Pathologic Telephone Speech. In: Proceedings of the Automatic Speech Recognition and Understanding Workshop (ASRU), Kyoto, Japan, pp. 717–722. IEEE Computer Society Press, Los Alamitos (2007)
13. Wahlster, W. (ed.): Verbmobil: Foundations of Speech-to-Speech Translation. Springer, Berlin (2000)
14. Stemmer, G., Hacker, C., Steidl, S., Nöth, E.: Acoustic Normalization of Children's Speech. In: Proc. European Conf. on Speech Communication and Technology, Geneva, Switzerland, vol. 2, pp. 1313–1316 (2003)
15. Gales, M., Pye, D., Woodland, P.: Variance compensation within the MLLR framework for robust speech recognition and speaker adaptation. In: Proceedings of the International Conference on Speech Communication and Technology (Interspeech), Philadelphia, USA, ISCA, vol. 3, pp. 1832–1835 (1996)
16. Maier, A., Haderlein, T., Nöth, E.: Environmental adaptation with a small data set of the target domain. In: Sojka, P., Kopeček, I., Pala, K. (eds.) TSD 2006. LNCS (LNAI), vol. 4188, pp. 431–437. Springer, Heidelberg (2006)
17. Kießling, A.: Extraktion und Klassifikation prosodischer Merkmale in der automatischen Sprachverarbeitung. Berichte aus der Informatik. Shaker, Aachen, Germany (1997)
18. Wahlster, W. (ed.): Verbmobil: Foundations of Speech-to-Speech Translation. Springer, Berlin (2000)
19. Nöth, E., Batliner, A., Kießling, A., Kompe, R., Niemann, H.: Verbmobil: The Use of Prosody in the Linguistic Components of a Speech Understanding System. IEEE Trans. on Speech and Audio Processing 8, 519–532 (2000)
20. Batliner, A., Buckow, A., Niemann, H., Nöth, E., Warnke, V.: The Prosody Module. In: [18], pp. 106–121
21. Fawcett, A.: An introduction to ROC analysis. Pattern Recognition Letters 27, 861–874 (2006)
22. Freund, Y., Schapire, R.E.: Experiments with a new boosting algorithm. In: Thirteenth International Conference on Machine Learning, pp. 148–156. Morgan Kaufmann, San Francisco (1996)
23. Hacker, C., Cincarek, T., Maier, A., Heßler, A., Nöth, E.: Boosting of Prosodic and Pronunciation Features to Detect Mispronunciations of Non-Native Children. In: Proceedings of the International Conference on Acoustics, Speech, and Signal Processing (ICASSP), Hawaii, USA, vol. 4, pp. 197–200. IEEE Computer Society Press, Los Alamitos (2007)

A Comparison of Acoustic Models Based on Neural Networks and Gaussian Mixtures

Tomáš Pavelka and Kamil Ekštein

University of West Bohemia, Faculty of Applied Sciences,
Deptartment of Computer Science and Engineering
{tpavelka,kekstein}@kiv.zcu.cz

Abstract. This article tries to compare the performance of neural network and Gaussian mixture acoustic models (GMMs). We have carried out tests which match up various models in terms of speed and achieved recognition accuracy. Since the speed-accuracy trade-off is not only dependent on the acoustic model itself, but also on the settings of decoder parameters, we have suggested a comparison based on equal number of active states during the decoding search. Statistical significance measures are also discussed and a new method for confidence interval computation is introduced.

1 Introduction

The most widely used mathematical framework for automatic speech recognition are the continuous density hidden Markov Models (CDHMMS). Despite of their success these models make various assumptions that are not true for speech data (see [4]). There are attempts to solve some of the drawbacks of the CDHMM paradigm by employing neural networks. The research into the so called *hybrid systems* (see e.g. [1], [8]) has shown that it can be advantageous to use neural networks (instead of the more traditional Gaussian mixtures) as emission probability estimators for hidden Markov model based automatic speech recognizers. Our results presented in [5] demonstrate that there are two main benefits:

- The application of neural networks to emission probability estimation does not place any constraints on the form of its inputs (as opposed to GMM models with diagonal covariance matrices which add delta and acceleration coefficients to the input vector because the elements of the final composed vector are loosely uncorrelated). This is usually exploited by presenting several subsequent speech frames to the input of the neural network and thus allowing the network to "see" a larger context of the speech signal.
- When compared to Gaussian mixture based acoustic models the neural networks need less trainable parameters to achieve similar or better recognition accuracy. As we will show, this can lead to faster recognition speed.

While the above stated can be said about context independent phonetic units (monophones) the experiments presented in [2] make clear that much better results can be gained with adding context dependency (e.g. by using triphones). In [7] a solution is

V. Matoušek and P. Mautner (Eds.): TSD 2009, LNAI 5729, pp. 291–298, 2009.

proposed that attempts to use decision tree clustering for the reduction of the number of physical models in order to solve the sparse data problem and also in order to limit the number of output neurons and thus decrease the time needed for training. In this article we would like to further explore the difficulties in comparing acoustic models based on neural networks and GMMs.

2 Speech Corpora

All the available speech data is in Czech language, recorded in quiet environment at 16 KHz sampling rate and 16 bits per sample. The corpora are divided into sentences; each sentence is stored in a separate file. The training set consists of three parts:

- **Train Schedule Queries.** This corpus consists of questions about train schedules and related information. An example of such question would be "When does the train for Plzeň leave".
- **LAC-HP Chess.** Stands for LASER Audiocorpus High Precision. The corpus was recorded in an audio studio; all the audio files have been verified during the recording. This set consists of voice commands for a chess game. The commands could be either chess moves (e.g. "Move the king to b5") or miscellaneous commands like "I want to start a new game".
- **LAC-HP Phonetic.** This is a set of nonsense sentences with words containing infrequent phonetic units.

The testing corpus for the train schedules is a subset taken out from the original corpus. The testing corpus for the chess game contains only move commands because we have found out that other commands can skew the recognition results (the move commands are much harder to recognize). This means that if other commands are present in the training data the resulting accuracy is highly correlated with moves / other commands ratio. Table 1 shows statistics for all the speech data used in our experiments.

Table 1. Training and testing corpora

Training Corpus	Speakers	No. of sentences	Vocabulary size [words]	Total Length [hours]
Train Schedules	81 (48M, 33F)	11270	1490	11:28:06
LAC-HP Chess	81 (31M, 50F)	2050	96	1:51:50
LAC-HP Phonetic	81 (31M, 50F)	1200	96	1:33:02
Testing Corpus	Speakers	No. of sentences	Vocabulary size [words]	Total Length [hours]
Train Schedules	4 (2M, 2F)	400	1490	0:31:34
Chess Moves	20 (10M, 10F)	2000	96	1:18:28

3 Gaussian Mixture Acoustic Models

Gaussian mixture models (GMMs) were trained by the Hidden Markov Toolkit (HTK, [9]). Only models with diagonal covariance matrices were tested. The parameter estimation was done by a flat start embedded training which only requires the phonetic transcriptions of the training utterances to be available. On the other hand the neural network needs exact locations of phonetic units in the training data. This leads to the second reason for having a set of trained GMM models: the GMM based recognizer can be used to label the training data for the neural network (by the means of forced Viterbi alignment).

In the case of the GMMs the whole set consists of 36 (35 context independent units + silence) phonetic unit models. Each phonetic unit is a three state HMM, each state has its own mixture of Gaussians. The training starts with one Gaussian per state. In each training cycle embedded re-estimation is performed four times (our tests show that the error decrease after four iterations is negligible). After the cycle is completed the number of Gaussians for each state is increased twofold. We have trained models with up to 128 Gaussians per state.

The training of triphone GMM models is described in detail in [2]. The process is similar to the training of the monophones, the difference is that after the models with single Gaussian per state are trained a decision tree clustering of all the states is performed. The result is that the triphone models which do not have sufficient amount of training data available are tied together with all the other models in their respective clusters. The clustering provides a mapping between the logical models (i.e. any triphone) and the physical models (actual Gaussian mixtures).

The clustering algorithm uses yes-no questions about triphone context which are used to split the models into two parts. A measure exists (see [3]) that evaluates the information gained by the split and thus the best question can be chosen. After that the same is done for each of the newly created model and a tree structure is created. This process is eventually stopped based on a threshold which sets the minimum information gain needed for a split to be allowed. By setting the value of this threshold one can control the total number of physical states (see Table 2).

Table 2. Accuracy achieved for different values of decision tree clustering threshold

Threshold	Number of state models	Best results	
		Train Schedules	Chess Moves
1000	1733	84.46 @ 16mix	97.55 @ 16mix
2000	1070	83.93 @ 32mix	97.51 @ 32mix
5000	517	82.46 @ 32mix	97.52 @ 32mix

4   Neural Network Acoustic Models

All networks discussed in this paper are multi layer perceptrons with nine consecutive speech frames on input (altogether 177 neurons). Various numbers of hidden neurons were tested. The number of output vectors corresponds to the number of phonetic units (36 for the monophone case) or the number of physical states (for the triphone case).

Hidden Markov models work with likelihoods $p(\text{input}|\text{class})$ instead of the class posteriors $p(\text{class}|\text{input})$ that we get on the output of the neural network. These can be converted using the Bayes theorem:

$$p(\text{input}|\text{class}) = \frac{p(\text{class}|\text{input}) \cdot p(\text{input})}{p(\text{class})} \ . \tag{1}$$

Since the probability of an input P(input) is the same for all HMM states examined in a given frame it can be discarded from the equation without affecting the result. Our experiments (details in [7]) have shown that the division by priors is not always beneficial and depends on the testing corpus. In the case of the train schedule corpus the division by priors increases accuracy but in the case of the chess corpus its use is actually harmful. For this reason the division by priors was only done in tests with the train schedule corpus.

The GMMs with the lowest number of physical states (517, see Table 2) were used to generate training vector labels. The resulting triphone neural network had 2000 hidden neurons and 517 output neurons.

5   Confidence Intervals

The problem with the results of speech recognition experiments is that accuracies obtained by tests with various models can be close together and random errors (i.e. caused by the choice of the testing corpus which should be representative of the domain but is always limited in size) should be taken into account. A possible solution would be to treat the recognition results as a random variable and compute confidence intervals for its distribution.

However, the standard binomial distribution confidence interval cannot be used, because it assumes statistical independence of individual errors of which the proportion (i.e. error rate or accuracy) is computed. This is clearly not true in speech recognition where each error can have influence on subsequent errors. We can, however, assume independence of errors in different sentences since all search variables are discarded after a sentence is recognized and there is no way a recognition of one sentence can affect recognition of the next one.

The statistical properties of a sample of independent observations is investigated in [10]: Let $(r_1, s_1), (r_2, s_2), \ldots, (r_n, s_n)$ be the sample where (in the case of speech recognition) r_i is the number of correctly recognized words in sentence i and s_i is the number of incorrectly recognized words. The distribution function of the variable

$$\xi_n = \frac{\sum_{i=1}^{n} r_i}{\sum_{i=1}^{n} r_i + \sum_{i=1}^{n} s_i} \tag{2}$$

is derived in [10] which can be used to find the boundaries of a confidence interval. If we know the expected value estimations

$$e_r = \frac{1}{n} \sum_{i=1}^{n} r_i, \quad e_s = \frac{1}{n} \sum_{i=1}^{n} s_i \tag{3}$$

as well as the variance estimations

$$\sigma_r^2 = \frac{1}{n} \sum_{i=1}^{n} (r_i - e_r)^2, \quad \sigma_s^2 = \frac{1}{n} \sum_{i=1}^{n} (s_i - e_r)^2 \tag{4}$$

and the correlation coefficient

$$\rho_{r,s}\sigma_r\sigma_s = \frac{1}{n}\sum_{i=1}^{n}(r_i - e_r)(s_i - e_s). \tag{5}$$

then the asymptotic distribution function (for sufficiently large n) of the random variable ξ_n can be expressed (see [10] to see how it is inferred) as

$$F_{\xi_n} = 1 - \Phi\left(-\sqrt{n}\frac{e_s - \left(\frac{1}{x} - 1\right)e_r}{\sqrt{\sigma_s^2 + \left(\frac{1}{x} - 1\right)^2\sigma_r^2 - 2\left(\frac{1}{x} - 1\right)\rho_{r,s}\sigma_r\sigma_s}}\right) \Leftrightarrow 0 < x < 1. \tag{6}$$

We are looking for a confidence interval $(x_L; x_U)$ that satisfies the condition

$$P(x_L \leq \xi_n \leq x_U) = 1 - \alpha. \tag{7}$$

To uniquely determine the bounds we chose α_1 and α_2 so that $\alpha = \alpha_1 + \alpha_2; 0 < \alpha, \alpha_1, \alpha_2 < 1$. The interval boundaries can then be found by solving the equations

$$F_{\xi_n}(x_U) = 1 - \alpha_1 \tag{8}$$

and

$$F_{\xi_n}(x_L) = \alpha_2. \tag{9}$$

In all our experiments the value of α was set to 0.005 which corresponds to 99% confidence. See [10] for details of the interval computation.

6 Experimental Results

Even though the training of the GMM models was done by the HTK software the testing of both the GMM and MLP acoustic models was carried out with the JLASER [6] recognizer. For both acoustic models Mel-frequency cepstral coefficients (MFCCs) served as input.

There was a grammar representing all the possible utterances in the chess moves test corpus but the tests with the train schedule corpus were run without any language model. We do not consider the lack of language model to be a problem since our main goal is to compare the two kinds of acoustic models.

For all tests pruning was performed during the decoding phase. For the train schedule corpus a word insertion penalty was applied. Both the pruning threshold and the word insertion penalty were tuned for each acoustic model in order to achieve the highest possible speed while maintaining the highest recognition accuracy. Decoding with pruning means that the emission probabilities are usually not needed only for all phonetic units (this is especially true for triphones). Some computation can be avoided by computing only those emission probabilities that are requested by the decoding algorithm. This is quite straightforward in the case of GMM models. In the case of neural networks the

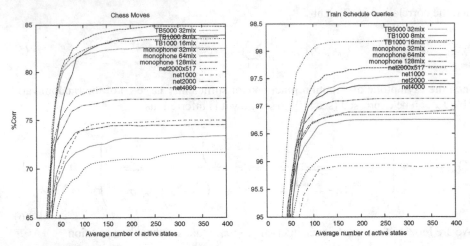

Fig. 1. The relationship between the average number of active states (controlled by the beam threshold) and the resulting accuracy

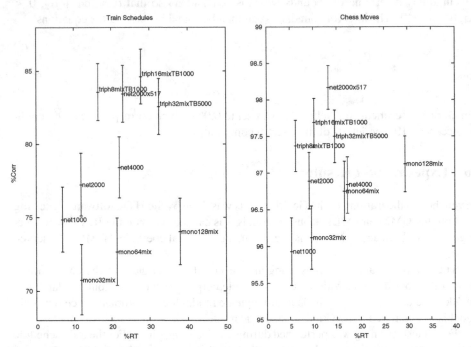

Fig. 2. Comparison of recognition speed and percentage of correct results for all tested acoustic models. The error bars represent the confidence intervals for 99% probability. Neural networks are specified by two numbers: the first denotes the number of hidden neurons and the second the number of output neurons. GMMs are specified by a prefix ("mono" for monophones, "triph" for triphones), the number of mixtures and, in the case of triphones the clustering threshold.

activations of all the hidden neurons need to be computed for every speech frame. But the computation of the output layer neuron activations can be delayed until those are requested by the decoder.

Another problem with pruning is that the choice of pruning threshold allows one to make trade-off between speed and accuracy which makes the comparison in terms of speed and accuracy more complicated. One thing we have observed is that accuracy rises with less stricter threshold but eventually stabilizes and more loosening of the threshold only decreases speed. So our solution is to compare the models at this point. But, the stabilizing point is different for different corpora and different models. We have found out that it is best to view the relationship between the average number of active states (as opposed to the actual threshold value) and the resulting accuracy. It can be seen from Figure 1 that this stabilization point is around 200 average active states regardless of corpus and model choice. For this reason, all the models are tested with a beam threshold that leads to an average of 200 active states during decoding.

Figure 2 shows the results for models with different numbers of trainable parameters (controlled by the number of hidden neurons or the number of mixtures and clustering threshold). Besides showing the recognition accuracy the figure also shows the recognition speed measured as a percentage of real time processing power on a referential machine needed for the recognition.

7 Conclusions

The tests reported in [7] were favorable towards neural networks and showed that neural network based acoustic models usually led to higher recognition speeds. In this article we have tested GMM based models with different values for the decision tree clustering threshold which, in some cases (see Figure 2) led to higher recognition speeds. But, there is still an outlying result achieved for the chess corpus where the neural network triphone acoustic model clearly outperforms all other models. This may suggest that triphone neural networks may be suitable for small vocabulary grammar based applications (such as our voice controlled chess game). For LVCSR tasks the Gaussian mixture models provide more flexibility without compromising speed.

We have also suggested an approach to speed vs. accuracy testing based on controlling of the number of active states during the decoding phase. A novel method for confidence interval computation based on findings in [10] has been shown in this article.

Acknowledgement

This work was supported by grant No. 2C06009 Cot-Sewing.

References

1. Bourlard, H., Morgan, N.: Hybrid HMM/ANN Systems for Speech Recognition: Overview and New Research Directions, Summer School on Neural Networks (1997)
2. Hejtmánek, J., Pavelka, T.: Use of context-dependent units in Czech speech. In: Proc. of PhD Workshop 2007, Balatonfüred, Hungary (2007)

3. Odell, J.J.: The Use of Context in Large Vocabulary Speech Recognition, PhD Thesis, Cambridge University Engineering Dept. (1995)
4. Rabiner, L.R.: A Tutorial on Hidden Markov Models and Selected Applications in Speech Recognition. Proceedings of the IEEE 77(2) (1989)
5. Pavelka, T., Ekštein, K.: Neural Network Acoustic Model for Recognition of Czech Speech. In: Proc. of PhD Workshop Systems & Control, Izola, Slovenia (2005)
6. Pavelka, T., Ekštein, K.: JLASER: An Automatic Speech Recognizer Written in Java. In: Proc. of XII International Conference Speech and Computer (SPECOM 2007), Moscow, Russia (2007)
7. Pavelka, T., Král, P.: Neural Network Acoustic Model with Decision Tree Clustered Triphones. In: Proceedings of 2008 IEEE International Workshop on Machine Learning for Signal Processing, Cancún, Mexico (2008)
8. Tebelskis, J.: Speech Recognition using Neural Networks, PhD Thesis, Carnegie Mellon University (1995)
9. Young, S., et al.: The HTK Book (for HTK v. 3.3), Cambridge University Engineering Dept. (2002)
10. Vávra, F., Pavelka, T., Šedivá, B., Vokáčová, K., Marek, P., Neumanová, M.: Ratio Statistics. In: Proceedings of JČMF ROBUST 2008, Pribylina, Slovakia (2008)

Semantic Annotation for the LingvoSemantics Project

Ivan Habernal and Miloslav Konopík

University of West Bohemia, Department of Computer Sciences,
Univerzitní 22, CZ - 306 14 Plzeň, Czech Republic
{habernal,konopik}@kiv.zcu.cz

Abstract. In this paper, a methodology of semantic annotation of the LingvoSemantic corpus is presented. Semantic annotation is usually a time consuming and
expensive process. We thus developed a methodology that significantly reduces
the demands of the process. The methodology consists of a set of techniques and
computer tools designed to simplify the process as much as possible. We claim
that in this way it is possible to obtain sufficient amount of annotated data in a reasonable time frame. The LingvoSemantic project focuses on semantic analysis of
user questions to an Internet information retrieval system. The semantic representation approach is based on abstract semantic annotation methodology. However,
we advanced the annotation process. The bootstrapping method was used during
the corpus annotation. The resulting annotated corpus consists of 20292 annotated sentences. In comparison to the straight-forward style of annotation, our
approach significantly improved the efficiency of the annotation. The results, as
well as a set of recommendations for creating the annotated data, are presented at
the end of the paper.

1 Introduction

The purpose of a semantic analysis system is to obtain a context-independent semantic
representation from a given input sentence. Most of today's semantic analysis system
are based on statistics. Therefore, the corpora are required for the system training and
performance testing. A corpus consists of annotated data. Annotated data are data enriched with semantic information. The annotation formalism depends on the particular
semantic representation of the developed system.

Since the annotations are created by human annotators, the annotation process is very
time consuming. Given a certain annotation methodology, the annotator uses his or her
semantic and domain knowledge and associates the annotation with each sentence from
the training data. Thus, a reliable annotation methodology can lead to better results in
sense of inter-annotator agreement (see section 4), as well as speed up the annotation
process. Some annotation guidelines will be presented later in this paper.

Our semantic representation is based on abstract semantic annotation, similar to the
one proposed in [1]. This is a tree based representation, where the tree nodes represent
semantic concepts. The leaves of the tree are called *lexical classes* and they are bound
to words or to groups of words from the sentence. In opposite to the abstract semantic
annotation, our model requires the exact assignment of lexical class and words (also
called *lexical model*). An example of semantic annotation is shown in Fig 1.

V. Matoušek and P. Mautner (Eds.): TSD 2009, LNAI 5729, pp. 299–306, 2009.

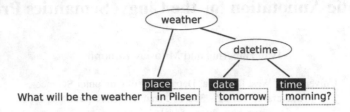

Fig. 1. An example of semantic annotation; the black boxes represent the lexical classes

1.1 Related Work

Although various systems for semantic annotation exist, they are not suitable for our work. For example, the ATIS corpus in [4] was annotated automatically first, by using the Phoenix syntactic parser. Another approach based on patterns is used in [5]. The usage of Conditional Random Fields is described e.g. in [6].

2 The LingvoSemantic Corpus

For collecting the data, we created a system that simulated a real application for search-ing on Internet. The user queries are natural language questions in Czech language. Every query put into the system was stored in the database. Totally, we have collected 20 292 user queries in this way. The collection of the sentences took half a year. Ap-proximately 450 people participated in the collection of the data. The statistic values of the collected corpus are given in Table 1.

Table 1. LingvoSemantics corpus statistics

Unique Words	19 125
Total Word Forms	145 183
Unique Sentences	20 292
Total Count of Sentences	21 161
Word Perplexity	19,6

Here are some examples of the selected areas:

- weather forecast
- public transport
- local city transport
- accommodation

3 Semantic Annotation Methodology

To crate the annotated data, a team of 4 annotators was assembled. Each sentence in the corpus was annotated by two annotators. If the annotations differ from each other, the conflict is resolved by the annotation coordinator (a person in charge of coordinat-ing annotators). Semantic annotations always reflect the point of view of the annotator.

Having annotated every sentence twice allows us to make the semantic annotation more objective. Then the *inter-annotator agreement* can be computed. The inter-annotator agreement is a percentage of how many times annotators agreed on the annotation of each sentence. The inter-annotator agreement gives us some information about the ambiguity contained in the semantic annotation formalism that is used to describe the meaning in annotated sentences.

A commonly used formalism of semantic representation and annotation is the *bracketing notation*. The following example shows the semantic annotation of the sentence *"What will be the weather in Pilsen tomorrow morning?"*:

```
WEATHER(PLACE(Pilsen), DATETIME(DATE(tomorrow), TIME(morning)))
```

Apparently, the bracket notation is a formalism for describing a tree structure. The previous example is equivalent to the tree representation from Fig. 1. Although creating such annotation for the sentence might appear as a straight-forward solution, it still requires a lot of annotator effort (ie. checking the bracket pairing etc.). This annotation methodology is used in our baseline system.

3.1 Annotation Scheme

The semantic representation is defined by an *annotation scheme*. This is a hierarchical structure (a tree) that defines the dominance relationship among concepts, theme and lexical classes. It says which concepts can be associated with which super-concepts, which lexical classes belong to which concepts and so on.

An example is shown in Fig. 2. The figure shows the annotation scheme for the *Weather forecast* theme. The scheme defines that the *Weather forecast* theme can contain the *Meteorological event*, the *Place* and the *Date and Time* concepts. Apparently, this hierarchy is valid for the rest of the concepts, according to the tree in Fig. 2.

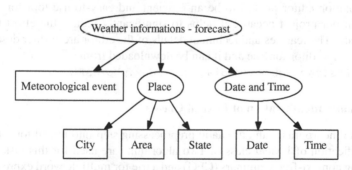

Fig. 2. An annotation scheme example

4 The Basic Annotation Methodology

Members of the team were trained for the semantic annotations, according to the in-depth annotation manual. The training was done on the first 300 sentences under the supervision of the annotator coordinator to check the errors and mistakes made by the annotators.

After the initial team training, another 1000 sentences were annotated. For the annotation, a text editor with bracket completion was used. However, there was still a considerable amount of errors in the annotated sentences. The types of errors and its counts are shown in table 2.

Table 2. Errors after the annotation of first 1000 sentences

Error type	# of sentences	in %
Unpaired brackets	12	0.01
Invalid concept name (e.g. typing error)	89	8.9
Invalid structure	63	6.3
No errors	835	83.5

As we can see, almost 17% of the annotated data contained errors and thus they were unusable for further processing. This number of errors was reached even after training the team and with supporting detailed manual and annotation schemes. Moreover, the time demands of the annotation was enormous. Therefore, we developed a more efficient methodology.

5   The Advanced Annotation Approach

To avoid errors shown in Table 2 (invalid concepts, structure, etc.), a supporting annotation editor was developed. This annotation editor ensures that annotations are syntactically correct and conform to a predefined annotation scheme (see section 3.1). The editor also helps to remember the names of concepts because they are given to the annotator to choose from. The annotator does not need to consult the annotation manual so often.

The annotation editor proved to be an efficient and easy-to-use tool for semantic annotation in our project because it helps to significantly reduce the effort to create annotated data. The features and technical details of the editor are further described in [2]. We offer this editor for free and it can be downloaded from :
https://liks.fav.zcu.cz/mediawiki/index.php/JAEE.

5.1   Automatic Identification of Lexical Classes

We have also incorporated an automatic preprocessing step into the editor – the automatic identification of lexical classes. Two algorithms are used for this task: shallow parsing using context-free grammars (CFG) and a trie for multiple word expressions. In the data, we found two main types of lexical classes:

1. proper names, multi word expressions, enumerations
2. structures (date, time, postal addresses, etc.)

The first type can be usually learned from a database. Therefore we have created a huge database of all proper names and enumerations (cities, names of stations, types of means of transport, etc.) from free available Internet sources. The database contains string expressions and corresponding lexical classes. Since a lexical class can consist of more than one word, it is necessary to look for multiple word expressions (MWEs). To find MWEs in a sentence, it is necessary to try to look for all possible combinations of words that can create a MWE in the database. A sentence contains $\frac{n^2+n}{2}$ of possible MWEs, where n is the number of words. We decreased the complexity from $O(n^2 * m)$, where m is number of the database items, to $O(n)$ using the *trie*[1] structure in the algorithm. The lexical classes are stored in the leaves of the trie.

The second type of lexical classes can be analyzed with a specialized parser, the LINGVOParser. This is an implementation of context-free grammar parser with some additional features. More details can be found in [3].

We created a set of grammars that describe the following lexical classes:

date – specific date (e.g. on Monday, tomorrow, 26.3.1981; March 26th, 1981; etc.), date interval (e.g. next week, between Monday and Friday, during Christmas, etc.)
time – specific time (e.g. twelve o'clock, 12:30, etc.), time interval (e.g. in the evening, before three o'clock, etc.)
currency – specific currency (e.g. 5200 crowns, etc.), currency interval (e.g. under 5200 crowns, etc.)

Since this tool is a part of the annotation editor, the annotator does not have to select the lexical classes manually. The lexical classes are identified and proposed automatically, however, they can be corrected by the annotator.

5.2 Semantic Annotation of the Corpus

Having the supporting tools, the *bootstrapping* methodology was adapted for semantic annotation. It is an iterative process of training unsupervised or partially supervised models. At the beginning, only a small part of the data is annotated by the human. Using this data, the initial model is trained. Then the trained model runs over the remaining data, from which a part is corrected by human annotators. The model is then re-trained again and the process continues. An example is shown in Fig. 3.

In the first iteration, 1500 sentences was chosen. They were annotated manually, using the preselected lexical classes. Thereafter, the semantic analysis system [7] was trained using this data. The rest of unannotated data was split into four parts, one part for each iteration process. Then the system analysed another approx. 4700 sentences. In the next step, these data were hand corrected by annotators. The annotator checks whether the suggested annotation tree is appropriate, otherwise a hand correction is performed. The asset of the annotation methodology can bee seen in table 3. Instead of annotating the whole corpus, in each step the number of corrected sentences decreases. However, the efficiency of the automatic annotation is directly influenced by the system used for semantic parsing.

The so-called *"sleeping annotator"* effect was also taken into account. This means, that in the later iterations if the annotations proposed by the automatic annotation are mostly accurate, the annotator does not pay sufficient attention and accepts even the wrong annotations (*"sleeps on the enter key"*). Thus, we randomly put a randomly

---

[1] A tree for storing strings in which there is one node for every common prefix.

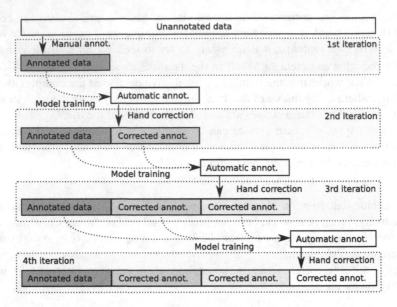

Fig. 3. An example of bootstrapping method

Table 3. Annotation iterations and number of corrected sentences in each step

Iteration	Sentences annot. aut.	Corrected sent.	Corrected sent. (%)
2	4698	3320	70,6
3	4698	1870	39,8
4	4698	1409	29,9
5	4698	1450	30,8

chosen sentence with an invalid annotation into the data for correction to ensure, that the annotator makes the task responsibly. These sentences were discarded later.

6   Results

The inter-annotation agreement after the first attempt described in section 4 was only 55%. The main reason of different annotations among the annotators was the different assignment of the lexical classes. Although the annotation manual was very detailed in this problematic, the results were quite poor.

After using the semi-automatic approach described in section 5 with automatically proposed lexical classes, the inter-annotator agreement increased to 72%. Totally 20292 sentences from the LingvoCorpus were annotated. Furthermore, a time demands between these two approaches are very significant. Using the bootstrapping method, only a part of the data must be fully annotated from scratch.

6.1 Abstract Annotation Recommendation Rules

After our experience with long lasting annotation process, we propose some recommendations for semantic annotation. Some of the rules are specific for chosen semantic representation, other are more general.

- The annotation schemes have to be as simple as possible. The semantic annotation is very subjective process. The more complicated semantic description the more space for subjective point of view.
- To obtain consistent annotations a very detailed annotation manual has to be provided. There should be a lot of examples and counter-examples in the manual.
- A sentence should be annotated twice by different annotators. In that way, the inter-annotator agreement can be computed and many mistakes can be quickly revealed.
- Lexical classes should be annotated automatically in advance. Most of the lexical classes are known in advance (cities, countries, station names, etc.). Such lexical classes should be annotated automatically to reduce the work of annotators and most of all to reduce the ambiguities in the annotations. Obviously all lexical classes cannot be covered automatically in advance. Some shortcuts and colloquial expressions of lexical classes as well as infrequent variations of general lexical classes (date, time) may remain unannotated. Anyway, annotators will have many examples in other classes. Then it is easy for them to capture the few remaining.
- The annotations should be done in a bootstrapping manner. Again, it reduces the work of annotators and the possibility of subjective results. It can be seen from the results that even with small amount of training data significant portion of the annotation is done correctly by the parsing algorithm.

7 Conclusion

We have developed and evaluated a methodology for semantic annotation. It is based on a set of computer tools and techniques. The methodology proved to be efficient for annotating the corpora of 20k sentences. Moreover, the time demands have been rapidly decreased. We provide the tools for free at our Internet pages[2].

Acknowledgement

This work was supported by grant No. 2C06009 Cot-Sewing.

References

1. He, Y., Young, S.: Semantic processing using the Hidden Vector State model. Computer Speech and Language 19(1), 85–106 (2005)
2. Habernal, I., Konopík, M.: JAAE: the Java Abstract Annotation Editor. In: INTERSPEECH 2007, pp. 1298–1301 (2007)

---

[2] http://liks.fav.zcu.cz/

3. Habernal, I., Konopík, M.: Active tags for semantic analysis. In: Sojka, P., Horák, A., Kopeček, I., Pala, K. (eds.) TSD 2008. LNCS (LNAI), vol. 5246, pp. 69–76. Springer, Heidelberg (2008)
4. Zhou, D., He, Y.: Discriminative Training of the Hidden Vector State Model for Semantic Parsing. IEEE Trans. on Knowl. and Data Eng. 21(1), 66–77 (2009)
5. Meurs, M.J., Duvert, F., Bechet, F., Lefevre, F., Mori, R.D.: Semantic Frame Annotation on the French MEDIA corpus. In: Proc. Language Resources and Evaluation (LREC), Marrakech, Morocco (2008)
6. Rodriguez, K., Raymond, C., Riccardi, G.: Active Annotation in the LUNA Italian Corpus of Spontaneous Dialogues. In: Proc. Language Resources and Evaluation (LREC), Marrakech, Morocco (2008)
7. Konopík, M.: Hybrid Semantic Analysis, PhD thesis, University of West Bohemia (2009)

Hybrid Semantic Analysis

Miloslav Konopík and Ivan Habernal

University of West Bohemia, Department of Computer Sciences,
Univerzitní 22, CZ - 306 14 Plzeň, Czech Republic
{habernal,konopik}@kiv.zcu.cz

Abstract. This article is focused on the problem of meaning recognition in written utterances. The goal is to find a computer algorithm capable to construct the meaning description of a given utterance. An original system for meaning recognition is described in this paper. The key idea of the system is the hybrid combination of expert and machine-learning approaches to meaning recognition. The system utilizes a novel algorithm for semantic parsing. The algorithm is based upon extended context-free grammars. The grammars are automatically inferred from the data.

1 Introduction

Recent achievements in the area of automatic speech recognition started the development of speech enabled applications. Currently it starts to be insufficient to merely recognize an utterance. The applications demand to understand the meaning. Semantic analysis is a process whereby the computer representation of the sentence meaning is automatically assigned to an analyzed sentence.

This article is focused on the development of a semantic analysis system. The system is developed in the context of information retrieval on the Internet. The used data are written in the Czech language.

Our approach to semantic analysis is based upon a combination of expert methods and stochastic methods (that is why we call our approach a hybrid semantic analysis). We show that a robust system for semantic analysis can be created in this way. During the development of the system an original algorithm for semantic parsing was created. The algorithm is based upon chart parsing and extended context-free grammars. Some of the ideas used in the thesis come from the Chronus system [8] and the HVS model [3].

2 Semantic Representation

There are several ways how to represent semantic information contained in a sentence. In our work we use tree structures (see Figure 1) with the so-called *concepts* and *lexical classes*. The theme of the sentence is placed on the top of the tree. The inner nodes are called concepts. Concepts describe some portion of semantic information contained in the sentence. They can contain other sub-concepts that specify the semantic information more precisely or they can contain the so-called lexical classes. Lexical classes are the leaves of the tree. A lexical class covers certain phrases that contain the same type of

V. Matoušek and P. Mautner (Eds.): TSD 2009, LNAI 5729, pp. 307–314, 2009.

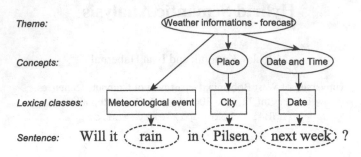

Fig. 1. The example of an annotation tree

information. For example a lexical class "date" covers phrases "tomorrow", "Monday", "next week" or "25th December" etc.

The semantic representation formalism used in this paper is taken from [3]. The advantage of this approach is that it does not require the annotation of all words of a sentence.

3 Data

The data used during the development are the questions to an intelligent Internet search engine. The questions are in the form of whole sentences because the engine can operate on whole sentences rather than just on keywords as is usual. The questions were obtained during a system simulation. We asked users to put some question into a system that looked like a real system. In this way we obtained 20 292 unique sentences. The sentences are annotated with the aforementioned semantic representation (Section 2).

The following box shows an example of the data:

WEATHER(Jaká EVENT(teplota) vzduchu bude DATETIME(DATE(pozítří))?)
How warm will it be the day after tomorrow?

BUSTRAIN(Jede něco DEPART(TIME(po půlnoci)) do DESTIN(CITY(Nýrska)) ?)
Is there any line to Nýrsko after midnight?

4 System Description

The system consists of three main blocks (see Figure 2). The *preprocessing* phase prepares the system for semantic analysis. It involves sentence normalization, tokenization and morphological processing. The *lexical class analysis* is explained in Section 5 and the *probabilistic parsing* is explained in Section 6.

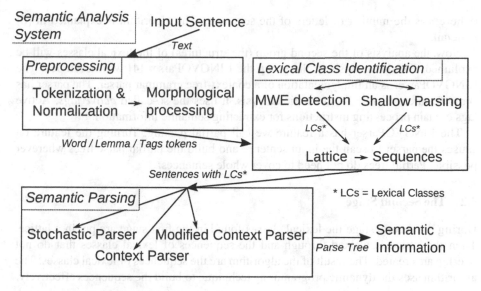

Fig. 2. The structure of our semantic analysis system

5   Lexical Class Identification

The lexical class identification is the first phase of the semantic analysis. During this phase the lexical classes (see Section 2) are being found in the input sentence.

The lexical class identification consists of two stages. First, several dedicated parsers are run in parallel. During this stage a set of lexical classes are found. In the second stage the lexical classes are stored in a lattice. Then the lattice is walked through and a set of possible sequences of lexical classes are created. Only the sequences that contain no overlapping classes are created.

5.1   The First Stage

During the first step the lexical classes are being found as individual units. We found in our data two groups of lexical classes:

1. Proper names, multi word expressions, enumerations.
2. Structures (date, time, postal addresses, ...).

To analyze the first group we created a database of proper names and enumerations (cities, names of stations, types of means of transport etc). Since a lexical class can consist of more than one word it is necessary to look for multiple word expressions (MWEs). To solve the search problem effectively a specialized searching algorithm was developed.

The main ideas of the algorithm are organizing the lexical classes in the trie structure [7] and using parallel searching. The algorithm ensures that all lexical classes (possibly consisting of more words) are found during one pass with a linear complexity $O(n)$

(where n is the number of letters of the sentence). The algorithm is explained in [6] in detail.

Now, the analysis of the second group (the structures) of the lexical classes will be explained. For the analysis of the group the LINGVOParser [4] was developed. The LINGVOParser is an implementation of a context-free grammar parser. The parser has some features suitable for semantic analysis. It uses the so-called *active tags*. Active tags contain processing instructions for extracting semantic information.

The LINGVOParser has a feature we call partial parsing. Turning the feature on causes the parser to scan the input sentence and build the partial parse trees wherever possible. Partial trees do not need to cover whole sentences.

5.2   The Second Stage

During the second stage the lexical classes found in the first stage are put in a lattice. Then the lattice is walked through and the sequences of lexical classes that do not overlap are created. The result of the algorithm are the sequences of lexical classes. The algorithm uses the dynamic programming technique to build the sequences effectively.

5.3   Performance Testing

The performance is measured for both stages. We use precision, recall and F-measure (see [5]). To measure the first stage we compare the generated lexical classes with the lexical classes annotated by human annotators. For the second stage we compare the whole sequences. The results are shown in Table 1.

Table 1. The performance of both stages

	First Phase	Second Phase
Precision	0.79	0.68
Recall	0.96	0.83
F-measure	0.86	0.74

6   Semantic Parsing

The second phase of semantic analysis will be described in this section. In the previous section the process of finding lexical classes was described. In this section we will presume that the lexical classes are known and we have to build the tree. The structure of the tree is shown in Figure 1. The task of the parsers described here is to create the same trees as in the data section (see Section 3).

6.1   Stochastic Parser

This parser works in two modes: training and analysis. The training phase requires aforementioned annotated data (see Section 3). During the training the annotation trees are transformed to context free grammar rules in the following way. Every node is transformed to one rule. The node name makes the left side of the rule and the children of the node make the right side of the rule (for example see node "Place" in Figure 1, this node is transformed into the rule `Place -> City`). In this way all the nodes of the annotation tree are processed and transformed into grammar rules. Naturally, identical rules are created during this process. The rules are counted and conditional probabilities of rule transcriptions are estimated:

$$P(N \rightarrow \alpha | N) = \frac{\text{Count}(N \rightarrow \alpha)}{\sum_\gamma \text{Count}(N \rightarrow \gamma)} \tag{1}$$

where N is a nonterminal, α and γ are strings of terminal and nonterminal symbols.

The analysis phase is in no way different from standard stochastic context-free parsing. The sentence is passed to the parsing algorithm. The stochastic variant of the active chart parsing algorithm (see e.g. [1]) is used. The lexical classes identified in the sentence are treated as terminal symbols. The words that are not members of any lexical class are ignored. The result of parsing – the parse tree – is directly in the form of the result tree we need.

During parsing the probability of the so far created tree $P(T)$ is computed by:

$$P(T) = P(N \rightarrow A_1 A_2 ... A_k | N) \prod_i P(T_i) \tag{2}$$

where N is the top nonterminal of the subtree T, A_i are the terminals or non-terminals to which the N is being transcribed and T_i is the subtree having the A_i nonterminal on the top.

When the parsing is finished a probability is assigned to all resulting parse trees. The probability is then weighted by the prior probability of the theme and the maximum probability is chosen as the result:

$$\hat{T} = \arg\max_i P(S_i) P(T_i) \tag{3}$$

where $\hat{T}$ is the most likely parse tree and $P(S_i)$ is the probability of the starting symbol of the parse tree T_i.

6.2 Context Parser

This parser looks at other words of the sentence rather than looking at lexical classes only. For this purpose it was necessary to extend both the training algorithm and the analysis algorithm.

The training phase shares the same steps with the training of the previous parser in Section 6.1. The node is thus transformed into the grammar rule and the frequency of the rule occurrence is counted. However instead of going to the next node, the context of the node is examined. Every node that is not a leaf has a subtree beneath. The subtree spans across some terminals. The context of the node is defined as the words before and after the span of the subtree. During training the frequency of the context and a nonterminal (Count($word, nonterminal$)) are counted. The probability of a context given a nonterminal is computed via MLE as follows:

$$P(w|N) = \frac{\text{Count}(w, N) + \lambda}{\sum_i \text{Count}(w_i, N) + \lambda V} \qquad (4)$$

where λ is the smoothing constant, V is the estimate of the vocabulary size, w is the actual context word and w_i are all the context words of nonterminal N.

Additionally, to improve the estimate of the prior probability of the theme (the root node of the annotation) we add words to the estimate as well:

$$P(w|S) = \frac{\text{Count}(w, S) + \kappa}{\sum_i \text{Count}(w_i, S) + \kappa V} \qquad (5)$$

where κ is the smoothing constant, w_i are the words of the sentence and S is the theme of the sentence (the theme constitutes the starting symbol after annotation tree transformation).

The analysis algorithm is the same as in the previous parser but the probability from formula 2 is reformulated to consider the context:

$$P(T) = \sum_i P(w_i|N)P(N \to A_1 A_2 ... A_k|N) \prod_j P(T_j) \qquad (6)$$

Then the best parse is selected using context sensitive prior probability:

$$P(\hat{T}) = \arg\max_i P(S_i) \prod_j (P(w_j|S)P(T_i)) \qquad (7)$$

where S_i is the starting symbol of the parse tree T_i and w_j are the words of the analyzed sentence.

6.3 Modifications for Morphologically Rich Languages

We tried to further improve the performance of parsing algorithms by incorporating features that consider the specific properties of the Czech language. The Czech language is a morphologically rich language [2] and it has also a more flexible word order than for example English or German. To deal with specific properties of the Czech language *lemmatization* and *ignoring word order* features were incorporated to the Context Parser.

The rich morphology increases the size of the dictionary. Hence we tried to lemmatize input sentences used for both training and testing. During lemmatization all word forms in the sentences are converted to their basic forms.

The flexible word order also decreases the robustness of probability estimations. Different word order creates different events in the probability estimations. Therefore we incorporated the ability to ignore the word order in the Context Parser.

6.4 Results of Semantic Parsing

This section describes the performance of the introduced semantic parsers. The testing is limited to the semantic parsers only. The performance of previous phases of semantic analysis plays no role in this testing.

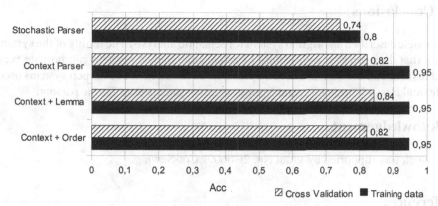

Fig. 3. Results of Semantic Parsers

The performance is given by the accuracy measure. The accuracy (Acc) is computed as the ratio between the number of correct results (C) and the number of tested sentences (T): $Acc = \frac{C}{T}$. The result is considered to be correct only if the result is identical to the semantic annotation contained in the data.

There are two results given in Figure 3. First, all sentences were used for both training and testing and thus the testing data were seen during training. Second, results are computed by the *cross validation* technique. Cross validation ensures that no testing data were seen during training.

The context parser shows a significant improvement over the baseline stochastic parser. The results imply that introducing some sort of context to context-free grammars is beneficial. The results of the context parser are very promising.

The results of the modifications for the Czech language have not met our expectations. Lemmatization slightly improved the performance however ignoring the word order was not beneficial at all. The explanation can be found in the nature of our data. The questions are oriented to some given topics. It is natural that the topic orientation determines the word order of the sentences. The fixed structure also to some extent determines the word forms that are used in the sentence. The effect of the lemmatization then does not show.

7   Overall Results

Table 2 shows the results for both steps (lexical class identification and semantic parsing) together. The results are lower than in Section 6.4 since the results shown here are affected by the presence of the error in the lexical class identification phase.

Table 2. Results of Semantic Parsers

	Stochastic Parser	Context Parser	Context + Lemma	Context + Order
Acc	0.63	0.70	0.71	0.68

8 Conclusions

This article described an original system for semantic analysis. The testing of the system shows that the proposed architecture of a semantic analysis system can provide very good results. The architecture benefits from the advantages of both expert systems used for lexical class identification and from stochastic systems used for the parsing.

Acknowledgement

This work was supported by grant No. 2C06009 Cot-Sewing.

References

1. Allen, J.: Natural Language Understanding. Benjamin/Cummings Publ. Comp. Inc., Redwood City (1995)
2. Grepl, M., Karlík, P.: Skladba éeštiny, Olomouc, Czech Republic (1998)
3. He, Y., Young, S.: Semantic processing using the Hidden Vector State model. Computer Speech and Language 19(1), 85–106 (2005)
4. Habernal, I., Konopík, M.: Active tags for semantic analysis. In: Sojka, P., Horák, A., Kopeček, I., Pala, K. (eds.) TSD 2008. LNCS (LNAI), vol. 5246, pp. 69–76. Springer, Heidelberg (2008)
5. Jurafsky, D., Martin, J.: Speech and Language Processing. Prentice-Hall, Englewood Cliffs (2000)
6. Konopík, M.: Hybrid Semantic Analysis, PhD thesis, University of West Bohemia (2009)
7. Knuth, D.: The Art of Computer Programming, 2nd edn. Sorting and Searching, vol. 3. Addison-Wesley, Reading (1997)
8. Pieraccini, R., Tzoukermann, E., Gorelov, Z., Levin, E., Lee, C.-H., Gauvain, J.-L.: Progress Report on the Chronus System: ATIS Benchmark Results. In: Proc. of the workshop on Speech and Natural Language (1992)

On a Computational Model for Language Acquisition: Modeling Cross-Speaker Generalisation

Louis ten Bosch, Joris Driesen, Hugo Van hamme, and Lou Boves

Dept Language and Speech, Radboud University Nijmegen, NL
ESAT, Katholieke Universiteit Leuven, Belgium
{l.tenbosch,boves}@let.ru.nl
{joris.driesen,hugo.vanhamme}@esat.kuleuven.be
http://lands.let.ru.nl

Abstract. The discovery of words by young infants involves two interrelated processes: (a) the detection of recurrent word-like acoustic patterns in the speech signal, and (b) cross-modal association between auditory and visual information. This paper describes experimental results obtained by a computational model that simulates these two processes. The model is able to build word-like representations on the basis of multimodal input data (stimuli) without the help of an a priori specified lexicon. Each input stimulus consists of a speech signal accompanied by an abstract visual representation of the concepts referred to in the speech signal. In this paper we investigate how internal representations generalize across speakers. In doing so, we also analyze the cognitive plausibility of the model.

1 Introduction

Learning the sound patterns of new words is an essential part of learning language. In the literature on language learning, two fundamental processes are discussed that are needed to accomplish this task [13][20].

The first process is the word discovery and recognition from an acoustic point of view. Infants (until about six months old) are truly 'universal' learners without preference for a particular language or sound structure. Recent research strongly suggests that infants store a great deal of acoustic information about words [12][17]. For infants of 6-7 months old, the reliance on acoustic detail can even hamper the recognition of the same word when it is spoken by another speaker or in another speaking style ([11][19]). When infants are about one year old, they become better in generalizing across speaker and gender.

The second process in word discovery is cross-situational and cross-modal learning. Infants are mostly confronted with *multimodal* stimuli: they hear speech in the context of tactile or visual information that is associated with the information in the auditory channel (*word-referent pairs*). In *individual* word-scene pairs, the relation between word and referent may be ambiguous, and it may only be revealed by accumulation of statistical evidence across many situational examples. [20] provides arguments for the hypothesis that young learners make use of *statistical* cues to associate acoustic patterns and referents.

V. Matoušek and P. Mautner (Eds.): TSD 2009, LNAI 5729, pp. 315–322, 2009.
© Springer-Verlag Berlin Heidelberg 2009

How can word learning by young infants be modeled computationally? Currently, automatic speech recognizers (ASR) are the most elaborate computational models of speech recognition. Contrary to virtually all psycholinguistic models of speech processing, an ASR-based model is able to handle the entire chain from speech signal to a sequence of words. However, current ASR algorithms certainly cannot claim cognitive or ecological plausibility. Moreover, ASR systems perform substantially worse than humans [15] [21]. In [2][4][7] we have described a novel computational model that is designed in the framework of the ACORNS project [23]. The model bears some similarity with the CELL model developed by [18], but unlike CELL it does not assume (unrealistic) phonemic input representations for the speech signal. The ACORNS model is able to learn words in a cognitively plausible analogy of the way in which infants acquire their native language. Following [20], we model the detection of words by searching recurrent patterns in the speech modality, in such a way that hypothesized word-like units statistically correspond with visual information. In designing the model we have made an attempt to make it as cognitively plausible as possible with respect to processing and representations in memory. Our research pursues two goals: to better understand human speech recognition and to improve ASR by including essential knowledge from human speech recognition.

More often than not infants hear speech from a small number of speakers (their caregivers). This raises the question to what extent initial representations of word-like units are speaker dependent. Behavioral experiments suggest that infants do not readily generalize 'words' learned from one speaker to other speakers [17]. However, there are indications that recognizing other speakers is improved if infants learn from speech of several different speakers. In this paper, we investigate these issues by means of a number of simulation experiments.

The structure of this paper is as follows. In the next section, we will discuss the main components of the computational model. The following sections describe the experiment and results in more detail. The final section contains a discussion and conclusions.

2   Brief Description of the Model

An input stimulus in our model consists of a spoken utterance in combination with a visual representation of (some of) the objects referred to in the speech. For the time being, the visual representation of a referent (e.g. 'car') is represented by a low-dimensional vector containing visual features. This is provided in synchrony with the speech signal. The acoustic input consists of continuous speech; there is no word segmentation and there is no orthographic representation available for the learner. It is the task of the learner to find a coherent relation between acoustic forms (word-like units) and the referent.

The learning is incremental and takes place in a communicative loop between the learner and the 'caregiver'. An interaction takes place as follows. The (virtual) 'caregiver' presents one multimodal stimulus to the (virtual) learner. After the learner receives the input stimulus, structure discovery techniques are applied to hypothesize new and/or adapt existing internal representations for word-like units. This process is based on the stimulus and the information stored in the learner's memory (see e.g. [2][4][5][7] for more detail). During *training*, the learner uses *both* modalities of an input stimulus.

In the *test*, only the auditory part of the stimulus is processed, and the learner responds with the hypothesized concept(s) that match(es) best with the utterance.

In the ACORNS project we are experimenting with several different structure discovery techniques [23]. In this paper we only report results obtained with Non-negative Matrix Factorization (NMF) [10][14][22]. NMF is member of a family of computational approaches that represent input data in a (large) observation matrix V and uses linear algebra to decompose V into much smaller matrices W and H such that $V \approx W \cdot H$. After this decomposition, W can be interpreted as a set of representations of speech units, while H contains the associated internal activations. In combination, W and H represent the information in V in a more condensed form ('reconstruction'). The number of columns in W (and rows in H) is equal to the number of different internal representations. This number is a model parameter (see next section). The other dimension of W is specified by the dimension of the input. In our experiments, an input utterance is coded in the form of counts of co-occurrences of Vector Quantization labels. This allows us to represent utterances of arbitrary length in the form of a fixed-length vector (this is the acoustic part of the stimulus). The *visual* representation of the stimulus is appended to the acoustic part to obtain its full vectorial representation.

NMF is reminiscent of Latent Semantic Analysis, e.g. [1]. It can be shown that NMF provides a clear interpretation of concepts such as *abstraction* in terms of linear algebraic operations [23]. Prior to a training, both W and H are initialized randomly, and during training both matrices are updated on the basis of the stimuli. In its conventional form NMF is a technique for decomposing matrices that comprise a large number (possible all) of input stimuli [14]. To make the approach more cognitively plausible, the decomposition algorithm has been adapted such that the influence of past stimuli decays exponentially over time. The decay is determined by a model parameter γ (set to 0.99 – the closer to 1, the smaller the forgetting decay). Apart from details, the adapted NMF-update has the following form (Θ denoting an auxiliary matrix) ([8], cf [14]):

$$
\begin{aligned}
W_{new} &= W_{old} \cdot ((V/W_{old}H_{old}) * H_{old}^{tr}) + \gamma\Theta_{old} \\
\Theta_{new} &= W_{new} \cdot ((V/W_{new}H_{new}) * H_{new}^{tr}) + \gamma\Theta_{old}
\end{aligned}
\qquad (1)
$$

It is left to the learner to decide how many different internal representations will be built. Also, we exercise no control over the 'contents' of the internal representations: they may represent phrases, words, sub-word units, etc. Input utterances are not kept in memory after the update of the W matrix is complete. Therefore, in psycholinguistic terms our present model is not purely episodic. However, it is possible to mix abstractionist and episodic representations by deferring the update of W until after a certain number of input utterances have been received.

The virtual learner is endowed with the *intention* to learn words in order to maximize the appreciation (s)he receives from the caregiver. Here it is assumed that appreciation is a function of the proportion of utterances that are correctly understood. In addition, the learner tries to minimize the stress between the internal representations of speech units and the representation of a new utterance. This optimization boils down to the minimization of the Kullback-Leibler distance between the original input V and the reconstructed $\tilde{V} = WH$ [10].

3 Experiment

In this section we describe the design of the experiments with which we tested the hypothesis that if the observed variation during training is small, the ability to generalize will be small as well, while a larger amount of variation observed during training will lead to a larger ability to generalize.

3.1 Data

A pre-recorded database of Dutch utterances was applied for constructing specific training and test sets ([4]). It consists of utterances with a simple syntactic structure, in analogy to infant-directed speech. For this experiment, the data was narrowed down to utterances with exactly one concept (target word). For example, one of the stimuli is 'daar is een auto' (English: 'there is a car'), in which 'car' is one of the target words. In total, there are 13 different target words, all inspired on the basis of child language inventories [24].

3.2 Procedure

Table 1 presents a summary of the experiments that have been carried out. Each row represents a separate experiment. The column 'training' indicates which speaker(s) are used by the model for learning the internal word representations. The column 'test' indicates the test speaker(s). Each experiment consists of one incremental training starting with blank memory, using 600 stimuli presented in randomized order. Each test set consists of 400 stimuli. Experiments 1 and 2 assume a single speaker as the primary caregiver for training *and* test. The other experiments represent different possibilities for speaker generalization. Experiments 3 and 4 deal with cross-speaker, within-gender generalization. Experiments 5 and 6 show cross-gender performance. The final two experiments 7 and 8 can be compared to experiment 1 and 3, but investigate the effect of mixing in other speakers during training. In experiment 7, the 600 training stimuli are a randomized combination of 300 training stimuli from F1 and another 300 from F2. In experiment 8, the training consisted of the *entire* combined set from F1 and F2 (1200 utterances). The number of columns in W was set to 70, which is more than enough to build speaker-dependent representations of the 10 concepts presented to the model.

Table 1. Overview of the experiments on generalization across speakers

experiment	training (600)	held-out test (400)	accuracy (%)	conf. intervals (5% 1%)
1	female 1 (F1)	F1	99–100	- - - -
2	male 1 (M1)	M1	99–100	- - - -
3	F1	F2	85.25	82.33–88.17 81.13–89.37
4	M1	M2	87.30	84.56–90.04 83.43–91.17
5	F1	M1	61.80	56.24–65.90 55.36–66.76
6	M1	F1	67.10	63.57–70.82 61.51–72.44
7	F1+F2 (600)	F1	89.00	86.43–91.57 85.36–92.64
8	F1+F2 (1200)	F1	95.25	93.50–97.00 92.78–97.72

4   Results

The results are shown in the column 'accuracy' of table 1. The accuracy is given as the percentage correctly recognized keywords in a test with 400 utterances. The confidence intervals (based on the student's t-test and a binomially distributed success-rate, both for the $p = 0.05$ level and the $p = 0.01$ level) are provided in the last column.

Not surprisingly, the model performs best in case of speaker-matched training-test conditions. This is true for both the female and the male speaker. Rows 1 and 2 in table 1 show that speaker-dependent learning is almost perfect. Rows 3 and 4 show that generalization to a new speaker of the same gender, however, already provides significantly worse results on the $p = 0.01$ level. Generalization across gender (experiments 5 and 6) is significantly worse than the performance in the same gender cases.

Experiments 7 and 8 show that learning is affected by the number of training speakers, and not necessarily in a positive way. If the model learns from speakers F1 and F2, recognition of F1 deteriorates compared to the situation (in experiment 1) where there is only one training speaker. The accuracy of 89% in experiment 7 cannot be attributed to lack of training data. Fig. 1 shows how learning proceeds for speaker F1 if she is the only person from whom the model learns. It can be seen that recognition performance is already close to 100% after some 200 utterances. In the presence of speaker F2 that performance level is not even reached after learning from 600 utterances.

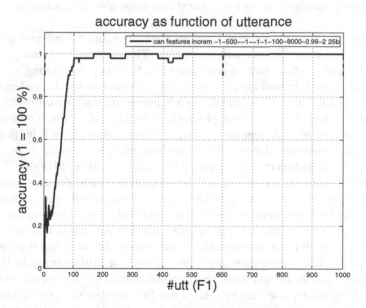

Fig. 1. Example of the learning performance of the computational model. The stimuli are taken from a single (female) speaker (F1). The horizontal axis indicates the stimulus index, while the accuracy of the learner (in terms of fraction of correctly recognized key words) is displayed along the vertical axis. The first 600 utterances are processed in incremental training mode, and the remaining 400 utterances are tested without any further training. Within about 200 utterances, that is, about 20–25 tokens per key word, the learner is able to build useful word representations.

5   Discussion

Most of the results in table 1 are unsurprising when considered from an Automatic Speech Recognition (ASR) point-of-view. The table shows that test results are best for within-speaker conditions, and deteriorate for the different-speaker same-gender condition, and are worst for different-gender conditions. However, we did not perform a conventional ASR training-test experiment. It was left to the learner to figure out how many different representations to build from the input utterances, and how to link these to the keywords. In our experiments the word-referent pairs were quite strongly tied. While this may seem to enhance learning, it might actually have had the opposite effect. From experiments 7 and 8 it appears that the learner tried to build speaker independent representations for the referents and that this slows down learning and results in sub-optimal performance.

A possible explanation for this behavior of the learner is the use of memory in the present NMF-based structure discovery approach. In all experiments reported above, a 70-column W matrix was initialized randomly (more than 5 times the number of concepts to be learned). The close to perfect recognition performance on *independent* test sets observed in experiments 1 and 2 shows that the ratio between the number of concepts to be learned and the number of representations that can be built is not the culprit. More likely, the interference between two speakers is due to the fact that the learner did not try to build (seemingly inefficient) speaker dependent representations, which would then only (much) later be reorganized into more abstract speaker independent representations.

For adults, the recognition of new speakers seems easy, but the underlying learning and adaptation processes are not well understood. The literature on language acquisition provides evidence that children must *learn to adapt*: they start storing a great deal of phonetic detail and gradually learn to generalize towards new speakers when they are between 7 and 12 months old. Recent theories in psycholinguistics assume that adaptation (by adults) can probably be best explained by hybrid models in which both so-called 'episodes' (detailed acoustic representations) and abstractions play a role ([9][16]). If we would exactly understand the processes involved, the 'gap' between human and automatic speech recognition could be narrowed in many realistic conditions.

When humans adapt to a new speaker, this adaptation process does not come at the cost of 'forgetting' older information that appeared to be useful in the past. Instead, learning to understand new speakers involves adapting existing internal representations if this does not destroy older information, and by creating additional representations if the new input differs too much from what was already in memory. In our model this can be accomplished if we change the present update procedure: instead of updating the W matrix after every utterance the learner should estimate the degree of fit between the new utterance and the internal representations. Only if that fit is good enough, the update should proceed. In all other cases a new representation should be added to the W matrix. However, implementing such a conditional update procedure is not straightforward. So far, we have not been able to define a reliable distance measure for the fit between an utterance and the internal representations. In [6] we have shown that reorganization of a large set of over-detailed representations (formed by enforcing the learner to create fully speaker dependent representations) can be implemented by means of clustering procedures.

So far, the ACORNS project has been more successful in applying knowledge from ASR and machine learning to elucidating models of human speech processing. However, we are confident that the underlying idea, viz. that the learning system should build its representations on the basis of its experience with input data rather than building models of pre-defined units, will result in automatic speech recognition systems that will rival human performance.

Acknowledgment

This research was funded by the European Commission, under contract number FP6-034362, in the ACORNS project [23].

References

1. Bellegarda, J.R.: Exploiting Latent Semantic Information for Statistical Language Modeling. Proc. IEEE 88, 1279–1296 (2000)
2. Van hamme, H.: Integration of Asynchronous Knowledge Sources in a Novel Speech Recognition Framework, ISCA ITRW, Speech Analysis and Processing for Knowledge Discovery (2008)
3. ten Bosch, L., Van Hamme, H., Boves, L.: Unsupervised detection of words - questioning the relevance of segmentation. In: ISCA ITRW, Speech Analysis and Processing for Knowledge Discovery (2008)
4. ten Bosch, L., Boves, L.: Language acquisition: The emergence of words from multimodal input. In: Sojka, P., Horák, A., Kopeček, I., Pala, K. (eds.) TSD 2008. LNCS (LNAI), vol. 5246, pp. 261–268. Springer, Heidelberg (2008)
5. ten Bosch, L., Van Hamme, H., Boves, L.: Discovery of words: Towards a computational model of language acquisition. In: Mihelic, F., Zibert, J. (eds.) Speech Recogition: Technologies and Applications, pp. 205–224. I-Tech Education and Publishing KG, Vienna (2008)
6. ten Bosch, L., Van Hamme, H., Boves, L.: A computational model of language acquisition: focus on word discovery. In: Proc. Interspeech 2008, pp. 2570–2573 (2008)
7. Boves, L., ten Bosch, L., Moore, R.: ACORNS - towards computational modeling of communication and recognition skills. In: Proceedings IEEE-ICCI 2007 (2007)
8. Driesen, J., Van Hamme, H.: personal communication
9. Goldinger, S.D.: Echoes of echoes? An episodic theory of lexical access. Psychological Review 105, 251–279 (1998)
10. Hoyer, P.O.: Non-negative matrix factorization with sparseness constraints. Journal of Machine Learning Research 5, 1457–1469 (2004)
11. Houston, D.M., Jusczyk, P.W.: The role of talker-specific information in word segmentation by infants. Journal of Experimental Psychology: Human Perception & Performance 26, 1570–1582 (2000)
12. Jusczyk, P.W., Aslin, R.N.: Infants' detection of the sound patterns of words in fluent speech. Cognitive Psychology 29, 1–23 (1995)
13. Kuhl, P.K.: Early language acquisition: cracking the speech code. Nat. Rev. Neuroscience 5, 831–843 (2004)
14. Lee, D.D., Seung, H.S.: Algorithms for non-negative matrix factorization. In: Advances in Neural Information Processing Systems, vol. 13 (2001)
15. Lippmann, R.: Speech Recognition by Human and Machines. Speech Communication 22, 1–14 (1997)

16. McQueen, J.M., Cutler, A., Norris, D.: Phonological abstraction in the mental lexicon. Cognitive Science 30, 1113–1126 (2006)
17. Newman, R.S.: The level of detail in infants' word learning. Current directions in Psychological Science 17(3), 229–232 (2008)
18. Roy, D.K., Pentland, A.P.: Learning words from sights and sounds: a computational model. Cognitive Science 26, 113–146 (2002)
19. Singh, L., Morgan, J.L., White, K.S.: Preference and processing: The role of speech affect in early spoken word recognition. Journal of Memory and Language 51, 173–189 (2004)
20. Smith, L., Yu, C.: Infants rapidly learn word-referent mappings via cross-situational statistics. Cognition 106(2008), 1558–1568 (2008)
21. Sroka, J.J., Braida, L.D.: Human and machine consonant recognition. Speech Communication 44, 401–423 (2005)
22. Stouten, V., Demuynck, K.: Van hamme, H.: Automatically Learning the Units of Speech by Non-negative Matrix Factorisation. In: Interspeech 2007, Antwerp, Belgium (2007)
23. http://www.acorns-project.org
24. http://www.sci.sdsu.edu/cdi/

Efficient Parsing of Romanian Language for Text-to-Speech Purposes

Andrei Şaupe[1], Lucian Radu Teodorescu[1], Mihai Alexandru Ordean[1],
Răzvan Boldizsar[1], Mihaela Ordean[1], and Gheorghe Cosmin Silaghi[2]

[1] iQuest Technologies, Cluj-Napoca, Romania
{Andrei.Saupe,Lucian.Teodorescu,Mihai.Ordean,
Razvan.Boldizsar,Mihaela.Ordean}@iquestint.com
[2] Babeş-Bolyai University of Cluj-Napoca, Romania
Gheorghe.Silaghi@econ.ubbcluj.ro

Abstract. This paper presents the design of the text analysis component of a TTS system for the Romanian language. Our text analysis is performed in two steps: document structure detection and text normalization. The output is a tree-based representation of the processed data. Parsing is made efficient with the help of the Boost Spirit LL parser [1], the usage of this tool allowing for a greater flexibility in the source code and in the output representation.

1 Introduction

Converting words from written form into speakable forms is a non-trivial process and it strongly influences the performance of a text-to-speech (TTS) system [2]. The text analysis component of a TTS system is generally responsible for determining the document structure, the conversion of non-orthographic symbols and the parsing of the language structure and meaning. It indicates all the knowledge about the text and reveals out the message that is not specifically phonetic or prosodic in nature and encodes it in a format easy to use in speech synthesis. The core element of the text analysis component is the text parser which translates and disambiguates the human language and conveys the meaning among a potentially unlimited range of possibilities.

In this paper the focus is on the text analysis component of a TTS system tailored for the Romanian language. As we are on the early stages of the development of a TTS system for the Romanian language, our goals is to produce an efficient parser of the written text and proper data structures to allow the tuning of the core unit selection component of our TTS system.

For the Romanian language, Burileanu et al. [3] presents a text preprocessor based on *lex/flex* and *yacc/bison* lexer and parser generators, including facilities like conversion of anomalous symbol strings into orthographic characters and interpretation of certain punctuation marks. Buza et al. [4] presents another text parser focused toward syllable detection, using the same *lex* tool as the basic lexer. R. Sproat [5] indicates the weighted finite-state transducer approach as being applied also for Romanian, but without entering the implementation details of the Romanian language. We go beyond this previous research by enhancing our parser with further computational efficiency and better output representation to allow a unit selection approach for the core speech synthesis.

V. Matoušek and P. Mautner (Eds.): TSD 2009, LNAI 5729, pp. 323–330, 2009.

Our parser uses the power of the Boost Spirit LL parser [1] from the C++ Boost Framework[1] [6], targeting rather ambiguous EBNF grammars, completely embedded in the source C++ code. Further, we resolve more text ambiguities than Burileanu et al. [3].

This paper is organized as follows. Section 2 details the text analysis component, emphasizing the document structure detection and text normalization. Section 3 presents the performance of our text processing component and discuss some specific features. Section 4 concludes the paper.

2   The Text Processing Component

In this section we enter the details of the Text Processing component of our TTS system, depicted in figure 1. This component works directly on the raw text and produce a suitable internal representation. It is further structured in several smaller subcomponents: the Dictionary, containing the common words and exceptions; the Parsers, containing the parsing logic, taking as input the text and constructing the internal data structures representation of the text; the Data Structure, representing the storage of the processed text; and the Prosody component which analyzes the data structure and adds prosodic information accordingly.

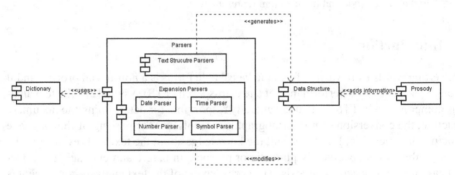

Fig. 1. The structure of the Text Preprocessing component

The Text Processing component intensively uses the Boost Spirit LL parser [1]. Boost Spirit is an object-oriented recursive descent parser implemented using template meta-programming techniques in C++. It allows the programmer to describe the grammar in way similar with EBNF [7], while still writing C++ code. We design EBNF-like grammar specifications (presented further in this section) for matching the input text. When such a specification matches different predicates (word, phrases, paragraphs etc.), different methods are activated for saving the matched structure and for building the internal tree structure representation of the text.

Building a big grammar specification for the Romanian language like Burileanu et al. [3] allows to process all the text in only one traversal, but with the cost of a high coupling, high maintenance and low extensibility of the code. We improve this by employing a two phase parsing described in subsections 2.1 and 2.2, which better structures the data representation. In the first phase, we identify the text structure and in the

---

[1] http://www.boost.org (consulted on 20 March 2009).

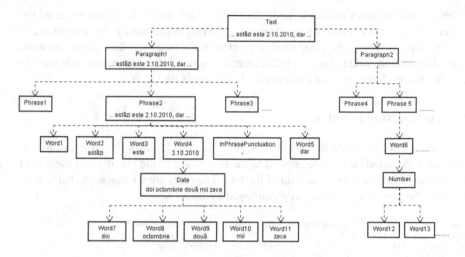

Fig. 2. The tree representation of a text

second, we enter the ambiguities details and normalize the text. Thus, for almost similar processing costs, we succeed to write independent grammar rules for text structure and for normalization, which overcome the drawback of high coupling inside the big grammar of the Romanian language.

We construct a tree representation of the text, facilitating ease traversal and other general functionalities. We design two sorts of nodes: *structure nodes* and *expansion nodes*. The structure nodes are used to model the input text in paragraphs, phrases, words, etc. The expansion nodes are used in text normalization, to replace new text for the original input: E.g. if the raw input text is "2 e un număr", number '2' is an expansion node of type Number for which we will create the replacement text "doi". Figure 2 represents the tree structure produced by the parsing, together with the tree representation of the following Romanian sequence: "... *astăzi este 2.10.2010, dar* ...". To construct this tree structure representation, we basically employ two types of parsers: for text segmentation and for expansion of words. The text segmentation parser identifies the text structure while the expansion parsers are used on the structure nodes to identify special cases like numbers, time, date, currency, etc. They also attach an expansion node to the structure node when necessary.

The structure nodes are the following ones: (i) Text: is the root node in our structure. It contains all the raw text that needs to be processed, being a container for this text. All other nodes contain only references to portions of the text held in this node. This allows us to optimize the data storage; (ii) Paragraph: identifies paragraphs in the text; (iii) Phrase: identifies a phrase in the text and holds properties like the type of the phrase (interrogative, affirmative etc.) and other related data; (iv) Word: represents a word in the text. A Word might be expanded if it can not be pronounced as it is i.e. needs normalization: for example the date, number or time words. It also stores the phonetic representation, and other intonation properties; (v) InPhrasePunctuation: represents the punctuation used inside a phrase like ", : ;". These nodes are used by the Prosody component to determine phrase breaks and other related prosody issues;

The expansion nodes includes: node `Number` which holds the literal value of the number after normalization; nodes `Date` and `Time`, representing the expansion of a date or a time respectively; node `Abbreviation` holding the whole expanded word for an abbreviation; node `SpecialCharacter` which is an expansion node used to translate special characters like currency \$, € or others like #, @.

2.1 Text Structure Detection

In this section we describe the EBNF-like rules used to identify the structure nodes. These rules, depicted below, are applied in the first phase of the parsing and the output is a non-expanded tree representation of the text. The rules are written such that to conform with the greedy style of parsing employed by Boost Spirit.

```
text ::= {WS | PAS}, {paragraph, PAS+, {WS | PAS}}, {WS | PAS}
paragraph ::= phrase { phrase }
phrase ::= (PHS, {PHS | SS}) | ((word | IPP+ | SC+),
{word | IPP+ | SC+ | WS+}, {PHS | SS} )
    word ::= ^( WS | PAS | PHS | IPP | SC) , { { ^( WS | SC ) },
^( WS | PAS | PHS | IPP | SC ) }
    WS ::= " " | "\t" | "\n"
    PAS ::= "\n"
    SS ::= "\t"
    IPP ::= "," | ";" | ":"
    SC ::= "~"|"@"|"#"|"$"|"&"|"%"|"^"|">"|"<"|"*"|"/"|"+"|"="|"\"|"""|"'"
    PHS ::="." | "!" | "?"
```

Basically the text is formed by zero or more paragraphs separated by at least one paragraph separator PAS. Paragraphs are among the easiest to detect, character \n (new line) being reliable clue for paragraph ends. When reading the text, the {WS | PAS} structure helps to jump over the spaces that can be inserted by mistake or intentionally at the beginning, end of the text or between paragraphs. The WS (word separators) and PAS (paragraph separators) are used this way for defining a less restrictive way of writing the input text.

A paragraph is composed by a succession of phrases. Phrase boundaries (PHS) are signaled by terminal punctuation from the set: {., !, ?} followed by white spaces.

A phrase is a set of words, IPP (in phrase punctuation), SC (special characters) and PHS (phrase separators). The phrase could start with a word, several IPP or SC, could contain one or more words, IPP and SC separated by one or more WS (word separators) and could end with zero or more PHS separated by SS (sign separators). If the phrase starts with phrase separators, we encounter the case in which the paragraph is formed only by phrase separators. Another exception is the one where the first phrase starts with some phrase separators. In this situation, an empty phrase is made containing only that phrase separator. Because the parsing is performed in a greedy descend manner, any other phrase separators that are in front of a phrase will be taken by the phrase separators predicate of the previous phrase. If more phrase separators are encountered at the end of the phrase, they will be all considered to depend on that phrase.

A word could contain any character except special characters and word separators but could not start or end with any of the characters relevant for the other predicates: WS, PAS, PHS, IPP, SC. Special characters (SP) like '@' are considered as individual non words that will be expanded in the second phase of the text analysis. In-phrase punctuation (IPP) are also considered independently of words and they will modify the previous and next word token properties. In this way, a web page (e.g.) "www.dlalex.com" will constitute one word and will allow a later processing module to identify it as a web page.

2.2 Text Normalization

In this subsection we describe how the text analysis component further expands and normalizes each identified expansion node. Our text normalization deals with abbreviations, acronyms, number formats, phone numbers, dates, times, money and currency, account numbers, ordinal numbers, cardinal numbers, domain specific tags like mathematical expressions and chemical formulas and miscellaneous formats. No ambiguous text remains unexpanded after the text normalization. Even if the text does not match any normalization rule, it is expanded up to its letter decomposition. We will detail only the abbreviation, the number formats and the date and time, as they are the most important text normalization issues. After text normalization, we further perform the phonetic transcription and syllable construction for those words which are not identified in the main dictionary. These later steps are not in the scope of this paper.

An abbreviation is marked by a final dot "." . When a "." is encountered by the text parser, the previous word is looked-up in the dictionary for an expansion. If the word is found to be an abbreviation, it is expanded accordingly with the dictionary and the phrase ended by the dot of the abbreviation is linked with the next phrase in the case there is no additional punctuation mark after.

Next, we depict the EBNF-like syntax for numbers, date and time.

```
number ::= numberWS | numberWNS
numberWNS ::= (digit, {digit}) | (digit, {digit}, DES, digit, {digit})
numberWS ::= (digit | digit, digit | digit, digit, digit, {DIS, digit,
digit, digit}) | (digit | digit, digit | digit, digit, digit, {DIS, digit,
digit, digit}, DES, digit {digit})
DES ::= ","
DIS ::= "."
date ::= day, DMYS, month, (DMYS, year) | month, MDYS, day,
(MDYS, year) | day, DS, month | month, DS, year
day ::= ("0"|"1"|"2"|"3"), digit   where day is less then 32 and greater then 0
month ::= (("0"| "1"), digit ) | writtenMonth  where month is less then 13 and greater then 0
year ::= (digit - "0", digit) | (digit - "0", digit, digit, digit)
DMYS ::= "."
MDYS ::= "/"
DS ::= DMYS | MDYS
time ::= (hour, TS, minute) | (hour, TS, minute, TS, second)
hour ::= ("0"|"1"|"2"), digit  where hour is less then 25
minute ::= ("0"|"1"|"2"|"3"|"4"|"5"), digit  where minute is less then 60
second ::= ("0"|"1"|"2"|"3"|"4"|"5"), digit  where second is less then 60
TS ::= ":"
```

Generic number formats or cardinal numbers are mainly used in simple counting or the statement of amounts. A complex number format can be divided in several cardinal number tokens and treated individually. This is usually the case for most complex number formats but not all; for example the month of a date is better to be expanded in the word specifying the month than in a number. Generic number formats in the Romanian language are fitting in two categories: number formats that use the digit separator '.' for grouping groups of tree digits (numberWS) and the ones that do not use this kind of

---

Algorithm 1. Expanding the generic numbers

---

1. Check number type
if number token n has DES separator **then**
    separate number token n in integer token i and decimal token d
    perform decimal expansion
else
    set number token n as integer token i
    perform integer expansion
end if
2. Decimal expansion
if decimal token d starts with digit 0 **then**
    perform decimal expansion digit by digit
else
    perform decimal expansion from right to left in groups of tree digits
end if
3. Integer expansion
if integer token i starts with digit 0 **then**
    perform integer expansion digit by digit
else if integer token i has DIS separator **then**
    apply integer expansion from right to left in groups of tree digits separated by DIS
else
    perform integer expansion from right to left in groups of tree digits
end if
4. Advance to next number and **GOTO** step 1.

---

separators (numberWNS). For both categories the numbers can be real numbers which means that they can have decimals. A decimal is separated from the rest of the number by the decimal separator ','. Algorithm 1 presents the verifications performed when expanding the generic numbers.

The rule for the date numbers includes two formats: the US form (month day year) and the Romanian form (day month year). We also treat the case when only the day and month are specified or only the month and year. The time format is recognized by the hour and minutes separated by the time separator (TS) and eventually followed by the seconds.

3 Evaluation

In this section we evaluate our Text Processing component against various inputs. We designed three test cases, covering spoken and scientific language. The first text case consists of full page excerpt from a Romanian literature book. The second test case is selected from a scientific paper from biology. To further stress the Text Processing component, the third test case represents the selected text from the biological scientific paper appended with several paragraphs full of terms to be disambiguated. Table 1 sumarizes the results. We counted how many paragraphs, phrases and words were correctly identified or disambiguated by our parsers. Table 1 depicts both the percentage and the succesfull hits.

Table 1. Performance evaluation of the `Text Processing` component

Test case	Paragraphs	Phrases	Words
1. Book	100%, (5/5)	90.56%, (48/53)	97.45%, (996/1022)
2. Scientific paper	100%, (10/10)	100%, (32/32),	99.30% (713/718)
3. Scientific paper plus extra	100%, (12/12)	100% (37/37)	96.85% (863/891)

We note that the parser gives a good performance in identifying the text structure. For specific excertps (test cases 2 and 3), paragraphs and phrases were fully identified. A worse performance is reported only when identifying phrases on the book excepts (spoken language). But, even in this case, words were identified with a pretty good accuracy. As properly identifying the words is our main target, because speech synthesis will be done mainly at the word level,we can conclude that our Text Processing component fulfill the requirements of a TTS system.

Further, we should note that using the Boost Spirit technology allowed us to write small pieces of grammar specification, easily embedded into the C++ source code. Thus, we opted for a higher degree of transparency and idenpendence between the source code and the grammar specification. As opposed to Burileanu et al. [3] which implemented a highly coupled parser, our approach allows us to quickly write new pieces of grammar specification and embed them into the system, when some exceptions are encountered during the TTS system usage.

4   Conclusion

In this paper we present the design of the text analysis component of a concatenative TTS system for the Romanian language. As our overall goal of the TTS system is a high speech accuracy through unit selection, we employed a two phases text analysis. First, we present the design of a parser that identifies the text structure. Next, we enter the details of the text normalization component that performs abbreviation disambiguation, number, date and time expansion and other normalization specific issues. Text processing is made efficient with the help of the Boost Spirit parser [1], targeting a tree-based data representation.

As a further work, we will describe the letter-to-sound conversion and the main unit selection component of our TTS system, emphasizing the use of intelligent tools to achieve a good speech accuracy.

Acknowledgement

iQuest TTS system is partially supported by the Romanian Authority for Scientific Research under contract INOVARE no. 186/2008.

References

1. de Guznam, J., Kaiser, H., Nuffer, D.: Spirit, http://spirit.sourceforge.net (consulted on March 18, 2009)
2. Huang, X., Acero, A., Hsiao-Wuen, H.: Spoken Language Processing: A Guide to Theory, Algorithms, and System Development. Prentice Hall, Englewood Cliffs (2001)

3. Burileanu, D., Dan, C., Sima, M., Burileanu, C.: A Parser-Based Text Preprocessor for Romanian Language TTS Synthesis. In: Proceedings of the 6th European Conference on Speech Communication and Technology (Eurospeech 1999), Budapest, Hungary, September 1999, vol. 5, pp. 2063–2066 (1999)
4. Buza, O., Toderean, G., Bodo, A.Z.: Syllable detection for romanian text-to-speech synthesis. In: Sixth International Conference on Communications COMM 2006, Bucharest, June 2006, pp. 135–138 (2006)
5. Sproat, R.: Multilingual text analysis for text-to-speech synthesis. Natural Language Engineering 2(4), 369–380 (1996)
6. Karlsson, B.: Beyond the C++ Standard Library: An Introduction to Boost. Addison-Wesley, Reading (2005)
7. ISO/IEC 14977: 1996(E), Extended BNF. ISO (1996)

Discriminative Training of Gender-Dependent Acoustic Models

Jan Vaněk, Josef V. Psutka, Jan Zelinka, Aleš Pražák, and Josef Psutka

Department of Cybernetics, University of West Bohemia in Pilsen, Czech Republic
{vanekyj,psutka_j,aprazak,zelinka,psutka}@kky.zcu.cz
http://www.kky.zcu.cz

Abstract. The main goal of this paper is to explore the methods of gender-dependent acoustic modeling that would take the possibly of imperfect function of a gender detector into consideration. Such methods will be beneficial in real-time recognition tasks (eg. real-time subtitling of meetings) when the automatic gender detection is delayed or incorrect. The goal is to minimize an impact to the correct function of the recognizer. The paper also describes a technique of unsupervised splitting of training data, which can improve gender-dependent acoustic models trained on the basis of manual markers (male/female). The idea of this approach is grounded on the fact that a significant amount of "masculine" female and "feminine" male voices occurring in training corpora and also on frequent errors in manual markers.

1 Introduction

The gender-dependent acoustic modeling is a very efficient way how to increase the accuracy in LVCSR systems. The training of acoustic models is usually based on manual markers connected with each utterance stored in a corpus. Such training of male/female acoustic models usually ignores diametrically different types of voices, e.g. "masculine" female and "feminine" male voices, whose occurrence in the corpus is not negligible. Also a problem with frequent errors in manual markers (male/female) connected with individual utterances is not solved. We proposed an unsupervised clustering algorithm which can reclassify training voices into more acoustically homogeneous classes. The clustering procedure starts from gender-dependent splitting and finishes in somewhat refined distribution which yields higher accuracy score. This approach is discussed in more detail in Section 2.1.

In the following part of the paper we discuss discriminative training (DT) of gender-dependent acoustic models. All the discussed methods come from frame-based discriminative training that seeks such solution (such acoustic models) which yield on one hand favorable quality (increased accuracy) of DT models, on the other hand these DT models should not be overly sensitive to imperfect function of a gender detector. A profit of this solution can be observed in real-time recognition tasks (e.g. real-time subtitling of meetings) when a reaction of the gender detector to the changes of speakers is not immediate or the detector evaluates changes incorrectly. The goal is to minimize an impact to the correct function of the recognizer. Let us mention that the Discriminative Training (DT) or Frame-Discriminative training (FD) are described in Section 2.2 and

V. Matoušek and P. Mautner (Eds.): TSD 2009, LNAI 5729, pp. 331–338, 2009.

incorporating DT to a gender-dependent training procedure is discussed in Section 4.4. Obtained results are presented in Section 5.

2 Methods

2.1 Automatic Clustering

Training of gender-dependent models is the most popular method how to split training data into two more acoustically homogeneous classes [1]. But for particular corpora, it should be verified that the gender-based clusters are the optimal way, i.e. the criterion $L = \prod_u P(u|M(u))$, where u is an utterance in a corpus and $M(u)$ is a relevant acoustic model of its reference transcription, is maximal. Because of some male/female "mishmash" voices contained in corpora we proposed an unsupervised clustering algorithm which can reclassify training voices into more acoustically homogeneous classes. The clustering procedure starts from gender-dependent splitting and finishes in somewhat refined distribution which yields higher accuracy score. [2].

The algorithm is based on similar criterion like the main training algorithm – maximize likelihood L of the training data with reference transcription and models. The result of the algorithm is a set of trained acoustic models and a set of lists where all utterances are assigned to exactly one cluster. Number of clusters (classes) n has to be set in advance and for gender-dependent modeling naturally $n = 2$. The process is a modification of the Expectation-Maximization (EM) algorithm. The unmodified EM algorithm is applied for estimation of acoustic model parameters. The clustering algorithm goes as follows:

1. Initial splitting of training utterances into n clusters. The clusters should have similar size. In case of two initial classes it is reasonable to start the algorithm from gender-based clusters/lists. In general case it should be a random splitting.
2. Train (retrain) acoustic models for all clusters.
3. Posterior probability density $P(u|M)$ of each utterance u with its reference transcription is computed for all models M (so-called forced-alignment).
4. Each utterance is assorted to the cluster with the maximal evaluation $P(u|M)$ computed in the previous step:

$$M_{t+1}(u) = \arg\max_M P(u|M). \tag{1}$$

5. If clusters changed then go back to step 2. Otherwise the algorithm is terminated.

Optimality of results of the clustering algorithm is not guaranteed. Besides, the algorithm depends on initial clustering. Furthermore, even convergence of the algorithm is not guaranteed, because there can be a few utterances which are reassigned all the time. Therefore, it is suitable to apply a little threshold as a final stopping condition or to use a fixed number of iterations. Thus, if we want to verify that the gender-dependent splitting is "optimal" we use this initial male/female distribution and start the algorithm. The intention is that the algorithm finishes with more refined clusters, in which "masculine" female and "feminine" male voices and also errors in manual male/female annotations will be reclassified which should ensure better performance of a recognizer.

2.2 Discriminative Training

Discriminative training (DT) was developed in a recent decade and provides better recognition results than classical training based on Maximum Likelihood criterion (ML) [3,4,5,6]. In principle, ML based training is a machine learning method from positive examples only. DT on the contrary uses both positive and negative examples in learning and can be based on various objective functions, e.g. Maximum Mutual Information (MMI) [7], Minimum Classification Error (MCE) [5], Minimum Word/Phone Error (MWE/MPE) [3]. Most of them require generation of lattices or many-hypotheses recognition run with appropriate language model. The lattices generation is highly time consuming. Furthermore, these methods require good correspondence between training and testing dictionary and language model. If the correspondence is weak, e.g. there are many words which are only in the test dictionary then the results of these methods are not good. In this case, we can employ Frame-Discriminative training (FD), which is independent on a used dictionary and language model [8]. In addition, this approach is much faster.

2.3 Frame-Discriminative Training

In lattice based method with MMI objective function the training algorithm seeks to maximize the posterior probability of the correct utterance given the used models [7]:

$$\mathcal{F}_{MMI}(\lambda) = \sum_{r=1}^{R} \log \frac{P_\lambda(O_r|s_r)^\kappa P(s_r)^\kappa}{\sum_S P_\lambda(O_r|s)^\kappa P(s)^\kappa}, \tag{2}$$

where λ represents the acoustic model parameters, O_r is the training utterance feature set, s_r is the correct transcription for the r'th utterance, κ is the acoustic scale which is used to amply confusions and herewith increases the test-set performance. $P(s)$ is a language model part.

Optimization of the MMI objective function uses Extended Baum-Welch update equations and it requires two sets of statistics. The first set, corresponding to the numerator (num) of the equation (2), is the correct transcription. The second one corresponds to the denominator (den) and it is a recognition/lattice model containing all possible words. An accumulation of statistics is done by forward-backward algorithm on reference transcriptions (numerator) as well as generated lattices (denominator). The Gaussian means and variances are updated as follows [8]:

$$\hat{\mu}_{jm} = \frac{\Theta_{jm}^{num}(O) - \Theta_{jm}^{den}(O) + D_{jm}\mu'_{jm}}{\gamma_{jm}^{num} - \gamma_{jm}^{den} + D_{jm}} \tag{3}$$

$$\hat{\sigma}_{jm}^2 = \frac{\Theta_{jm}^{num}(O^2) - \Theta_{jm}^{den}(O^2) + D_{jm}(\sigma'^2_{jm} + \mu'^2_{jm})}{\gamma_{jm}^{num} - \gamma_{jm}^{den} + D_{jm}} - \mu_{jm}^2, \tag{4}$$

where j and m are the HMM-state and Gaussian index, respectively, γ_{jm} is the accumulated occupancy of the Gaussian, $\Theta_{jm}(O)$ and $\Theta_{jm}(O^2)$ are a posteriori probability weighted by the first and the second order accumulated statistics, respectively. The Gaussian-specific stabilization constants D_{jm} are set to maximum of (i) double of the smallest value which ensures positive estimated variances, and (ii) value $E\gamma_{jm}^{den}$, where constant E determines the stability/learning-rate and it is a compromise between stability and number of iteration which is needed for well-trained models [9]. In Frame-Discriminative case, the denominator lattices generation and its forward-backward processing is not needed. The denominator posterior probability is calculated from a set of all states in HMM. This very general denominator model leads to good generalization to test data. Furthermore, statistics of only a few major Gaussians are need to be updated and their probability has to be exactly calculated in each time. It can lead to very time-efficient algorithm [10]. Optimization of the model parameters uses the same two equations (3) and (4), the computation of $\Theta_{jm}^{den}(O)$ and γ_{jm}^{den} is modified only.

2.4   Frame-Discriminative Adaptation

In case that only limited data are available, maximum a posteriori probability method (MAP) [11] can be used even for discriminative training [12]. It works in the same manner as the standard MAP, only the input HMM has to be discriminatively trained with the same objective function. For discriminative adaptation it is strongly recommended to use I-smoothing method to boost stability of new estimates [13].

3   Train Data Description

For training of acoustic models a microphone-based high-quality speech corpus was used. The high-quality speech corpus of read-speech consists of the speech of 800 speakers (384 males and 416 females). Each speaker read 170 sentences. The database of text prompts from which the sentences were selected was obtained in an electronic form from the web pages of Czech newspaper publishers[14]. Special consideration was given to the sentences selection, since they provide a representative distribution of the more frequent triphone sequences (reflecting their relative occurrences in natural speech). The corpus was recorded in the office where only the speaker was present. Sentences were recorded by a close-talking microphone (Sennheisser HMD410-6). The recording sessions yielded totally about 220 hours of speech.

4   Experimental Setup

4.1   Acoustic Processing

The digitization of an analogue signal is provided at 22.05 kHz sample rate and 16-bit resolution format. The aim of the front-end processor is to convert continuous speech into a sequence of feature vectors. Several tests were performed in order to determine the best parameterization settings of the acoustic data (see [15] for methodology). The best results were achieved using PLP parameterization [16] with 27 filters and 12 PLP cepstral coefficients with both delta and delta-delta sub-features (see [17] for details). Therefore one feature vector contains 36 coefficients. Feature vectors are computed each 10 milliseconds (100 frames per second).

4.2 Acoustic Model

The individual basic speech unit in all our experiments was represented by a three-state HMM with a continuous output probability density function assigned to each state. As the number of Czech triphones is too large, phonetic decision trees were used to tie states of Czech triphones. Several experiments were performed to determine the best recognition results according to the number of clustered states and also to the number of mixtures. In all presented experiments, we used 16 mixtures of multivariate Gaussians for each of 4922 states. The baseline acoustic model was made using HTK-Toolkit v.3.4 [18].

4.3 Two Class Splitting

As was presented above, the whole training corpus was split into two acoustically homogeneous classes (gender-based). Initial splitting was achieved via manual markers. However, due to several "masculine" female and "feminine" male voices occurring in the training corpora and also because of possible errors in manual annotations we applied algorithm introduced in Subsection 2.1 to refine initial "gender-based" training subcorpora. The whole set of sentences (109.5k) was split into male (52.4k) and female (57.1k) parts based on manually assigned markers. The percentage of sentences which were moved from the male to female ($M_{i-1} \rightarrow F_i$) cluster as well as from the female to male ($F_{i-1} \rightarrow M_i$) cluster in two following iteration steps $(i, i-1)$ is given in Table 1.

Table 1. The shift between male and female clusters

Iteration	[%]			
step (i)	$M_{i-1} \rightarrow M_i$	$M_{i-1} \rightarrow F_i$	$F_{i-1} \rightarrow F_i$	$F_{i-1} \rightarrow M_i$
1	96.81	4.81	95.76	2.63
2	99.37	0.90	99.21	0.52
3	99.34	0.37	99.93	0.36
4	99.75	0.25	99.69	0.31
5	99.65	0.25	99.78	0.32
6	99.90	0.06	99.94	0.10
7	99.97	0.01	99.99	0.03

4.4 Discriminative Training of Two-Class Models

Our next attention was to explore a suitable way of discriminative training of gender-dependent acoustic models which would yield on one hand favorable characteristics of DT models but on the other hand developed models should not be overly sensitive to imperfect function of a gender detector, e.g. a negative impact of reversely selected (male/female) acoustic model. Such situation could happen for instance in real-time recognition tasks in case that the reaction of a gender detector to the change of speaker is not immediate and/or the detector evaluates the change incorrectly. We performed a set of experiments in which an impact of speaker independent and gender-dependent acoustic models were tested in combination with a technique of frame-based discriminative training. In case when only single acoustic model is trained, the situation is simple. The model is trained from all data under ML approach or some DT objective function. Nevertheless some parameters could be tuned, for example a number of tied-states and a number of Gaussians per state. In DT case, the number of tuned parameters is higher but it is still an optimization task. In our experiments corresponding models are marked as SI (Speaker Independent) and SI_DT for ML and DT, respectively. The DT model was developed from SI via two training iterations based on FD-MMI objective function. The E constant was set to one. Furthermore, the I-smoothing was applied and smoothing constant τ^I was set to 100. If the training data is split into more than one class, the situation is a bit complicated because of more training strategies that we have in our disposal. Naturally the same training procedure can be used for each part of data. This is concluded by a set of independent models. For a real application this approach is not a good option because final models have different topology which is generated during tied-states clustering and therefore obtained models cannot be simply switched/replaced in the recognizer. The better strategy is to split the training procedure just after state clustering. In our experiments such model sets are marked as $ClusterGD$ and $ClusterGD\_DT$ for ML and DT, respectively. Secondly, the ML or DT adaptation can be applied. In our experiments the adaptation starts from SI or SI_DT and two iterations were done via MAP or DT-MAP with parameter τ equal to 25. Two models developed by these techniques are marked as $SI\_MLAdapt$ and $SI\_DT Adapt$.

4.5 Tests Description

The test set consists of 100 minutes of speech from 10 male and 10 female speakers (5 minutes from each) which were not included in training data. In all recognition experiments a language model based on zerograms was applied in order to judge a quality of developed acoustic models. In all experiments the perplexity of the task was 2190, there were no OOV words.

5 Results

As can be seen from Table 2 we achieved a significant gain in terms of recognition results for all gender-dependent acoustic models ($ClusterGD$, $ClusterGD\_DT$, $SI\_MLAdapt$ and $SI\_DT Adapt$) against speaker independent acoustic models (SI and SI_DT). Moreover the automatic clustering procedure decreases the word error rate more than 1.7% relatively, see the row GD with recognition results for manually marked training data and $ClusterGD$ with recognition results after the automatic re-clustering procedure of training data was performed. This gain is more than 11% relatively for $ClusterGD\_DT$ (discriminative re-training of gender dependent ML model) when the information on speaker gender is correct. But on the other hand the recognition results are considerably worse when the speaker gender information is not correct. From this point of view the best tradeoff between recognition results of gender-dependent acoustic model with correct and non-correct gender information is $SI\_DT Adapt$ ($SI\_DT Adapt$ is SI_DT after two iterations via DT-MAP). In this case the recognition results are slightly worse (improvement 8% relatively to SI) than in case of $ClusterGD\_DT$ but the non-correct gender information decreases the recognition results only slightly comparing with the original SI acoustic model.

Table 2. The results of recognition experiments

	WER [%]	
SI	40.19	
SI_DT	39.02	
Gender identification	correct	non correct
GD	37.50	64.08
ClusterGD	36.89	63.57
ClusterGD_DT	35.81	61.92
SI_MLAdapt	38.08	52.18
SI_DTAdapt	36.99	46.60

6 Conclusion

The goal of our work was to build the gender-dependent acoustic model which is more robust to the incorrect decisions of gender detector. We tried several methods based on combination of gender-based data and discriminative training procedures. In all experiments a zero-gram language model was applied in order to better judge the quality of developed acoustic model. The best gender-dependent training procedure depends on the performance of gender detection. If the gender detector works perfectly the GD_DT model is the best solution. But if the gender detector works incorrectly, e.g. a change of speaker is not detected in time or is evaluated sometimes wrongly then $SI\_DT\,Adapt$ acoustic model seems to be a good trade off.

Acknowledgement

This research was supported by the Grant Agency of Academy of Sciences of the Czech Republic, project No. 1QS101470516.

References

1. Stolcke, A., Bratt, H., Butzberger, J., Franco, H., Gadde, V.R., Rao, P.M., Rickey, C., Shriberg, E., Sonmez, K., Weng, F., Zheng, J.: The SRI Hub-5 Conversational Speech Transcription System. In: Proc. NIST Speech Transcription Workshop, College Park, MD (March 2000)
2. Zelinka, J.: Audio-visual speech recognition. PhD. thesis, University of West Bohemia, Department of Cybernetics (2009) (in Czech)
3. Povey, D.: Discriminative Training for Large Vocabulary Speech Recognition. Ph.D. thesis, Cambridge University, Department of Engineering (2003)
4. Yu, D., Deng, L., He, X., Acero, A.: Use of incrementally regulated discriminative margins in MCE training for speech recognition. In: Proc. Interspeech 2006 (2006)
5. McDermott, E., Hazen, T., Roux, J.L., Nakamura, A., Katagiri, S.: Discriminative training for large vocabulary speech recognition using minimum classification error. IEEE Trans. Speech and Audio Proc. 14(2) (2006)
6. Reichl, W., Ruske, G.: Discriminative Training for Continuous Speech Recognition. In: Proc. 1995 Europ. Conf. on Speech Communication and Technology, Madrid, September 1995, vol. 1, pp. 537–540 (1995)

7. Bahl, L.R., Brown, P.F., de Souza, P.V., Mercer, L.R.: Maximum Mutual Information Estimation of Hidden Markov Model Parameters for Speech Recognition. In: ICASSP (1986)
8. Kapadia, S.: Discriminative Training of Hidden Markov Models. Ph.D. thesis, Cambridge University, Department of Engineering (1998)
9. Povey, D., Woodland, P.C.: Improved discriminative training techniques for large vocabulary continuous speech recognition. In: IEEE international Conference on Acoustics Speech and Signal Processing, Salt Lake City, Utah, May 7-11 (2001)
10. Povey, D., Woodland, P.C.: Frame discrimination training for HMMs for large vocabulary speechrecognition. In: Proceedings of the ICASSP, Phoenix, USA (1999)
11. Gauvain, L., Lee, C.H.: Maximum A-Posteriori Estimation for Multivariate Gaussian Mixture Observations of Markov Chains. In: IEEE Transactions SAP (1994)
12. Povey, D., Gales, M.J.F., Kim, D.Y., Woodland, P.C.: MMI-MAP and MPE-MAP for acoustic model adaptation. In: EUROSPEECH, pp. 1981–1984 (2003)
13. Povey, D., Woodland, P.: Minimum phone error and I-smoothing for improved discriminative training. In: Proceedings of the ICASSP, Orlando, USA (2002)
14. Radová, V., Psutka, J.: UWB-S01 Corpus: A Czech Read-Speech Corpus. In: Proceedings of the 6th International Conference on Spoken Language Processing ICSLP2000, Beijing, China (2000)
15. Psutka, J., Müller, L., Psutka, J.V.: Comparison of MFCC and PLP Parameterization in the Speaker Independent Continuous Speech Recognition Task. In: 7th European Conference on Speech Communication and Technology (EUROSPEECH 2001), Aalborg, Denmark (2001)
16. Hermansky, H.: Perceptual linear predictive (PLP) analysis of speech. J. Acoustic. Soc. Am. 87 (1990)
17. Psutka, J.: Robust PLP-Based Parameterization for ASR Systems. In: SPECOM 2007 Proceedings. Moscow State Linguistic University, Moscow (2007)
18. Young, s., et al.: The HTK Book (for HTK Version 3.4), Cambridge (2006)
19. Stolcke, A.: SRILM - An Extensible Language Modeling Toolkit. In: International Conference on Spoken Language Processing (ICSLP 2002), Denver, USA (2002)

Design of the Test Stimuli for the Evaluation of Concatenation Cost Functions*

Milan Legát and Jindřich Matoušek

University of West Bohemia in Pilsen, Faculty of Applied Sciences
Department of Cybernetics, Univerzitní 8, 306 14, Plzeň, Czech Republic
{legatm,jmatouse}@kky.zcu.cz

Abstract. A large number of methods for measuring of audible discontinuities, which occur at concatenation points in synthesized speech, have been proposed in recent years. However, none of them proved to be comparatively better than others across all languages and recording conditions and the presented results have sometimes even been in contradiction. What is more, none of the tested concatenation cost functions seem to be reliably reflecting the human perception of such discontinuities. Thus, the design of the concatenation cost functions is still an open issue, and there is a lot of work remaining to be done. In this paper, we deal with the problem of preparing the test stimuli for evaluating the performance of these functions, which is, in our opinion, one of the key aspects in this field.

1 Introduction

Unit selection based concatenative speech synthesis currently represents an approach that, without question, produces synthetic speech of the highest naturalness. The idea of this method is to have more than one instance of each unit stored in a large speech database and to search at runtime for the best sequence of units to generate the desired utterance. Ideally, no smoothing is required, which results in the high naturalness and intelligibility of the synthesized speech. In order to select the best sequence of units, two cost functions are calculated – *target cost* and *concatenation (join) cost* [1]. The task of the target cost function is to estimate the perceptual difference between the target and the candidate unit, and the concatenation cost function should reflect the level of the perceived discontinuity between two consecutive units. While the problem of the designing of the target cost functions has been more or less solved, many concatenation cost functions have been tested in the last decade with results often being in contradiction [2], [3].

The concatenation cost consists mostly of a set of sub–components associated with the difference in pitch, energy and spectra of adjacent segments of the concatenated units. The weak point of the concatenation cost functions is the spectral component as no objective measure seems to correlate well with human perception of discontinuities in spectra. Many spectral parametrizations in combination with various distance measures have been tested, including mel–frequency cepstral coefficients (MFCCs),

---

* This research was supported by the Ministry of Education of the Czech Republic, project No. 2C06020 and the Grant Agency of the Czech Republic, project No. GACR 102/09/0989.

V. Matoušek and P. Mautner (Eds.): TSD 2009, LNAI 5729, pp. 339–346, 2009.

linear prediction coefficients, line-spectrum frequencies, bispectrum [2], Wigner–Ville distribution–based cepstrum [5], FFT–based cepstra, perceptual linear prediction coefficients, multiple–centroid analysis coefficients [6] or formant frequencies, in combination with the Euclidean, Mahalanobis or symmetrical Kullback–Leibler distances, to name but a few. Besides these traditional approaches, some other techniques have been proposed, e.g. the application of the Latent Semantic Mapping [7], Kalman–filters [6], or Auditory Modelling [8].

Generally, there are two ways of evaluating the concatenation cost functions. The first one is to have a set of concatenation cost functions, synthesize the same sentences using each of them separately, and then ask listeners to choose the best version or to compare the synthesized versions with the natural forms of the same sentences. The other and more preferred option is to simply concatenate some units, let the listeners assess the quality of the concatenation points and then calculate the correlation between the values obtained by discontinuity measure and listeners' scores. Since the listeners' responses may vary, the Mean Opinion Score (MOS) is typically used as reference.

The crucial point for the latter approach is to have appropriate test stimuli for the listening tests and to collect the reliable results based on the listeners' answers. One of the issues related to the designing of the test stimuli for the evaluation of the naturalness of the synthetic speech is the length of the stimuli presented to listeners. It can vary from isolated phonemes [9] to whole sentences [5]. Shorter stimuli (e.g. isolated phonemes) could seem to be appropriate for evaluation of the concatenation cost functions as they can be synthesized containing only one concatenation point, which is important for avoiding the influence of other joins on listeners' assessments. On the other hand, if the test stimuli are too short, it is difficult for the listeners to perceive discontinuities [9]. Thus, monosyllabic words containing only one concatenation point in the middle seem to be an optimal choice.

However, there is another point of view. The general task of TTS is to synthesize sentences or longer texts (in most cases). If we use isolated words for evaluation of join cost functions, we can expect lower variability of the quality of the concatenated sounds than in real cases. In addition, in isolated words the discontinuities may be masked or difficult to perceive, compared to whole sentences. What we mean is that if the listener is presented with a whole sentence which is absolutely natural except one concatenation point, a possible discontinuity may be more salient than in the case of being presented in a short monosyllabic word.

In this work, we propose a framework for the design of the test stimuli for the listening tests for evaluating the concatenation cost functions. Note that the initial focus of our experiments is on vowels as they are the sounds which can be characterized as highly energetic and having rich spectral content.

2   Preliminary Listening Test Preparation

We have recorded a large database of short Czech sentences containing three words each, e.g. /kra:lofski: **kat** konal/ (SAMPA notation), where the middle word is of special interest as it is a mono–syllabic word containing a vowel in the middle surrounded by consonants, i.e. the word in the form CVC. All five short Czech vowels were

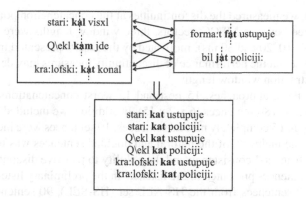

Fig. 1. Construction of the set of sentences containing only one concatenation point (SAMPA notation). The sentences were synthesized combining initial and final halves of the recorded sentences so that they contain one of the five design words in the middle (kat in this case).

taken into consideration. The middle mono–syllabic words contain all possible combinations of initial and final consonants. The sentences were uttered by both female and male speaker. This database is planned to be used for future concatenation costs evaluation and design, but first we need to find a reasonable way to exploit this database for that purpose.

We have divided all the sentences in the middle, i.e. at mid–vowels of the central mono–syllabic words. Then, the left and right halves of the sentences were concatenated in order to obtain sentences which are completely natural except one concatenation point in the middle (see Fig. 1). Note that no smoothing techniques were applied, and the concatenation was performed at the middle pitch mark position; this set is henceforth referred to as HS-ALL. The objective of synthesizing the HS-ALL set in this way was to present listeners with stimuli long enough for consistent assessment of the audible discontinuities while containing only one concatenation point at the same time.

In addition to the HS-ALL set, we have also included into the test stimuli for the preliminary listening test a set of sentences (henceforth referred to as DI-ALL), in which the middle words were synthesized using diphones taken from the diphone inventory created from the whole set of the recorded sentences. Thus, the synthesized sentences were natural except three concatenation points in the middle word. The inclusion of these sentences was motivated by questioning whether the range of possible discontinuities is not too narrowed by taking only the middle words into account. However, these sentences were handled with special care in the evaluation procedures because there were two other concatenation points present in the surrounding of the middle concatenation point, which was of interest. To limit the effect of surrounding joins on listeners' scores, the spectrograms and also waveforms at the surrounding concatenation points were checked, and the sentences containing some visible discontinuities at these points were removed.

Since the number of the sentences in both HS-ALL and DI-ALL sets was very large, there was a need to make a limited selection for the preliminary listening test. For that purpose, two approaches were used. First, a discontinuity metric proposed by Belle-garda [7] was implemented, using three extraction window lengths ($K = 3, 4, 5$ pitch periods). Second, we measured the discontinuities at the concatenation points using the Euclidean distance between MFCC vectors, three window lengths were used for the feature extraction (10, 20 and 30 ms), motivated by the results presented in [10], where the authors suggest that the performance of discontinuity measures may depend heavily on the feature extraction window length.

With each of these approaches, 15 best and 15 worst concatenations were found resulting in the set of 180 sentences (6 x 2 x 15). In addition, we included 15 sentences chosen randomly and 5 completely natural sentences, 10 sentences were included twice. The objective of the inclusion of the natural and doubled sentences was to have a tool for checking the listeners' consistency and their ability to perceive discontinuities. The total number of sentences presented to the listeners in the preliminary listening test was 210, including 90 sentences from the HS-ALL set (HS-SEL), 90 sentences from the DI-ALL set (DI-SEL), 15 sentences chosen randomly (mixture of both DI and HS), 5 natural sentences and 10 sentences included twice.

The task of the listeners was to assess the concatenations in the middle vowel of the central word of each sentence on both the five point scale (*no join at all* – 1, *unnatural but not disturbing* – 2, *slightly perceived join* – 3, *highly perceived join* – 4, and *highly disturbing join* – 5), as well as the binary scale (*perceived join* or *not perceived join*). To make the task even easier for listeners, the natural versions of the middle words were also played to them prior to the whole sentences. The overall number of participants in this preliminary listening test was 20.

3   Evaluation of the Preliminary Listening Test

3.1   Checking the Listeners' Consistency and Reliability

After collecting all the answers, the evaluation of the listeners was performed. Firstly, the ability of listeners to identify the natural sentences was estimated. The listeners, who assessed the natural sentences as containing audible joins, were given one minus point for each such decision. Based on this penalization, one listener was excluded. Secondly, the consistency of the listeners comparing their answers for the sentences included twice in the listening test was evaluated. Again, for each inconsistent decision on the binary scale the listener was given one minus point. All the listeners were found to be consistent using this measure as one mistake was allowed.

We also assessed the consistency of the listeners' answers on the five-point scale. See the penalization scheme in Tab. 3.1. If a listener, for instance, assessed the same sentence using *unnatural but not disturbing* and *highly disturbing join*, the penalization was -0.1 point. Based on this consistency measure, two listeners were excluded obtaining a score of -0.201 and -0.140, respectively. The average score of the remaining listeners was -0.008.

The answers of the 17 remaining listeners were used to calculate the Mean Opinion Score (MOS). We found the correlation between listeners' scores and MOS, and two listeners were excluded at this point as their answers correlated poorly with MOS (0.24, 0.51 respectively). The average correlation between the scores of the remaining 15 listeners and MOS was 0.76, resulting in a reliable set of listeners' scores.

Table 1. Penalization scheme based on the listeners' answers on the five–point scale. "Diff" stands for the difference in a listener's scores given to the same sentence.

Diff	Penalty
0	0
1	-0.001
2	-0.01
3	-0.1
4	-1

The next step of our evaluation procedure was the formulation of "facts". By "fact" we mean a sentence which was assessed by 80% listeners in the same way on the binary scale, either as containing an audible join or as being completely natural. In the HS-SEL set we found 39 "facts" (14 continuous and 25 discontinuous), which is about 42% of HS-SEL sentences. In the DI-SEL, 9 continuous "facts" were found. Before any discontinuous "facts" can be formulated, we need to be sure that any of the perceived discontinuities were not due to poor quality of the surrounding joins.

3.2   Checking the Surrounding Joins in DI-SEL Set

As mentioned above, the sentences in the DI-SEL set were evaluated in a special way, despite being checked visually before inclusion into the test stimuli, as the listeners' scores might have been affected by the presence of some discontinuities at surrounding joins. In order to analyze these unwanted joins, which are in fact introducing some noise into the results our listening test, we have calculated the energy and pitch differences at all concatenation points.

Since human loudness perception does not scale linearly with the intensity of the signal, the energy mismatches at the concatenation points were measured on the logarithmic scale (dB). The same applies to the measuring of differences in F_0, for which Mel scale was used. We found the maximum acceptable differences in both energy and pitch at all concatenation points in sentences which were found to be continuous "facts". These values were then used as thresholds for excluding sentences from the DI-SEL set based on checking the surrounding joins. Having these joins checked in this way, the discontinuous "facts" can also be found in the DI-SEL set. In our case, 17 discontinuous "facts" in DI-SEL set were found. The total number of the "facts" obtained from our listening test results was 65 (23 continuous and 42 discontinuous).

3.3   Energy and Pitch Differences at Concatenation Points

Since the objective of our work is to measure spectral discontinuities, we need to analyze the differences in pitch and also energy at these points as these are unquestionably the sources of audible discontinuities. Note that in related works such sources of discontinuities are very often eliminated by $F0$ and energy smoothing. Nevertheless, in our opinion, any signal modification can introduce some audible artefacts, which may also affect the listeners' discontinuity perception and that is why the unmodified concatenations are preferable for the design of the concatenation cost functions, especially building the test stimuli for the listening tests.

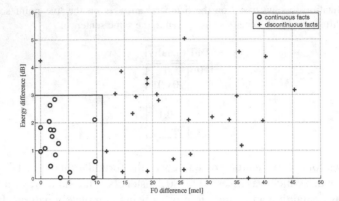

Fig. 2. Distribution of the continuous and discontinuous "facts" in the energy difference vs. $F0$ difference plain. The continuous "facts" can easily be separated from the discontinuous "facts".

In Fig. 2 the result of the analysis of the $F0$ and energy differences at concatenation points present in the "fact" sentences is shown. Surprisingly, there were no sentences among the discontinuous "facts" where the discontinuity in spectrum could be considered to be the only source of the perceived discontinuity. In fact, it was possible to classify all the sentences into continuous vs. discontinuous classes using only pitch and energy differences at the concatenation points as predictors. Obviously, there were some cases where some considerable spectral mismatches at concatenation points were observed. However, in all such cases this spectral mismatch was accompanied by either pitch or energy mismatch. Thus, there would be no need to measure spectral discontinuity to concatenate well at these points as the difference in $F0$ and energy would be sufficient components of the concatenation cost function.

4   Suggestions for Listening Test Stimuli Design

In this section, we summarize our observations related to the building of the listening test stimuli for the evaluation and design of concatenation cost functions.

One of the first questions we wanted to address was whether the concatenation points present in the synthesized sentences created from the halves of sentences as described in Sec. 2 contain enough discontinuities. Based on the obtained results, we could assume that this approach is a reasonable way as the number of perceived discontinuities follows the same trend as for the sentences in which diphones were used. The advantage of the synthesizing by halves is that we have stimuli long enough for reliable assessment by listeners, and containing only one concatenation point at the same time.

We also assume that the listeners were able to give consistent scores as the average correlation between the listeners' scores and MOS was 0.71. On the other hand, 42% "facts" do not seem to be enough, and this number needs to be taken into consideration for the design of the future listening test to obtain enough "facts" for the evaluation of the concatenation cost functions.

For the design of the larger listening test containing all the Czech vowels and utterances spoken by both of the speakers, the differences in pitch and energy at concatenation points need to be considered before the sentences are included into the listening test stimuli in order to find sentences where the spectral mismatch could be found as the only source of perceived discontinuity, which is crucial for the design of the spectral component of the concatenation costs.

It is also worth performing an analysis of listeners' scores as described in Sec. 3.1 in order to measure their ability to distinguish between continuous and discontinuous concatenation points, as well as the consistency of their answers.

5 Conclusion and Future Work

In this paper, we have addressed one of the key issues of evaluating the concatenation cost functions for the unit selection speech synthesis, the building of the listening test stimuli. To answer some questions related to this field, we have performed a preliminary listening test, the objective of which was to show that synthesizing sentences by halves could be one way of approaching the problem of "length vs. concatenation" of the test stimuli building. By the "length vs. concatenation" problem we mean that if the listening test stimuli are too short, it is difficult for the listeners to reliably score the concatenations, and having longer test stimuli requires some additional concatenation points, which might also affect the listeners' judgements.

We found that it is beneficial to check the listeners' for their ability to distinguish between smooth and discontinuous joins, and for the consistency of their answers. For this purpose, we proposed in this paper a simple method based on the inclusion of some natural and some doubled sentences into the test stimuli.

Special attention needs to be paid to the pitch and energy differences as the sources of perceived discontinuities. In our opinion, smoothing methods are not a reliable way of dealing with this problem as they may introduce some audible artefacts into the concatenation area, which might affect the listeners' scores.

In our preliminary listening test we have not observed any sentences where the spectral mismatch could be concluded to be the only source of perceived discontinuity as all discontinuous sentences contained either pitch or energy difference at the concatenation points. This particular issue is planned to be addressed in future listening tests containing all the short Czech vowels. For the evaluation and design of the spectral component of the concatenation cost functions, we need to find a set of sentences which are scored by the listeners as discontinuous and, at the same time, do not contain considerable pitch and energy discontinuities at the concatenation points.

References

1. Hunt, A., Black, A.: Unit selection in a concatenative speech synthesis system using a large speech database. In: ICASSP 1996, vol. 1, pp. 373–376 (1996)
2. Pantazis, Y., Stylianou, Y.: On the detection of discontinuities in concatenative speech synthesis. In: Stylianou, Y., Faundez-Zanuy, M., Esposito, A. (eds.) COST 277. LNCS, vol. 4391, pp. 89–100. Springer, Heidelberg (2007)

3. Vepa, J., King, S.: Join cost for unit selection speech synthesis. In: Alwan, A., Narayanan, S. (eds.) Speech Synthesis. Prentice Hall, Englewood Cliffs (2004)
4. Kawai, H., Tsuzaki, M.: Acoustic measures vs. phonetic features as predictors of audible discontinuity in concatenative speech synthesis. In: ICSLP 2002, pp. 2621–2624 (2002)
5. Chen, J., Campbell, N.: Objective distance measures for assessing concatenative speech synthesis. In: EUROSPEECH 1999, pp. 611–614 (1999)
6. Vepa, J.: Join cost for unit selection speech synthesis. PhD Thesis, University of Edinburgh (2004)
7. Bellegarda, J.R.: A novel discontinuity metric for unit selection text–to–speech synthesis. In: EUROSPEECH 1999, pp. 611–614 (1999)
8. Tsuzaki, M.: Feature extraction by auditory modelling for unit selection in concatenative speech synthesis. In: EUROSPEECH 2001, pp. 2223–2226 (2001)
9. Klabbers, E., Veldhuis, R.: Reducing audible spectral discontinuities. IEEE Transactions on Speech and Audio Processing 9, 39–51 (2001)
10. Kirkpatrick, B., O'Brien, D., Scaife, R.: Feature extraction for spectral continuity measures in concatenative speech synthesis. In: INTERSPEECH 2006 (2006)

Towards an Intelligent User Interface: Strategies of Giving and Receiving Phone Numbers

Tiit Hennoste, Olga Gerassimenko, Riina Kasterpalu, Mare Koit,
Andriela Rääbis, and Krista Strandson

University of Tartu, J. Liivi 2, 50409 Tartu, Estonia
{tiit.hennoste,riina.kasterpalu,mare.koit,
andriela.raabis,krista.strandson}@ut.ee
http://www.cl.ut.ee

Abstract. Strategies of giving and receiving phone numbers in Estonian institutional calls are considered with the further aim to develop a telephone-based user interface to data bases which enables interaction in natural Estonian language. The analysis is based on the Estonian dialogue corpus. Human operators give long phone numbers in several parts, making pauses after parts. Pitch contour works as a signal of continuation or finishing the process. Clients give feedback during the process (repetition, particles, and pauses). Special strategies are used by clients to finish receiving the number as well as to initiate repairs in the case of communication problems.

1 Introduction

Many telephone services are intermediated by spoken dialogue systems (DS) which include speech recognition, dialogue management, text-to-speech synthesis. Users are usually able to cope with limitations of a system [1]. That corresponds to the computer talk hypothesis which states that people behave differently when interacting with the computer as compared to communication between humans [2]. Still, the quality of such systems is improving quickly and users can expect that the computer as a dialogue partner will be even more human-like. Having a goal to develop such a DS which interacts with users like human officials do, we analyze human-human phone calls.

Directory inquiries are typical application of a spoken DS. In many cases, phone numbers are asked by clients calling a service. Giving a phone number may differ in different cultures. It is related to the length and the structure of numbers in use, to non-proclaimed agreements, and it depends on situation and on training of officials. On the other hand, the strategies of a hearer receiving a number may differ – s/he may keep silence or give a feedback during the process. Both participants are interested in fluent communication. Their common goal is that the client gets a correct number.

In the paper, we consider giving and receiving phone numbers in Estonian institutional phone calls. The results of our analysis will form the basis of the user interface which behaves like a human operator when giving phone numbers.

Empirical material is taken from the Estonian dialogue corpus [3] which includes about 1100 spoken human-human dialogues. Most of them are phone calls where a client

V. Matoušek and P. Mautner (Eds.): TSD 2009, LNAI 5729, pp. 347–354, 2009.

calls an institution in order to get some information. Clients are requesting phone numbers in 60% of dialogues. For the current analysis, we have chosen 170 directory inquiries and 60 calls to outpatients' offices.

In Estonia, phone numbers consist of seven or eight digits. Operators usually tell numbers in portions, each of which consisting of two or three digits. As a result, a number is typically told in 3–4 portions. The whole number is given in one block only in case of so-called short number consisting of four digits, or if the number begins with an invariable part and this part is known to the client (e.g an area code or a standard beginning for numbers of the same institution).

The paper has the following structure. In Section 2 we analyze how officials and clients indicate continuation. In Section 3 we consider which strategies they implement in order to indicate finishing. In Section 4 problems in telling and receiving phone numbers are studied. Section 5 discusses how the results can be implemented. In Section 6 we make conclusions.

2   Indicating Continuation

In the first part of giving/receiving a number, an operator (O) indicates after every portion that the number is not yet finished. A client (C) has to accept the portion, to fix it (remember, write down, enter it into a mobile phone, etc) and to give a signal if s/he is ready to listen to the next portion.

Our analysis has shown that O gives information about continuation of a phone number exclusively by pitch contour. Rising, smooth or weakly falling pitch contour indicates continuation (examples 1, 2 and 3).[1] Such pitch contour is clearly different from falling pitch contour which marks completeness. There are no deviations from that pitch contour model in our corpus.[2]

(1)
O: registra`tuuri `numbrilt sis saa[te `küsida,]
    you can ask from the registry number
C: [no kena,]   *OK*
O: `seitse kolm `üks,   *seven three one*
C: `seitse kolm `üks?   *seven three one*
O: öheksa `öheksa,   *nine nine*
C: öheksa `ööheksa?   *nine nine*
O: `kaheksa kaheksa.   *eight eight*
C: kaheksa `kaheksa.=   *eight eight*
O: =jah.   *yes*
C: aitäh?   *thank you*

---

[1] The transcription system of conversation analysis is used [4]. Special transcription marks are used to label unit-final pitch contour in dialogues (question mark – rising, comma – slightly falling, without a label – smooth pitch contour).

[2] This pitch contour model is not restricted to giving phone numbers only but is broadly used in Estonian for indicating the continuation.

(2)
O: .hh `Tallinna maantee `üks on kauplus `Kodumasinad.=
  Tallinn street one is a shop Kodumasinad
C: =mhmh *uhuh*
O: mt `number >kinnitamata andmetel< neli neli (.) kolm?
  unconfirmed number four four three
(.)
C: mhmh *uhuh*
O: üks null kaheksa kaheksa. *one zero eight eight*
(.)
C: mhmh, ai`täh teile? *uhuh, thank you*
O: palun. *you are welcome*

(3)
O: jaa? üks=`hetk (...) .hh ee kas `pubi toidu `kojuvedu või `kontor.
  yes, one moment, (would you like) pub's delivery service or office?
C: kontor. *office*
O: jah (.) `seitse kolm `null *yes, seven three zero*
(1.0)
kolm `neli, *three four*
(1.0)
`üks null. *one zero*
C: *suur aitäh* *thank you very much*

C can indicate his/her readiness to receive the next information portion in three different ways. C can repeat the portion told by O in the previous turn (Ex 1). Typically, s/he repeats the whole portion but in single cases only the last digit. C can use feedback particles *jah* (yes), *mhmh* (uhuh) or rarely *nii* (so) (Ex 2), or keep silence (Ex 3). Repetition and particles are characterized by rising, smooth or weakly falling pitch contour. Therefore, the pitch contour of C's speech turns also indicates continuation.

Repetition and particles signal that a part of a number is received and O may give the next portion. In addition, repetition informs O how C has fixed the portion and allows O to evaluate its correctness.

In 50% of cases, a pause precedes a repetition or a particle. There are only single cases where pause will follow the feedback. We can conclude that feedback is given only after C has finished his/her action (writing down etc) and its main role is to point out that O could and must take a turn.

If C keeps silence then O has to decide himself/herself, when s/he has to take a turn. The length of a silence is 0.5–1.0 seconds in our data. We can conclude that O knows, using his/her experience, that the length of the pause is sufficient for fixing the received portion.

The most frequent markers of receiving a phone number in our data are *jah* (yes) and repetition.

Repetition is also used in other functions in Estonian (e.g answers to yes/no questions [5]).

Jah (yes) is a multifunctional particle in Estonian. It is used as a positive answer to yes/no questions, offers and proposals, as agreement with assertions, opinions, as a reaction to partner's pre-closing, as a repair initiator and a means of giving the turn back to the partner [6]:1785–1787, [7].

Mhmh (uhuh) is also a feedback particle. It indicates active hearing and encourages the partner to continue. [6]:1787–1793.

Nii (so) is a particle that indicates that one sequence or action is finished and the user is ready to move to the next sequence or activity [6]:1799, [8].

In order to indicate continuation, C can use either one single strategy (Ex 1 to 3) or a combination of different strategies, e.g yes+silence, silence+yes, yes+repetition, so+yes, etc (Ex 4).

(4)
C: ahaa, ja mis ta `number on?   *oh and what is its number?*
O: seitse kolm neli?   *seven three four*
(.)
C: jah   *yes*
O: üheksa üheksa?   *nine nine*
C: mhmh   *uhuh*
O: üks kuus.   *one six*
C: üks kuus, ai`täh teile?   *one six, thank you*

How O can predict which strategy will be chosen by C? Our analysis shows that the strategy can not be determined by preceding context. Similarly, the previous turn or adjacency pair does not give a pre-knowledge about the reaction to the following portion of digits. The only relationship we have found is that C prefers repetition when there were problems with accepting the previous portion (not-hearing, inability to fix information, sound overlapping which disturbs receiving the previous portion).

3   Finishing a Phone Number

The second part of telling a phone number is a finishing sequence – telling the last portion of digits, its accepting and transition to the next part of conversation.

Our analysis shows that O tells the last portion of a phone number using falling pitch contour which indicates the end of a number (Ex 1 to 3).[3] The analysis of strategies of C is dividable into two groups.

First, our analysis shows that there is a clear interrelation between strategies of continuing and strategies of finishing used by C. If C has used particles or silence to indicate continuation then s/he almost always ends the call without any special feedback signal (Ex 3) or uses the same marker as in the previous turns (e.g *mhmh* uhuh, Ex 2). If s/he has used repetition or has changed continuation markers then s/he uses repetition to accept the last sequence (Ex 1, 4).

Second, in the most cases C finishes the call after getting the number. In rare cases, C asks for additional information or s/he expresses his/her misunderstanding. In such

---

[3] Such pitch pattern is marked with a dot in transcription.

cases, C uses an utterance which does not have finishing pitch contour or s/he asks a new question in the same turn.

There are two strategies used by C to finish the whole conversation. By the prevailing strategy C gives a closing signal in the same turn after accepting the last information portion (*aitäh/tänan/palju tänu* – thank you/many thanks). This strategy has three variants. C may provide a closing signal only (Ex 3), s/he may use a pause or a feedback marker before the ending signal. The most frequent markers are the particle *mhmh* (uhuh) (Ex 2) and repetition (Ex 4). *Jah* (yes) and *nii* (so) are used very rarely in this function.

The second strategy is used only if C repeats the last portion of digits. In this case C does not continue the turn him/herself but O reacts to his/her repetition with a confirming particle *jah* (yes) like in Ex 1, or with a particle *mhmh* (uhuh) in single cases. Then C gives a finishing signal in his/her next turn (*aitäh* - thank you) (Ex 1). In all analyzed cases *jah* (yes) fluently follows the repetition, without a pause or overlapping. We can suppose that O gets a sign from C, is his/her feedback expected or not.

Unfortunately, such signs are missing in most of calls. C may repeat the last portion using a usual finishing pitch contour, stretching some sounds, slowing or quickening the tempo. The confirmation of repetition is not influenced by a pause used sometimes by C after repetition which might indicate that C is expecting a reaction from the partner. Still, in most cases, C will continue after the pause himself/herself, giving a feedback and then a finishing signal, or asking a new question. Only in two cases, O has answered *jah* (yes) in such situations.

There are rare examples where C's *aitäh* (thank you) and O's confirming *jah* (yes) are overlapping. Usually, O's *palun* (you are welcome) follows C's *aitäh*. It indicates that O interprets *jah* as an entity before *aitäh* (thank you). Seldom, O says his/her *jah* before the end of C's repetition obviously predicting the correctness of echoing repetition.

The only clear relation with using of *jah* is the fact that this strategy is used in the case if there is overlapping or another communication problem in some previous turn (Ex 5).

(5)
C: tere? (.) ütelge palun kas `silmaarsti juurde pääsemiseks on `vaja `ette
   registreerida ka.
   hello. must I register beforehand to get to the eye doctor
O: .hh jah, kaheksakümmend (.) neli öeksa kaks viis telefonil.
   yes eighty four nine two phone
C: `kuidas see oli kaheksakend `neli?   *how it was eighty four*
O: öeksa kaks viis.   *nine two five*
(.)
C: öeksa kaks viis.   *nine two five*
O: jah   *yes*

4 Solving Communication Problems

Problems are rare when giving or receiving phone numbers. If a problem arises then the hearer initiates a repair which is performed by the partner who has caused the problem. There are two kinds of repairs.

First, C can initiate a repair if s/he did not hear or was unable to fix the previous portion. Repairs are initiated immediately after the turn where problem occured. The repair initiation typically consists of a repetition and a interrogative word *jah* (yes). Repairs can be initiated also using other markers (e.g, *üks hetk* – one moment; repetition of the first part of the problematic portion; Ex 5).

Otherwise, repair is initiated after the end of a number where the hearer repeats the whole number in order to check it. Words *kordan* (I repeat), turn-final *jah* (yes; Ex 6) are used as repair markers. Sometimes the repetition of the whole number itself is a sign of the need for correctness confirmation.

(6)
O: `kaks `kuus   *two six*
C: * mhmh? * (1.0) >kolm kuus kaks kuus jah?<   *uhuh three six two six yes*
O: .jah   *yes*
C: aitäh?=teile   *thank you*

5   Discussion

For developing a natural language interface that enables user get phone numbers from some data base, one has to study how human operators give phone numbers and how clients receive them. A user-friendly interface should not give a long phone number in one block but has to divide it into smaller parts – like human operators do – and stop after every part in order to make it possible for a user to fix it. Several strategies can be used by an intelligent agent for indicating continuation of giving information as well as its end. Signal of continuation is rising, smooth or weakly falling pitch contour. Signal of finishing is falling pitch contour. Text-to-speech synthesizer has to generate utterances using a suitable pitch contour.

On the other hand, the agent has to understand strategies used by clients when receiving phone numbers. Indicating a continuation, a client either repeats the information given by the agent in the previous turn, or uses feedback particles, or simply keeps silence. Responding to the final information portion, a client either gives a finishing signal in the same turn, after accepting the last information portion, or repeats the last portion. Then a client does not continue in the same turn himself/herself but mostly waits for the operator to confirm the repetition with *jah* (yes) and produces a finishing signal in the following turn.

If communication problems arise then the client initiates a repair and the agent has either to confirm information correctness or to provide correct information. User's strategies should be recognized by dialogue manager and speech recognizer.

A computational model of (communicative) feedback is worked out in [9]. An agent, C in our case, provides information to O about his/her processing of O's utterances. The feedback is considered positive if it reports that the processes observed have been completed without any problems and negative if it reports problems, uncertainties, or failure. Another form of feedback occurs when one agent, O, addresses the other agent's processing of his/her own (O's) utterances. Positive feedback is voluntary. That explains why a client sometimes keeps silence when receiving information. Negative feedback is necessary in cooperative interaction because both of participants are interested in giving/receiving the correct information.

A question-answer system has been developed which integrates speech recognition and text-to-speech synthesis for Estonian [10]. The system gives information about theatre performances in Estonia, using a special database. It uses a regular grammar for dialogue management and for understanding users' queries, and ready-made sentence patterns for information giving. We are oriented to transfer the system into other domains and to make it more human-like.

6 Conclusion and Future Work

Strategies of giving and receiving phone numbers are studied in Estonian institutional calls with the further aim to develop an intelligent user interface. Human operators give long phone numbers in several parts. Pitch contour gives the signal of continuation or finishing the process. Clients give feedback during the process (repetition, particles, and pauses). Special strategies are used by clients to finish receiving the number as well as to initiate repairs in the case of communication problems.

Our further work will concern implementation of the results of the current study for development an intelligent user interface.

Acknowledgments

This work was supported by the Estonian Science Foundation (grant 7503) and the Estonian Ministry of Education and Research (projects SF0180078s08, EKKTT09-57, and EKKTT09-61).

References

1. Möller, S.: Quality of telephone-based spoken dialogue systems. Springer, New York (2004)
2. Zoeppritz, M.: Computer Talk? Technical Report 85.05, IBM, WZH, Heidelberg (1985)
3. Hennoste, T., Gerassimenko, O., Kasterpalu, R., Koit, M., Rääbis, A., Strandson, K.: From Human Communication to Intelligent User Interfaces: Corpora of Spoken Estonian. In: Proceedings of the Sixth International Language Resources and Evaluation (LREC 2008), European Language Resources Association (ELRA), Marrakech, Morocco, May 28-30 (2008), http://www.lrec-conf.org/proceedings/lrec2008/
4. Hutchby, I., Wooffitt, R.: Conversation Analysis. Polity Press, Cambridge (1998)
5. Keevallik, L.: Üldküsimuse lihtvastuste funktsioonid. Keel ja Kirjandus 1, 33–53 (2009)
6. Hennoste, T.: Sissejuhatus suulisesse eesti keelde. Suulise kõne erisõnavara III. Partiklid. Akadeemia 8, 1779–1806 (2000)
7. Kasterpalu, R.: Partiklid jah, jaa ning jajaa naaberpaari järelliikmena müügiläbirääkimistes. Keel ja Kirjandus 11(12), 873–890, 996–1000 (2005)
8. Keevallik, L.: The deictic nii 'so, in this way' in interaction. In: Monticelli, D., Pajusalu, R., Treikelder, A. (eds.) From Utterance to Uttering and vice versa. Multidisciplinary views on deixis, Studia Romanica Tartuensia IVa, pp. 109–126. Tartu University (2005)

9. Bunt, H.: Dynamic Interpretation and Dialogue Theory. In: Taylor, M.M., Néel, F., Bouwhuis, D.G. (eds.) The Structure of Multimodal Dialogue II, pp. 139–166. John Benjamins, Philadelphia (1999)
10. Treumuth, M., Alumäe, T., Meister, E.: A natural language interface to a theater information database. In: Erjavec, T., Zganec Gros, J. (eds.) Language Technologies, IS-LTC 2006: Proceedings of 5th Slovenian and 1st International Conference, Slovenia, Ljubljana, pp. 27–30 (2006)

Error Resilient Speech Coding Using
Sub-band Hilbert Envelopes

Sriram Ganapathy[1], Petr Motlicek[2], and Hynek Hermansky[1,*]

[1]IDIAP Research Institute, Martigny, Switzerland
[2]École Polytechnique Fédérale de Lausanne (EPFL), Switzerland
{ganapathy,motlicek}@idiap.ch
hermansky@ieee.org

Abstract. Frequency Domain Linear Prediction (FDLP) represents a technique for auto-regressive modelling of Hilbert envelopes of a signal. In this paper, we propose a speech coding technique that uses FDLP in Quadrature Mirror Filter (QMF) sub-bands of short segments of the speech signal (25 ms). Line Spectral Frequency parameters related to autoregressive models and the spectral components of the residual signals are transmitted. For simulating the effects of lossy transmission channels, bit-packets are dropped randomly. In the objective and subjective quality evaluations, the proposed FDLP speech codec is judged to be more resilient to bit-packet losses compared to the state-of-the-art Adaptive Multi-Rate Wide-Band (AMR-WB) codec at 12 kbps.

1 Introduction

Conventional approaches to speech coding achieve compression with a linear source-filter model of speech production using the linear prediction (LP) [1]. The residual of this modeling process, which represents the source signal, is transmitted to reconstruct the speech signal at the receiver. On the other hand, modern perceptual codecs [2] typically used for multi-media coding applications are not as efficient for speech content. Furthermore, the reconstruction quality of the state-of-the-art speech codecs are significantly degraded in lossy channel conditions.

In this paper, we propose to exploit the predictability of sub-band spectral components for encoding speech signals. The proposed FDLP codec has been developed with low delay (about 50 ms) so that the codec is suitable for voice communications. Our approach is based on the assumption that speech signals in sub-bands can be represented as a product of a smoothed Hilbert envelope estimate and fine signal variations in the form of Hilbert carrier. The Hilbert envelopes are estimated using Frequency Domain Linear Prediction (FDLP) [3], which is an efficient technique for auto-regressive modelling of the temporal envelopes of the signal [4]. This idea was first applied for audio coding in the MPEG2-AAC (Advanced Audio Coding) [5], where it was primarily used for Temporal Noise Shaping (TNS).

* This work was partially supported by grants from ICSI Berkeley, USA; the Swiss National Center of Competence in Research (NCCR) on "Interactive Multi-modal Information Management (IM)2"; managed by the IDIAP Research Institute on behalf of the Swiss Federal Authorities.

V. Matoušek and P. Mautner (Eds.): TSD 2009, LNAI 5729, pp. 355–362, 2009.
© Springer-Verlag Berlin Heidelberg 2009

In the proposed speech codec, the technique of linear prediction in spectral domain is performed on sub-band signals. We use a non-uniform Quadrature Mirror Filter (QMF) bank to derive 5 critically sampled frequency sub-bands. FDLP is applied over short segments (25 ms) of speech signal to estimate Hilbert envelopes in each sub-band. The remaining residual signal (Hilbert carrier) is further processed and its frequency components are quantized. The bit-packets for the FDLP codec contain individual sub-band signals in the form of FDLP envelope parameters and spectral components of the residual signal. At the decoder, the sub-band signal is reconstructed by modulating the inverse quantized residual with the AR model of the Hilbert envelope. This is followed by a QMF synthesis to obtain the speech signal back. The current version of the FDLP speech codec operates at a bit-rate of 12 kbps.

For speech codecs operating in lossy channels, some bit-packets get distorted and hence, the reconstructed signal is degraded. The intelligibility of speech is also affected in cases of severe bit-packet loss (frame dropouts). In order to simulate these channel conditions in speech codecs, bit-packets are dropped randomly (with a uniform distribution) at the decoder.

For the proposed FDLP codec, the dropout of bit-packets corresponds to loss of sub-band signals at the decoder. The degraded sub-band signals are recovered from the adjacent sub-bands in time-frequency plane which are unaffected by the channel. The objective and subjective listening tests show that the proposed FDLP codec is more resilient to frame dropouts compared to the state-of-the-art AMR-WB codec at similar bit-rate.

The rest of the paper is organized as follows. Sec. 2 explains the FDLP technique for the auto-regressive modelling of Hilbert envelopes. Sec. 3 describes the proposed speech codec based on FDLP. The sub-band signal reconstruction technique in lossy channels is detailed in Sec. 4. The results of the objective and subjective evaluations are reported in Sec. 5.

2   Frequency Domain Linear Prediction

Typically, auto-regressive (AR) models have been used in speech applications for representing the envelope of the power spectrum of the signal by performing the operation of Time Domain Linear Prediction (TDLP) [6]. This paper utilizes AR models for obtaining smoothed, minimum phase, parametric models of temporal rather than spectral envelopes. The duality between the time and frequency domains means that AR modeling can be applied equally well to discrete spectral representations of the signal instead of time-domain signal samples [3].

For signals that are expected to consist of a fixed number of distinct transients, fitting an AR model constrains the temporal envelope to be a sequence of maxima, and the AR fitting procedure removes the finer-scale detail. This suppression of detail is particularly useful in speech coding applications, where the goal is to extract the general form of the signal by means of a parametric model and to characterize the finer details with a small number of bits.

The block schematic showing the steps involved in deriving the AR model of Hilbert envelope is shown in Fig. 1. The first step is to compute the analytic signal for the

Fig. 1. Steps involved in FDLP technique for AR modelling of Hilbert Envelopes

input signal. For a discrete time signal, the analytic signal can be obtained using the Fourier Transform [8]. Hilbert envelope (squared magnitude of the analytic signal) and spectral autocorrelation function form Fourier transform pairs [5]. This relation is used to derive the autocorrelation function of the spectral components of a signal which are then used for deriving the FDLP models (in manner similar to the computation of the TDLP models from temporal autocorrelations [6]).

For the FDLP technique, the squared magnitude response of the all-pole filter approximates the Hilbert envelope of the signal. This is in exact duality to the approximation of the power spectrum of the signal by the TDLP as shown in Fig. 2.

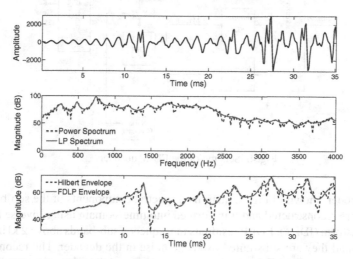

Fig. 2. Linear Prediction in time and frequency domains for a portion of speech signal

3 FDLP Based Speech Codec

Short temporal segments (25 ms) of the input speech signal are decomposed into 5 non-uniform QMF sub-bands. In each sub-band, FDLP is applied and Line Spectral Frequencies (LSFs) approximating the sub-band temporal envelopes are quantized using Vector Quantization (VQ). The residual signal (sub-band Hilbert carrier) is processed in spectral domain and the magnitude spectral parameters are quantized using VQ. Since a full-search VQ in this high dimensional space would be computationally infeasible, the split VQ approach is employed. Although this forms a suboptimal approach to VQ, it reduces computational complexity and memory requirements without severely affecting the VQ performance. Phase spectral components are scalar quantized (SQ) as they are found to have a uniform distribution. Graphical scheme of the FDLP encoder is given in Fig. 3.

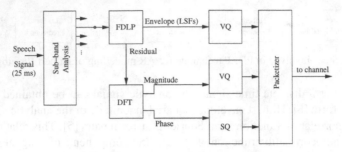

Fig. 3. Scheme of the FDLP encoder

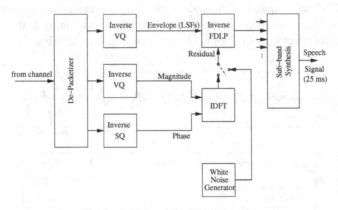

Fig. 4. Scheme of the FDLP decoder

In the decoder, shown in Fig. 4, quantized spectral components of the sub-band carrier signals are reconstructed and transformed into time-domain using Inverse Discrete Fourier Transform (IDFT). FDLP residuals in frequency sub-bands above 2 kHz are not transmitted, and they are substituted by white noise in the decoder. The reconstructed FDLP envelopes (from LSF parameters) are used to modulate the corresponding sub-band carriers. Finally, sub-band synthesis is applied to reconstruct the full-band signal. The final version of the FDLP codec operates at ∼ 12 kbps.

4   Signal Reconstruction in Lossy Channels

For the proposed FDLP codec, a bit-packet contains information about a single sub-band signal in the form of the Hilbert envelope parameters and the spectral components of the residual signal. Therefore, the loss of bit-packets in a lossy channel refers to the dropout of sub-band signals. Since short-term sub-band signals of speech are correlated with the neighboring sub-band signals (in the time-frequency plane), the sub-bands corresponding to degraded bit-packets can be reconstructed at the decoder using the adjacent sub-bands which are undistorted. Specifically, we estimate these sub-band signals as a time-frequency average of the contiguous sub-band signals which are unaffected by the channel.

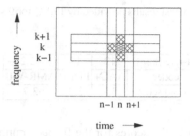

Fig. 5. Reconstruction of corrupted sub-band signals using adjacent time-frequency sub-bands

Let $x_{n,k}(t)$ denotes short-term signal in the kth sub-band for n^{th} time frame of a speech signal (Fig. 5). The neighboring sub-bands in the time-frequency plane are denoted as $x_{n,k-1}(t), x_{n,k+1}(t), x_{n-1,k}(t)$ and $x_{n+1,k}(t)$. If the bit-packet corresponding to $x_{n,k}(t)$ is distorted due to lossy channel, it is reconstructed by

$$x_{n,k}(t) = Av.\{x_{n,k-1}(t), x_{n,k+1}(t), x_{n-1,k}(t), x_{n+1,k}(t)\}, \tag{1}$$

where $Av.$ denotes operation of averaging. If one of the adjacent sub-bands used in the averaging is also degraded, then it is not included in the mean computation. It is found that such an averaging operation retains the intelligibility of speech although, it introduces colored noise in the reconstructed signal.

5   Experiments and Results

For the purpose of training the VQ codebooks, we use speech files from the subset of TIMIT database [9] which are sampled at 16 kHz. VQ codebooks for the magnitude spectral components of the sub-band residual signals and the FDLP envelope LSFs are trained using 400 speech files spoken by 20 male and 20 female speakers. The TIMIT database speech is alone used for training the VQ codebooks, whereas the evaluations are done with standard databases for speech coding. For the objective and subjective quality evaluations, 9 challenging speech files from [14,15] are used consisting of clean speech, radio speech, conversational speech and conference room recordings.

5.1   Objective Evaluations

Objective Evaluation is done with the Perceptual Evaluation of Speech Quality (ITU-T P.862 PESQ standard [10]). The quality estimated by PESQ corresponds to the average user perception of the speech sample under assessment PESQ – MOS (Mean Opinion Score). PESQ scores range from 1.0 (worst) up to 4.5 (best). The first set of experiments compare the PESQ scores for ideal channel conditions (without any bit-packet loss). The output quality for the following codecs are compared:

1. The proposed FDLP codec at 12 kbps denoted as FDLP.
2. Enhanced AAC plus [2] at 12 kbps denoted as AAC+.
3. AMR-WB standard [11] at 12.2 kbps denoted as AMR-WB.

Table 1. Mean objective quality test results provided by PESQ for 9 speech files in ideal channel conditions

bit-rate [kbps]	12	12	12.2
codec	AAC+	FDLP	AMR-WB
PESQ score	3.1	3.3	3.7

Table 1 shows the mean PESQ scores for the 9 speech utterances used for the evaluations. It can be seen that the proposed codec, without the use of standard bit-rate reduction techniques like Huffman coding and psycho-acoustic modules, provides better objective quality than the Enhanced AAC plus codec although the AMR-WB standard provides the best reconstruction speech quality for ideal channel conditions.

The lossy channel experiments are performed using the proposed FDLP codec and the AMR-WB codec. The recovery of the frame dropouts for AMR-WB codec is done as reported in [12]. Fig. 6 shows the mean PESQ score for the 9 speech files as function of the percentage of bit-packet loss.

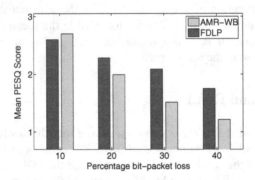

Fig. 6. Mean objective quality test results provided by PESQ for 9 speech files for lossy channel conditions

5.2 Subjective Evaluations

Formal subjective listening tests are performed on ideal channel conditions (0 % bit-packet loss) as well as for the reconstruction with 30% bit-packet loss conditions. We use the MUSHRA (MUltiple Stimuli with Hidden Reference and Anchor) methodology for subjective evaluation. It is defined by ITU-R recommendation BS.1534 documented in [16]. The mean MUSHRA scores for the subjective listening tests (with 95% confidence interval), using 9 speech files and 5 listeners is shown in Fig. 7. The results shown in this figure justify the objective quality evaluations from the PESQ scores (Table 1). Although the AMR codec performs well for ideal channel conditions, the proposed FDLP codec is more resilient to bit-packet losses.

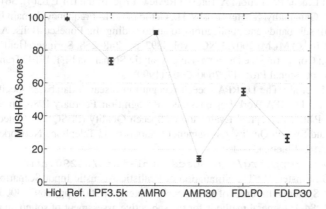

Fig. 7. MUSHRA scores for 9 speech files using 5 listeners at 12 kbps (FDLP codec for ideal channel conditions (FDLP0), FDLP codec for 30% bit-packet loss conditions (FDLP30), AMR-WB codec for ideal channel conditions (AMR0), AMR-WB codec for 30% bit-packet loss conditions (AMR30), hidden reference (Hid. Ref.) and 3.5 kHz low-pass filtered anchor (LPF3.5k)).

6   Conclusions

A technique for error resilient speech coding is proposed which uses auto-regressive modelling of the sub-band Hilbert envelopes. Specifically, the technique of linear prediction in the spectral domain is applied on short segments ($25ms$) of speech signals in QMF sub-bands. The sub-band signals which are degraded in lossy channel conditions are reconstructed by the time-frequency averaging of the adjacent undistorted sub-bands. The objective and subjective evaluations, performed with the current version of the FDLP codec, suggest that the FDLP codec operating at $\sim$ 12 kbps provides more error resilience compared to state-of-art AMR-WB codec at similar bit-rates. Furthermore, the current version of the FDLP codec does not utilize the standard modules for compression efficiency provided by entropy coding and simultaneous masking. These form part of the future work.

References

1. Schroeder, M.R., Atal, B.S.: Code-excited linear prediction (CELP): high-quality speech at very low bit rates. In: Proc. of the ICASSP, April 1985, vol. 10, pp. 937–940 (1985)
2. Enhanced aacPlus General Audio Codec, 3GPP TS 26.401
3. Athineos, M., Ellis, D.: Autoregressive Modeling of Temporal Envelopes. IEEE Trans. on Signal Proc. 55, 5237–5245 (2007)
4. Kumerasan, R., Rao, A.: Model-based approach to envelope and positive instantaneous frequency estimation of signals with speech applications. Journal of Acoustical Society of America 105(3), 1912–1924 (1999)
5. Herre, J., Johnston, J.D.: Enhancing the Performance of Perceptual Audio Coders by using Temporal Noise Shaping (TNS). In: Proc. of 101st AES Conv., Los Angeles, USA, pp. 1–24 (1996)

6. Makhoul, J.: Linear Prediction: A Tutorial Review. Proc. of the IEEE 63(4), 561–580 (1975)
7. Motlicek, P., Ganapathy, S., Hermansky, H., Garudadri, H.: Frequency domain linear prediction for QMF sub-bands and applications to audio coding. In: Popescu-Belis, A., Renals, S., Bourlard, H. (eds.) MLMI 2007. LNCS, vol. 4892, pp. 248–258. Springer, Heidelberg (2008)
8. Marple, L.S.: Computing the Discrete-Time Analytic Signal via FFT. IEEE Trans. on Acoustics, Speech and Signal Proc. 47, 2600–2603 (1999)
9. Fisher, W.M., et al.: The DARPA speech recognition research database: specifications and status. In: Proc. DARPA Workshop on Speech Recognition, February 1986, pp. 93–99 (1986)
10. ITU-T Rec. P.862: Perceptual Evaluation of Speech Quality (PESQ), an Objective Method for End-to-end Speech Quality Assessment of Narrowband Telephone Networks
11. Extended AMR Wideband codec,
    http://www.3gpp.org/ftp/Specs/html-info/26290.htm
12. Hirsch, H.G., Finster, H.: The Simulation of Realistic Acoustic Input Scenarios for Speech Recognition Systems. In: Proc. of Interspeech, September 2005, pp. 2697–3000 (2005)
13. ITU-R BS.1284-1: General methods for the subjective assessment of sound quality (2003)
14. ISO/IEC JTC1/SC29/WG11: Framework for Exploration of Speech and Audio Coding, MPEG2007/N9254, Lausanne, CH (July 2007)
15. Voice Age, http://www.voiceage.com/audiosamples.php
16. ITU-R Recommendation BS.1534: Method for the subjective assessment of intermediate audio quality (June 2001)

Prototyping Dialogues with Midiki:
An Information State Update Dialogue Manager

Lúcio M.M. Quintal[1] and Paulo N.M. Sampaio[2]

[1] Madeira Tecnopolo, 9020-105 Madeira, Portugal
lucio.q@gmail.com
[2] Laboratory for Usage-centered Software Engineering (LabUse),
Centro de Ciências Matemáticas (CCM), Department of Mathematics and Engineering,
University of Madeira (UMa), 9000-105 Madeira, Portugal
psampaio@uma.pt

Abstract. The context of the work presented in this paper is related to the design of systems and interfaces which enable natural language interaction between a person and a machine, through the use of Dialogue Managers (DM). Any practical DM must incorporate a model of a dialogue for the domain or task being addressed. Unfortunately, there are few methodologies and tools that allow authors to carry out an intuitive and easy modelling of such dialogues. This paper presents the proposal and application of a methodology and a tool for the authoring of generic dialogues for the MIDIKI DM, illustrated by the development of a concrete model of a dialogue.

1 Introduction

Access to a service (ex. banking) in the traditional manner requires a human mediator or the adaptation of the person to unnatural ways of interaction, such as WIMP () interfaces. Systems that provide a natural language intermediation between the human user and a back-end system are called Dialogue Systems (DS) [1], [2], at the core of which we find the so-called Dialogue Managers [3]. Therefore, the dialogue and its management represent a central aspect in this type of interaction. A dialogue system is designed to help the user to achieve a goal through a dialogue with the machine and therefore such a system must incorporate a good knowledge of the subject and show some communication skills.

In a typical DS we can identify a set of input interfaces; a component dedicated to the integration of inputs; a DM, which manages the interaction, at the core of the DS; a component for output planning (e.g. multimedia output) and a set of physical output interfaces. If the system supports more than one mode (or modality) of interaction, it is called a Multimodal DS (e.g. a system supporting gestures, touch and speech) [3].

A key aspect in the motivation for carrying out this work was the lack of tools for the rapid modelling/prototyping of generic dialogues. A tool of this kind must be supported by a methodology, in the sense of indicating the steps to follow in the modelling of dialogues [4]. Next section provides an overview of the adopted dialogue manager: The Midiki DM [5].

V. Matoušek and P. Mautner (Eds.): TSD 2009, LNAI 5729, pp. 363–370, 2009.
© Springer-Verlag Berlin Heidelberg 2009

2 Dialogue Managers and the Midiki DM

The main responsibilities of a DM are: i) to monitor the flow and the progress of the interaction with the user, managing the acquisition of information and making the interface with existing back-end services, and ii) to maintain a representation of the state of the dialogue [6], [7]. By comparing the main categories of existing approaches for modelling of dialogues (namely, Finite State Automata; Frames/Slots; Plans; Information State Update (ISU) and Agent/Collaborative approaches [8], [9]), we may conclude that the ISU approach, besides being not complex (in terms of modelling and implementing a dialogue), still provides a greater flexibility in modelling the dialogue, particularly by enabling natural language (NL) and/or multimodal interactions [10].

Midiki (MITRE Dialogue Kit) is a DM based on the ISU approach. One of the goals of its authors was to create a DM with a compromise between fidelity to linguistic aspects and implementation aspects, allowing the creation of relatively sophisticated dialogues being, at the same time, viable and functional on a practical standpoint [11]. The properties of Midiki provide a great flexibility in modelling a dialogue. It is therefore a good DM for carrying out experiments and research in this field. Nevertheless, there was no tool to support the modelling/authoring of dialogues for that DM.

Next section presents the main characteristics of the proposed methodology for the authoring of dialogues for Midiki.

3 Overview of the Methodology Proposed for Dialogue Modelling

The three main components that specify and structure a model of a dialogue in Midiki are the *Domain*, the *Lexicon* and the *Cells*. The proposed methodology [4] includes a declarative specification of components and the creation of procedural components, in a total of 13 steps. The *domain* encompasses the entities that may be managed or referenced in a dialogue and also includes the specification of the structure of the tasks to carry out, in this case through the construction/specification of *plans*. The specification of the domain corresponds to steps 1 to 3 of the methodology. The *lexical* component is mainly related to the specification of the vocabulary used by the system during the dialogue and corresponds to steps 4 and 5 of the methodology. *Cells* component includes the declaration of data structures and services accessed by the DM for each concrete dialogue and correspond to steps 6 to 9 of the methodology.

Table 1 identifies the 13 steps of the methodology and the main component to which they are related.

Note that the methodology was proposed in 13 steps in order to reflect faithfully the structure of the model of a dialogue in Midiki, including its final representation in Java. Next section describes the tool implemented to support this methodology.

Table 1. List of steps and related components in the methodology

Step	Component
Step 1 - Construction of plans with strategies	Domain
Step 2 - Declaration of attributes and synonyms	Domain
Step 3 - Association of strategies with types of attributes	Domain
Step 4 –Mapping of output matches	Lexicon
Step 5 - Mapping of input matches	Lexicon
Step 6 - Declaration of slots (strategy types)	Cells
Step 7 - Declaration of queries	Cells
Step 8 - Declaration of methods	Cells
Step 9 - Declaration of classes for handlers	Cells
Step 10 - Coding of handlers (in Java)	Procedural
Step 11 - Parameterization of Tasks class	Procedural
Step 12 - Parameterization of DomainAgent class	Procedural
Step 13 - Parameterization of "Dialogue Manager Executive" class	Procedural

4   The Web Tool Implemented to Support the Methodology

The tool (Midiki Authoring Tool, or MAT) was implemented on top of a Zope application server [12], following the MVC pattern [13], in a client-server model. The model layer is supported by a mySQL relational database. The controller layer was written in Python and Java. Java Emitter Templates (JET code) [14] was used to create the libraries which are called for the automatic generation of Java code. The view layer was implemented with Zope Page Templates [12] and structured in HTML. MAT automates the most relevant parts concerning the generation of the code (Java) for carrying out a dialogue managed by Midiki. Using MAT presents some clear advantages, such as: i) It defines a "project" as the central/aggregating element, which represents a model of a dialogue; ii) The author does not have to work with source code, but only with the relevant information that specifies each object, and; iii) The code generated by the tool always reflects the most recent changes/specifications, and most of the code necessary for a project/dialogue can be generated automatically by the tool.

   Next section illustrates a concrete dialogue, modelled using the implemented tool. Each step of the methodology is described accordingly.

5   Modelling a Concrete Dialogue with MAT for Natural Language Interacting with an ATM Machine

The dialogue presented in this paper consists in simulating, in a simplified form, the access to an Automatic Teller Machine (ATM), through the interaction in natural language (written or spoken), allowing the most common operations available on these terminal machines: i) Getting the balance of an account; ii) Carrying out a withdrawal, and; iii) Making a deposit. The requirements of this dialogue had to be simple and, simultaneously, had to address and highlight the main features of both the tool and Midiki.

In any case it isn't reasonable and/or practical to start applying the steps of the methodology without previously carrying out an analysis of requirements of the system to create/model. Table 2 lists the requirements to analyze and the corresponding components of the methodology. Analysing the requirements listed on the left column of Table 2 will provide the inputs to the specification of the components/elements of the methodology, identified on the right column.

Table 2. Mapping of requirements and related components in the methodology

Generic requirement	Related component
i) Identify the information needs of the system regarding its dialogue function	Lexical (*output matchers*) and Domain (*strategies and attributes*)
ii) Identify the back-end services required (services to be accessed)	Cells (*slots, queries, methods and handlers*)
iii) Anticipate possible responses of users	Lexical (*input matchers, synonyms*)
iv) Planning how to carry out the task	Domain (*plans and strategies*)

In the case of our ATM dialogue, this analysis of requirements is outlined / instantiated in Table 3.

Table 3. Summary of the analysis of the requirements for the ATM dialogue

Require-ment	Analysis (due to space restrictions, we include only a partial view of the analysis for each part, as an example)
i) Information needs	e.g. Operation "Making a deposit": In order to process a required deposit, the system needs an answer to the following questions: - Which account to access? / - What is the amount of the deposit?
ii) Back-end services	The list of operations and accounts available, as well as the amounts allowed per transaction, will all be defined as domain attributes. The "back end" system provides the following services: e.g. Update, in a database, of an account balance after a deposit.
iii) Possible responses	Typically, the client/user will apply expressions such as the following, for e.g. in the "consult balance" operation: - I want/would like to consult the balance of account <some_account_id>
iv) Planning	- In the main plan we identify which is the intended operation (asking to the user) and calling a sub-plan for each one of the three operations. For instance, sub-plan "making deposit" must: i) indicate the amounts that can be handled; ii) ask which account; iii) inform initial balance; iv) ask for the amount to deposit, and; v) inform the resulting final balance.

The application of the methodology to the ATM machine dialogue is exemplified in next paragraphs. Note that much of the work necessary in each step is assisted by the tool, avoiding unnecessary inputs or repetitions.

Step 1: Construction of plans with strategies - Plans are composed of a sequence of actions (called strategies in the context of our methodology), which enable the DM to "talk" with a user, presenting him questions or providing him with information in order to perform a certain task. The order/position of strategies in a plan should reflect a typical interaction scenario for the given task (or sub-dialogue). A plan may also include non-conversational strategies/actions, such as accesses to databases. As an example, the plan *consultBalance* consists of the strategies listed in Table 4:

Table 4. Strategies[1] in plan *consultBalance* (which is called from main plan)

Order of strategy in plan	Type of strategy	Name / Identifier	Short description
1st	*Inform*	*accountsAvailable*	Inform the user of the existing accounts.
2nd	*Findout*	*whichAccount-Consult*	Obtain from the user the name of the account to consult. A *Findout* strategy implies one or more questions being posed to the user.
3rd	*Query-Call*	*getAccount-Balance*	Access the account in a database to read/get its balance.
4th	*Inform*	*valueAccount-Balance*	Inform the user of the balance for the given account. In this case it takes/uses the value returned from previous query.

Step 2: Declaration of attributes and synonyms - In this step we declare attributes of the domain (attributes are similar to data types). For instance, an attribute "accounts" can be defined with the values {*checking, savings, certificates*} and, thus, an answer of this type must be resolved to one of these three values. For any of the values of an attribute we can define synonyms. For instance, "savings account" may also be referred to as "economy account" since we have defined "economy" as one of the synonyms of "savings".

Step 3: Association of strategies with types of attributes – This step involves associating an attribute type to the answers of each "question" of the plan (questions are generated by *Findout* strategies). As an example, the question *whichAccountConsult*, in plan *consultBalance* (illustrated in Table 4), was specified as being of the type "accounts" (which we described in step 2).

Step 4: Mapping of output matches – In this step, we must specify the outputs to be presented to users in terms of vocabulary, for instance, when posing them questions. The example illustrated in Table 5 is related to the mapping between the question *whichAccountConsult* and the text to present to the user in that case (or the equivalent speech, if spoken output is available). Besides the lexical content for the question itself, an *output matcher* includes another message, which the system will output in the case where the answer provided by the user to the question is not recognized.

Table 5. Mapping an output match for *Findout* strategy *whichAccountConsult*

Main output text (question)	*Which account do you want to consult?*
Text to output if user answer is not recognized	*Sorry, the given account was not recognized in order to provide it's balance.*

Step 5: Mapping of input matches – Specifying *input matches* allows the direct interpretation of certain terms or expressions from the user and the generation of a specific input to the system. For instance, if the user says "I need ten Euros and no receipt" he is indicating in advance that he wants to withdraw 10 Euros, but doesn't

---

[1] Strategies supported by Midiki are described in [12].

want a receipt of the transaction/operation and, thus, it is no longer necessary to present him later the related question.

Step 6: Declaration of slots (strategy types) – In this step we must declare storage space, i.e. *slots,* for information related to the strategies of the plans. Among other things, these slots will store the responses of the user to questions from the system. For instance, the slot for strategy *whichAccountConsult* will store a value of type "accounts", corresponding to the account the user wants to consult, which can be *checking, savings* or *certificates*.

Steps 7 & 8: Declaration of queries and methods – These two steps are identical. Typically, queries will be used to access databases and methods will be used to call/access other services. These two steps consist in the declaration of the queries (step 7) and methods (step 8) that are part of the existing plans. In the project described in this paper there are the following queries: *getAccountBalance* (to read the balance of an account in the database), *updateBalanceDeposit* (to update the balance of an account) *and checkBalanceOk* (to check if there's enough money in the account for the intended withdrawal). There are no methods in the plans of this project. The concrete implementation of *queries* and *methods* is carried via *handlers*, which are described in steps 9 and 10.

Step 9: Declaration of classes for handlers – This step involves the declaration of the external Java classes to be called to handle the queries and methods included in the plans. For e.g., the query handler of *updateBalanceDeposit* is the class *updateBalanceDepositHandler*.

Step 10: Coding of handlers (in Java) – This step consists in coding the "business logic". The functional algorithms that implement the processing, or handling, of queries and methods is specific of each service and must be implemented by means of Midiki handler classes. MAT can generate templates for each one of the handlers, however the author of the dialogue needs to write the business logic for each handler. For instance, in class *updateBalanceDepositHandler*, related to the query *updateBalanceDeposit,* it is necessary to perform the following actions: 1st) obtain the balance of the account specified; 2nd) add to it the amount of the deposit, and; 3rd) update the account balance with the new value.

Steps 11 to 13: Parameterization of classes Tasks, DomainAgent and "Dialogue Manager Executive" – Consists in the parameterization of three classes (Tasks, DomainAgent and "Dialogue Manager Executive") necessary to "run" the dialogue. Steps 11, 12 and 14 are fully automated in the tool.

6 Compiling and Testing the ATM Dialogue

Provided that the model of the dialogue is complete and all the classes required to its compilation have been obtained and/or written, the model can be compiled with the full set of Midiki classes in order to create a DM for the service. At least six classes are automatically generated by the tool: "domain", "lexicon", "cells", "Tasks", "DomainAgent" and "Dialogue Manager Executive", plus the templates for all handlers.

Once compiled, the dialogue can be executed and tested. The full set of Midiki classes may be imported to Eclipse [15] or other development environment (with the compilation, a JAR Java archive file is generated). The easiest way to test a dialogue is to use the interface available for written text based interaction (input and output).

The system can also be configured for spoken interaction. In order to evaluate spoken interaction, one possible basic configuration is to insert the text coming from a speech recognition application (e.g. Philips FreeSpeech 2000 [16]) in the text input field of the interface window provided by Midiki and carry the "Text to Speech" (TTS) output generated by the system, for e.g., via MSAPI [17]. This particular configuration is compatible with any dialogue created with MAT, providing a simple way of testing spoken interaction for any given dialogue created.

7 Conclusions

In this paper we presented the application of a methodology [4] and a tool (MAT) developed for the modeling/authoring of generic dialogues based on the Midiki DM. Specifically we presented their usage in the implementation of a simulation of the access to an ATM machine through a dialogue, supporting some operations common on ATM machines.

The proposed methodology and the tool implemented fill a gap felt in this area of research and contribute to the progress of the state of the art in this field. As for future work, we expect to: i) Extend Midiki Information State to support multimodal interaction; ii) Configure Midiki to allow access to the DM in distributed environments (e.g. from a PDA client); iii) Integrate dictionaries of synonyms with MAT; iv) Adapt the DM and MAT to provide the possibility of obtaining attributes dynamically from a database, and; v) Incorporate in MAT the support to XML representation of dialogues, as a way to decouple the generation of Java code from the project data which, currently, is stored in a database.

References

1. Allen, J.F., Byron, D.K., Dzikovska, M., Ferguson, G., Galescu, L., Stent, A.: Towards Conversational Human-Computer Interaction. AI Magazine 22, 27–37 (2001)
2. McTear, M.F.: Spoken dialogue technology: enabling the conversational user interface. ACM Computing Survey 34, 90–169 (2002)
3. Delgado, R., Araki, M.: Spoken, Multilingual and Multimodal Dialogue Systems – Development and Assessment. Wiley, Chichester (2005)
4. Quintal, L., Sampaio, P.: A methodology for domain dialogue engineering with the midiki dialogue manager. In: Matoušek, V., Mautner, P. (eds.) TSD 2007. LNCS (LNAI), vol. 4629, pp. 548–555. Springer, Heidelberg (2007)
5. Burke, C.: Midiki: MITRE Dialogue Kit (2005), http://midiki.sourceforge.net/ (Retrieved December 3, 2008)
6. Traum, D., Bos, J., Cooper, R., Larsson, S., Lewin, I., Matheson, C., Poesio, M.: A model of dialogue moves and information state revision (TRINDI project deliverable D2.1). Report: University of Gothenburg, Sweden (1999)

7. Traum, D., Larsson, S.: The Information State Approach to Dialogue Management. In: Smith, R., Kuppevelt, J. (eds.) Current and New Directions in Discourse & Dialogue, pp. 325–353. Kluwer Academic Publishers, Dordrecht (2003)
8. Trung, H.B.: Multimodal Dialogue Management (State of the art mini-report). University of Twente, Twente (2006)
9. Jurafsky, D., Martin, J.H.: Dialog and Conversational Agents. In: SPEECH and LANGUAGE PROCESSING - An Introduction to Natural Language Processing, Computational Linguistics, and Speech Recognition, ch. 19 (draft). Prentice Hall, Englewood Cliffs (2006)
10. Lemon, O.: Taking computer chat to a whole new level (TALK project). Research EU - Results Supplement, CORDIS Unit, UE Publications Office, p. 25 (2008)
11. Burke, C.: Midiki User's Manual, Version 1.0 Beta. The MITRE Corporation, pp. 1–58 (2005), http://midiki.sourceforge.net/
12. Fulton, J.: Introduction to Zope (1998), http://www.plope.com/Books/2_7Edition/IntroducingZope.stx (retrieved December 3, 2008)
13. Reenskaug, T.: Model View Controller (MVC) pattern (1979), http://heim.ifi.uio.no/~trygver/themes/mvc/mvc-index.html (retrieved January 20, 2009)
14. Popma, R.: JET Tutorial Part 1 (Introduction to JET) (2004), http://www.eclipse.org/articles/Article-JET/jet_tutorial1.html (December 9, 2008)
15. Eclipse: Eclipse 3.4 (2008), http://www.eclipse.org (retrieved November 1, 2008)
16. Guilhoto, P., Rosa, S.: Reconhecimento de voz. Departamewnto de Eng. Informática, Faculdade de Ciências e Tecnologia, Universidade de Coimbra, pp. 1–16 (2001)
17. Microsoft: Microsoft Speech API (SAPI) 5.3 (2009), http://msdn.microsoft.com/en-us/library/ms723627VS.85.aspx (retrieved January 20, 2009)

Experiments with Automatic Query Formulation in the Extended Boolean Model*

Lucie Skorkovská and Pavel Ircing

University of West Bohemia, Faculty of Applied Sciences, Dept. of Cybernetics
Univerzitní 8, 306 14 Plzeň, Czech Republic
{lskorkov,ircing}@kky.zcu.cz

Abstract. This paper concentrates on experiments with automatic creation of queries from natural language topics, suitable for use in the Extended Boolean information retrieval system. Because of the lack and/or inadequacy of the available methods, we propose a new method, based on pairing terms into a binary tree structure. The results of this method are compared with the results achieved by our implementation of the known method proposed by Salton and also with the results obtained with manually created queries. All experiments were performed on the same collection that was used in the CLEF 2007 campaign.

1 Introduction

The field of information retrieval (IR) has received a significant surge of attention in the recent two decades, mainly because of the development of World Wide Web and the rapidly increasing number of documents available in electronic form. As a result, new search paradigms that offer alternatives to (still widely-used) "classic" Boolean and vector space models has been introduced, such as the method employing language models [1] or the approach based on the concept of inference networks [2] that even allows the combination of multiple different IR approaches within a single framework, and many others.

Some of those paradigms are able to accommodate structured queries — for example the inference network framework mentioned above or the extended Boolean model, on which our paper focuses. However, the possibility of structured queries is rarely being exploited and the bag-of-word approach is used instead. This statement is certainly true for the experiments reported in the Czech task of the Cross-Language Speech Retrieval track that was organized within the CLEF 2007 evaluation campaign [3], in spite of the fact that the test collection used in this task contains example information requests in the format of rather richly structured TREC-like topics.

In our paper, we will concentrate on the possible ways of automatic creation of the structured queries from the topics formulated in natural language. All the presented methods of query formulation were developed to be used within the framework of the Extended Boolean model and were tested using the same collection that was used in the

---

* This work was supported by the Grant Agency of Academy of Sciences of the Czech Republic, project No. 1QS101470516, and by the Ministry of Education of the Czech Republic, project No. MŠMT LC536.

V. Matoušek and P. Mautner (Eds.): TSD 2009, LNAI 5729, pp. 371–378, 2009.

aforementioned CLEF 2007 campaign. This collection consists of automatically transcribed spontaneous interviews, segmented into "documents" [1] and, as was mentioned
above, the set of TREC-like topics. More information about the collection can be found
in [4].

2   Extended Boolean Model

Extended Boolean model P-Norm was proposed by Salton, Fox and Wu [5] in order
to take advantages of vector space retrieval model in Boolean informational retrieval.
This model preserves the structure included in Boolean queries, but uses document and
query term weights to compute query-document similarity, so the results can be ranked
according to decreasing similarity.

There are also many other models referred to as extended Boolean models such as
fuzzy set model, Waller-Kraft, Paice and Infinite-One model. All of them were designed
to improve the conventional Boolean model, but Lee [6] analyzed the aspects of these
models and found out that the P-Norm model is the most suitable for achieving high
retrieval effectiveness.

2.1   P-Norm Model

When we think of a simple query containing only two terms, connected with Boolean
operator (AND, OR), we can represent term assignment in two dimensional space,
where each term is assigned different axis. Then we can say, that for the AND query the
point $(1,1)$ (the case that both terms are present) is the most desirable location. On the
other hand, for the OR query is the point $(0,0)$ the least desirable location (both terms
are absent). The similarity of a document and a query is then computed as a distance
between these points and the document, according to the type of the query. For AND
query, the query-document similarity is represented by the complement of the distance
between the document and the $(1,1)$ point. For OR query the similarity measure is the
distance between the document and the point $(0,0)$.

The P-Norm model uses the L_p vector norm for measuring this distances. Conside-
ring a set of terms $t_1, t_2, t_3, ..., t_n$, we can denote w_{t_i, d_j} the weight of the term t_i in the
document d_j and $w_{t_i, q}$ the weight of the term t_i in a given query q. The query-document
similarity measure is then defined as

$$sim\left(q_{t_1 OR t_2}, d_j\right) = \left(\frac{w_{t_1,q}^p w_{t_1,d_j}^p + w_{t_2,q}^p w_{t_2,d_j}^p}{w_{t_1,q}^p + w_{t_2,q}^p}\right)^{1/p} \tag{1}$$

$$sim\left(q_{t_1 AND t_2}, d_j\right) = 1 - \left(\frac{w_{t_1,q}^p \left(1 - w_{t_1,d_j}\right)^p + w_{t_2,q}^p \left(1 - w_{t_2,d_j}\right)^p}{w_{t_1,q}^p + w_{t_2,q}^p}\right)^{1/p} \tag{2}$$

Changing the value of the parameter p from 1 to ∞ will modify the behavior of the
model from pure vector space model ($p = 1$) to conventional Boolean model ($p = \infty$,

---

[1] Those documents are created by sliding a fixed-size window across over a stream of text. Thus
they are not in any way topically coherent and that's why we put the term in quotation marks.

term weights only binary). For the values between 1 and ∞ we obtain an intermediate system between the vector space and Boolean model.

2.2 Term Weights

Since we have no special information about document terms importance, we use the combined $tf \cdot idf$ weight as w_{t_i,d_j}. The idf – inverse document frequency part of this weight represents the importance of a term in the whole collection of documents, the tf – term frequency part represents the occurrence of a term in a concrete document.

Suppose we have N documents in the collection, n_i of them contains the term t_i, $tf_{i,j}$ represents the number of occurrences of the term t_i in the document d_j. Then the weight w_{t_i,d_j} can be defined as

$$w_{t_i,d_j} = tf_{i,j} \cdot log\frac{N}{n_i} \tag{3}$$

To ensure, that $0 \leq w_{t_i,d_j} \leq 1$, the expression (3) is used in a normalized form

$$w_{t_i,d_j} = \frac{tf_{i,j}}{tf_{max_{term\ k\ in\ d_j}}} \cdot \frac{idf_i}{idf_{max_{term\ k}}} \tag{4}$$

For the $w_{t_i,q}$ weight we use simple idf_i weight.

3 Automatic Query Formulation

As mentioned before, the topics in our collection were in natural language (see Fig. 1 for example), so we needed to transform them into Boolean queries. It is not possible to always do that manually, because of the large number of topics and terms in them, so we need to create the queries automatically. We implemented Salton's original method for automatic creation of Boolean queries described in [7] and compared it with a new method we proposed, based on pairing the terms into a binary tree. For comparison, we tried simply ORing all terms and, on the other hand, manually creating some queries.

```
<top>
<num>1166
<title>Chasidismus
<desc>Chasidové a jejich nezlomná víra
<narr>Relevantní materiál by měl vypovídat o Chasidismu
v období před holokaustem, v průběhu holokaustu a po
něm. Informace o chasidských dynastiích a založených a
zničených geografických lokalitách.
</top>
```

Fig. 1. Example of a topic from collection

3.1 Simple Method

The simplest method for the automatic creation of a Boolean query is to take all words as they are in a natural language query and connect them with OR operators (we tried also only AND). Boolean query composed in that way takes no advantage of which Boolean model can give us.

3.2 Original Salton's Method

In this section the method for automatic creation of Boolean queries proposed by Salton is described in short. As an input, we need to have the natural language query and the desired number of retrieved documents wanted by the user (m). The algorithm of creating a Boolean query can be described as:

1. Take single terms t_i from the query that do not appear on stop words list. Set n_i as the number of items indexed by term t_i.
2. Generate term pairs with AND operator (t_i AND t_j) and estimate the number of items retrieved as $n_i \cdot n_j / N$. N is the total number of items.
3. Generate term triples (t_i AND t_j AND t_k) and compute $n_i \cdot n_j \cdot n_k / N^2$. When needed generate also quadruples and estimate $n_i \cdot n_j \cdot n_k \cdot n_l / N^3$.
4. Compute the weight for single terms, term pairs, triples, quadruples as an inverse term occurrence frequency (like idf value),computed as $1 - \frac{n_i}{N}, 1 - \frac{n_{ij}}{N}, 1 - \frac{n_{ijk}}{N}$ or $1 - \frac{n_{ijkl}}{N}$.
5. Formulate a broad OR query from single terms with the highest weight. Estimate the number of retrieved items by that query as $\sum n_i$ over all included terms.
6. If $\sum n_i > m$ modify the query by iteratively removing single terms of the lowest weight and adding term pairs (combinations of removed terms) . When needed start removing term pairs and adding term triples in the same way, possibly remove triples adding quadruples.
7. Stop the process when $\sum n_i + \sum n_{ij} + \sum n_{ijk} + \sum n_{ijkl}$ is approximately equal to the number of desired items m.

3.3 New "Tree-Growing" Method

The method described above is quite complicated and computationally intensive, so we tried to find out some new simpler method, which however should achieve the same or better results. The resulting query also should have some structure to take advantage of the Boolean query processing. The idea of this method is as follows:

1. Take all terms in a query (possibly remove that terms which appear on the stop words list). We tried some different ways of choosing only some terms from the topic, the comparison of these approaches is shown in Sect. 4.
2. Sort the terms by idf value (computed as described in Sect. 2.2) in decreasing order.
3. Make pairs from the terms by connecting two adjacent terms in the list with AND operator, starting from the terms with highest weights. The result of this step is a list of pairs of terms(t_i AND t_j). Compute the idf value of these pairs as an average of the idf weights of contained terms, i.e. $(t_i + t_j)/2$.

4. Sort the list of pairs of terms and connect these pairs further, but this time with the OR operator. Compute the *idf* weight of this pairs the same way as in step 3.
5. Continue this process (step 4) iteratively until you have the resulting pair (of pairs of pairs ....).

The process described above creates the binary tree structure of a query (see Fig. 2), where as leaves we have terms and as a root we have the final query.

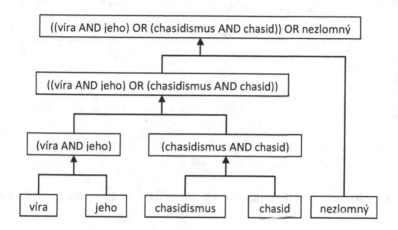

Fig. 2. Creation of the query from terms

3.4 Manual Construction

Queries were manually constructed from topics with respect to the structure of sentences and their meaning. For example, synonyms were connected with OR operator, term phrases with AND operator, then they were parenthesized according to the structure of the sentence. This way we can fully exploit the advantages of Boolean structure in the query.

4 Experiments

The segmented collection we used for experiments (as mentioned above) has 22581 "documents", methods were tested on 29 CLEF 2007 training topics. As an evaluation measure we used mean Generalized Average Precision (mGAP) as used in CLEF 2006 and 2007 Czech task[8] [3]. Both documents and topics were lemmatized in order to achieve better retrieval performance (as showed in [9]) and stop words were omitted. As terms for creating queries were used words from <title> (T) and <desc> (D) fields of the topics. The <narr> (N) field could not be used because then we would have too many terms for manually creating the query and even for the testing of Salton's method.

Salton's method looked promising at the first sight, but when we experimented with it we discovered many problems. This method was designed for queries with small

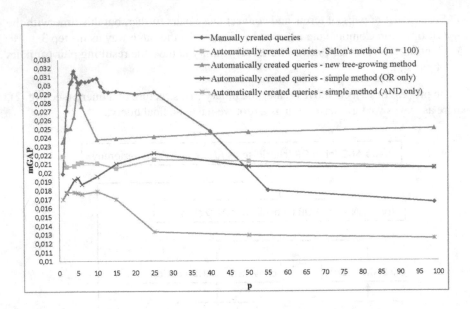

Fig. 3. Comparison of tested methods (p on the x-axis refers to the parameter of the P-Norm model as defined in Sect. 2.1)

number of terms and tested on a very small collection (1033 documents), but we have very long queries (sometimes 30 terms and even more with all TDN fields of the topic) and much bigger collection. For example the query used in original paper as example has 4 terms, which means that they can be combined into 6 pairs or 4 triples. That can be easily created and retrieved. When we take query with 10 terms, we have 45 pairs or 120 triples or 210 quadruples. With 30 terms in original query, we get 435 pairs, 4060 triples or 27405 quadruples. Creating and retrieving such a query is very time and memory consuming, so we had to use only terms with really high idf weight ($idf > 0.9$) in queries that had more than 10 terms in order to get any results at all.

The performance comparison of all tested methods is shown on Fig. 3. As expected, the simple method (both operator variants) had the worst results. This is not surprising, as it did not use any organization of the terms. All other methods had better results, although Salton's method only slightly. Salton's method was tested with different values for parameter m (30 - 1000), the value used in comparison ($m = 100$) had the best results. As can be seen, the method we proposed had better results, almost reaching the performance achieved for the manually created queries for $p = 5$ (which we suggest is the best setting – more experiments bellow). Manually created queries naturally yielded the best results as they were created with regard to the semantic properties of the terms (such as synonymy for example).

We have also done experiments with different ways of creating the query with the proposed method, such as using all terms or only terms with the highest weights, varying the number of levels with AND operator, etc. The results are shown in Fig. 4. As can be seen from the comparison, the best results are obtained when adding all terms

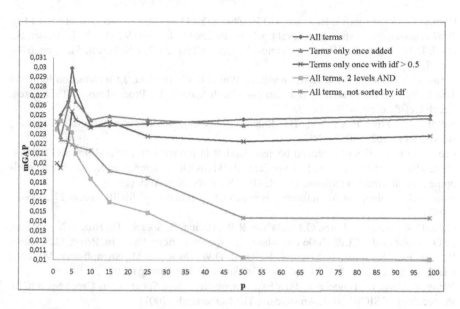

Fig. 4. Different ways of creating the query with the "tree-growing" method (p on the x-axis refers to the parameter of the P-Norm model as defined in Sect. 2.1)

from the topic as many times as they occur in the natural language query. It is also clear that sorting terms by decreasing idf is important.

5   Conclusions and Future Work

As our experiments have shown, our "tree-growing" method almost reaches the performance achieved when using manually created Boolean queries. This method is also much easier to implement and much less computationally intensive than the tested Salton's method. For the verification of our results, we have tested aforementioned methods on 42 CLEF 2007 evaluation topics. The results obtained have shown pretty similar course. The "tree-growing" method achieved even better results than manually created queries, on the other hand, Salton's method achieved slightly worse results than the simple method.

We suppose that using some semantic aspects of the terms in the "tree-growing" method would yield even better results (for example adding synonyms from some vocabulary or using word classes). This would, on the other hand, make the method far more complicated and computationally intensive.

References

1. Croft, W.B., Lafferty, J.: Language Modeling for Information Retrieval. Kluwer Academic Publishers, Norwell (2003)
2. Callan, J.P., Croft, W.B., Harding, S.M.: The INQUERY Retrieval System. In: Proceedings of the Third International Conference on Database and Expert Systems Applications, pp. 78–83 (1992)

3. Pecina, P., Hoffmannová, P., Jones, G.J.F., Zhang, Y., Oard, D.W.: Overview of the CLEF-2007 cross-language speech retrieval track. In: Peters, C., Jijkoun, V., Mandl, T., Müller, H., Oard, D.W., Peñas, A., Petras, V., Santos, D. (eds.) CLEF 2007. LNCS, vol. 5152, pp. 674–686. Springer, Heidelberg (2008)
4. Ircing, P., Pecina, P., Oard, D.W., Wang, J., White, R.W., Hoidekr, J.: Information Retrieval Test Collection for Searching Spontaneous Czech Speech. In: Proceedings of TSD, Plzeň, Czech Republic, pp. 439–446 (2007)
5. Salton, G., Fox, E.A., Wu, H.: Extended Boolean information retrieval. Commun. ACM 26(11), 1022–1036 (1983)
6. Lee, J.H.: Properties of extended boolean models in information retrieval. In: SIGIR 1994: Proceedings of the 17th annual international ACM SIGIR conference on Research and development in information retrieval, pp. 182–190. Springer, New York (1994)
7. Salton, G.: A blueprint for automatic Boolean query processing. SIGIR Forum 17(2), 6–24 (1982)
8. Oard, D.W., Wang, J., Jones, G.J.F., White, R.W., Pecina, P., Soergel, D., Huang, X., Shafran, I.: Overview of the CLEF-2006 cross-language speech retrieval track. In: Peters, C., Clough, P., Gey, F.C., Karlgren, J., Magnini, B., Oard, D.W., de Rijke, M., Stempfhuber, M. (eds.) CLEF 2006. LNCS, vol. 4730, pp. 744–758. Springer, Heidelberg (2007)
9. Ircing, P., Oard, D., Hoidekr, J.: First Experiments Searching Spontaneous Czech Speech. In: Proceedings of SIGIR 2007, Amsterdam, The Netherlands (2007)

Daisie: Information State Dialogues
for Situated Systems

Robert J. Ross and John Bateman

SFB/TR8 Spatial Cognition
Universität Bremen, Germany

Abstract. In this paper, we report on an information-state update (ISU) based dialogue management framework developed specifically for the class of situated systems. Advantages and limitations of the underlying ISU methodology and its supporting tools are first discussed, followed by an overview of the situated dialogue framework. Notable features of the new framework include its ISU basis, an assumption of agency in domain applications, a tightly-coupled, plugin-based integration mechanism, and a function-based contextualization process. In addition to reporting on these features, we also compare the framework to existing works both inside and outside of the situated dialogue domain.

1 Introduction

Just as many different grammatical formalisms continue to be investigated and debated, there is considerable variation in approaches to dialogue manager design. Different dialogue management models make varying assumptions about not only the dialogue manager's internal processes, but also about the nature of dialogue and domain-specific knowledge organization. Of the popular dialogue management model types, the so-called Information State Update (ISU) approach [1] is seen as a flexible dialogue management model based around the use of discourse objects (e.g., questions, beliefs) and rules which encode update-relationships between these objects. ISU-based systems may be viewed as practical instantiations of early agent-based theories (see [2] for an overview), where the complexities of language grounding processes are taken into account.

While the ISU methodology and the toolkits built upon it, e.g., TrindiKit [1] and Dipper [3], have been shown to provide a basis for flexible dialogic interaction, these existing works suffer a number of general limitations. From the perspective of system development, the ISU methodology and ISU toolkits have traditionally been based around a declarative update rule based design. Functionality both for integrating user dialogue contributions and planning system contributions is encoded as the firing of a complex update rules with complicated antecedents. While a rule-based approach is useful in working within an overtly declarative programing environment, the operation of resultant rules can become highly opaque. Moreover, in the classical ISU approach, the interface between application knowledge and dialogue management is described in terms of dialogue plans – which are procedural instructions regarding the parameters which have to be filled for a meaningful application update or query to be made. Thus,

V. Matoušek and P. Mautner (Eds.): TSD 2009, LNAI 5729, pp. 379–386, 2009.

dialogue plans can be seen as a procedural variant on traditional frame filling where the frame and the process of frame filling are conflated back into a single execution influencing unit. While this approach allows fast development, it deviates from the core advantage of frame-filling approaches, i.e., the separation of application knowledge from the general mechanisms of dialogue management.

More critically though, traditional ISU theories and implementations have been developed to the most part for phone-based information-seeking applications. However, we argue that the class of *situated* systems, i.e., computational applications which are embedded in a real or virtual environment, and which are typically capable of perceiving, reasoning on, and acting within that environment, are a more progressive domain for dialogue system development. This domain does however present notable research challenges to dialogue management in general, and to the ISU approach in particular. First, situated applications are fundamentally *agentive* in nature. Situated application have complex internal mental states, operate in a semi-autonomous manner, and perform actions that have clear temporal extent. Such agency features minimally require mixed-initiative and multi-threading in dialogues, but also a coupling of dialogue management with rational agency that recognizes the disparate, yet tightly coupled, nature of these elements. Second, in the situated domains, physical context dependencies are frequent as people make exophoric reference to objects in their environment, describe spatial relationships with respect to particular frames of reference and perspectives, and describe processes in ways which are in themselves related to the situational context. Understanding and handling these spatial language and situated dependences and the complications they add to the contextualization process becomes key to developing flexible interaction styles.

In this paper, we report on an ISU derived dialogue systems framework which we have developed. The framework, named Daisie (**Dia**Space's **A**daptive **I**nformation **S**tate **I**nteraction **E**xecutive), builds on what we see as the best elements of the ISU methodology, but rationalizes them on one hand to improve dialogue manager design, and also to addresses the issues of contextualization and agent-based domain design. We proceed by first stating the cornerstone modelling principles for the framework, before giving an overview of the framework itself, and then reporting on a particular class of situated dialogue that we have been investigating with the framework. Before concluding, we compare Daisie to a number of related works.

2   Modelling Principles

The six basic design principles of the Daisie framework are as follows:

- **Ginzburgean Dialogue Modelling:** Following [4], the dialogue management model assumes a Dialogue Game Board style structure rather than a *proposition bag* or a structural transition network as the primary organization of dialogue state at any given time. Moreover, following the rationalization of Ginzburg's theories, we assume distinct language integration and language planning processes which operate over the information state.
- **Modularity:** The dialogue processing architecture has been designed to be highly modular in terms of the solutions applied for particular processing needs. Here

a modularity of design is included not only in handling basic language processing resources such as speech recognition or language analysis interfaces, but is also included in the approach to dialogue manager design itself.

- **Two-Level Semantics:** In line with a requirement of modularity, the dialogue framework applies a two-level semantics methodology. In this methodology, we assume that the interface to reusable grammatical resources is defined in terms of a linguistic semantics. Domain specific knowledge on the other hand – including the representation of domain knowledge within the Information State – is expressed in terms of an application's own conceptual ontology. During runtime, mappings between these semantics layers must be maintained for both language analysis and production, but the benefit is a more re-usable approach to grammar development.

- **Embeddability:** Early dialogue system implementations typically placed dialogue functionality as the centre of a complete application that included one or more data sources that instantiate the application for a particular domain. However, in real-world applications, spoken dialogue elements are often regarded as an interface or view on a complete application. Thus, rather than being implemented as a stand-alone application, the Daisie toolkit has been implemented as a library of functionality that can be embedded within a third-party application.

- **Agent-Oriented Domain Modelling:** Although Daisie has been designed to be relatively information state theory independent through modularity, the core of the Daisie system implements an "agent-oriented dialogue management" strategy. This strategy views domain applications as agentive in nature. Daisie thus provides resources so that domain applications can be cast in agent terms. The agency features are however lightweight, and are comparable with current trends in the state of the art in agent-oriented design and programming.

- **Tight Integration:** Rather than applying middleware frameworks similar to the Open Agent Architecture [5] to the integration of dialogue systems, we reject such highly decentralised integration approaches in favour of a tightly coupled strategy. We argue that the future of spoken dialogue systems will require ever more tight integration between elements where contextual information is applied at increasingly early stages to resolve ambiguity in input. While constant communication between components could be achieved through a distributed architecture, we argue that a tighter coupling between components both improves efficiency at runtime, and also improves the development process since 'programming interface'-based design, rather than composing and interpreting messages, is in practice easier to implement. Moreover, we argue that although a multi-agent based approach to software integration is very useful in the case of dynamic systems, typical spoken dialogue systems are very static in organization, and thus little is actually gained from a fully distributed architecture.

3   Daisie Components

With these design principles in mind, Daisie consists of three functionality groups that are integrated into an application alongside linguistic resources. These three Daisie functionality groups are: (a) *The Daisie Plugins*, which provides a range of language technology solutions that have been pre-defined to populate Daisie with a minimum inventory of language parsers, speech synthesizers, and so forth; (b) *The Daisie Backbone*, which integrates language technology plugins with dedicated Daisie dialogue management code into a coherent dialogue system; and (c) *The Daisie Application Framework*, which is used directly by an application to provide the minimum interface to Daisie functionality.

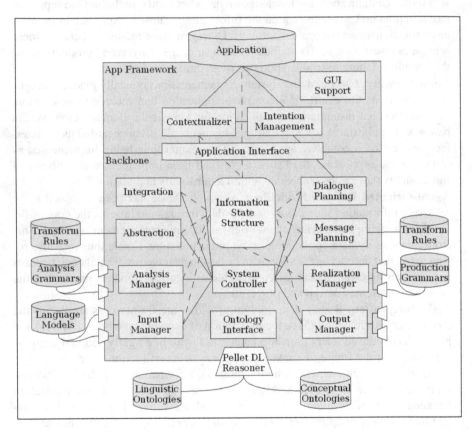

Fig. 1. Overview of the Daisie Dialogue System Framework

Figure 1 depicts Daisie's principle processing components. The Daisie Application Framework and Daisie Backbone are shown in the centre of the figure. The Daisie Plugins on the other hand are depicted as trapezoid's around the boundary of the Daisie Backbone. These plugins in turn interface with linguistic resources depicted as cylinders. In the following, the three Daisie functionality groups are described further.

3.1   The Daisie Backbone and Plugins

Following the ISU methodology, at the centre of the Daisie Backbone sits the Information State which provides a shared memory for each of the primary processing components (regular rectangles). The information state, like all Daisie components, can be specified in a modular way to support new dialogue strategies in conjunction with the language integration and dialogue planning components. Also at the centre of the Daisie backbone sits the system controller. Once again, based on a modular design, the system controller can be replaced to investigate specific control techniques in dialogue system design.

Moving outwards from the Information State and System Controller, the Daisie Backbone includes a range of language technology manager components and dialogue processing modules. The language technology manager components, i.e., *Input Manager*, *Analysis Manager*, *Realization Manager*, and *Output Manager* provide abstractions over language technology solutions. The language technology plugins, illustrated as trapezoids at the periphery of the Daisie Backbone, each implement a well-defined interface for that language technology component type. In addition to these language technology management modules, the Daisie backbone also contains four dialogue management modules:

- **Language Abstraction:** Transforms shallow semantics produced by language analysis grammars into domain-specific dialogue acts more suited to dialogic and situated reasoning. These dialogue acts are expressed in terms of the agent's own application ontology.
- **Language Integration:** Integrates abstracted language content into the agent's dialogue state through the construction and manipulation of input user dialogue acts and the information state.
- **Dialogue Planning:** Analyses the information state to determine the most appropriate dialogues acts to be made by the system in furthering user and/or system dialogue goals.
- **Message Planning:** Transforms planned system acts expressed in terms of abstract conceptual semantics into a shallow linguistic semantics suited to realization.

When taken together with the information state model, these processing modules constitute a *dialogue manager* in the traditional sense. In section 4, we briefly discuss a specific instantiation of Daisie for an agent-oriented dialogue management theory for the robotics domain.

Daisie includes a number of language technology plugins that can be used to construct a minimal operational spoken dialogue system. These plugin types are grouped according to Daisie's language technology manager classes. That is, four plugin types are assumed by Daisie: input plugins, analysis plugins, realization plugins, and output plugins. For each plugin type, an interface is provided which specifies required functional and ontological constraints that must be met by an implemented plugin solution. Plugins for example include an analysis plugin based on the OpenCCG system [6], and a language generation plugin based on the KPML language generator [7]. More complete information on the plugins available for use with Daisie, as well as descriptions of Daisie's application to concrete domains are provided in [8].

3.2 The Daisie Application Framework

The Daisie Application Framework provides functionality that is embedded within, and typically extended by, a given application. This potential for extension, or indeed replacement, is in contrast with the backbone modules which are stand-alone and not directly interacted with by an application. Referring again to Figure 1, such functionality for example includes Graphical User Interface support. However, since Daisie is not intended to act as a standalone dialogue engine, a complete GUI is eschewed in favour of functionality which allows the embedding of dialogue system and information state views directly within a developed application. In the following, we briefly comment on the three other main components of the Daisie Application Framework:

- **Intention Management:** As indicated, Daisie has been used to investigate the application of ISU style dialogue management in the situated domain – and in particular in the robotics domain. To achieve this, a tight coupling between core dialogue processing and an agentive or *intentional* view on an application is required. To support this tight coupling, the *Intention Management* module provides a lightweight behaviour management system based on the agent-oriented programming paradigm. This implementations is optional and may be replaced at development time with alternate theories or implementations.
- **Contextualization:** Also particular to the situated domain, contextualization of the discourse context is of critical importance. In the situated domain physical proximity, spatial structure, and the affordances offered by an interpreting agent's environment all critically effect language understanding. Daisie recognizes these factors through the inclusion of an explicit contextualization step in the normal dialogue control process. A generalized contextualization process based on resolution and augmentation functions associated with semantic roles is provided by Daisie as part of its default agent-oriented dialogue management model, but this may be replaced as necessary by specific applications.
- **The Application Interface:** The *Application Interface* provides a gateway between the Daisie Backbone and application functionality proper. As a true interface, this module must be instantiated by a given application to ensure functionality assumed by Daisie Backbone components is available.

4   Applying the Daisie Framework

The Daisie framework has been applied in the development of a dialogue interface for a class of real and simulated intelligent wheelchairs for the disabled and elderly [8]. This application of Daisie makes use of an *Agent-Oriented Dialogue Management Model* [9]. Features of this agent-oriented model include a Ginzburgean view on dialogue organization and grounding processes, the use of a dialogue move as a frame-like staging ground in information state, and the function-based contextualization process introduced in the last section.

5   Related Work

Rather than being developed as a replacement for existing, general-purpose ISU-based dialogue management frameworks, Daisie has been developed to support the investigation of knowledge representation, agency, and contextualization in dialogue processes. Despite this difference in motivation, it is useful to compare Daisie with both general purpose and specific ISU oriented dialogue frameworks. From an architectural perspective, the Daisie framework probably shares most similarities with the Allen's TRIPS architecture [10]. While TRIPS still provides what is arguably the most complete *neo* agent-based dialogue system, Daisie attempts to provide a more lightweight system and pays particular attention to the issue of situated contextualization. The Daisie framework is of course also comparable to long-standing, general purpose ISU toolkits such as TrindiKit [1] and Dipper [3]. Notable differences between Daisie and these works are Daisie's eschewing of a rule-based methodology in favour of a more transparent procedural method for language integration and dialogue planning. Of the specific ISU-based applications for the situated domain, Daisie, or rather the agent-oriented dialogue management model developed on top of it, is probably closest to Lemon's Conversational Intelligence Architecture [11] in that both models advocate a tight coupling between dialogue management and agency features – although in our work we have attempted to push towards issues of representation and contextualization in an ontologically modular architecture.

6   Conclusions and Future Work

Although maturing continuously, the nature of dialogue systems means that a framework like Daisie will always be under development. In immediate future work, we aim to improve the generalizability of Daisie by supporting rule-based, as well as procedural, specification of key dialogue processing components. We shall do this by building upon existing declarative programming constructs for Java-based systems already available in the agent frameworks community. We view some of the toughest tasks in semantics-rich dialogue systems as building a clear understanding of the complete dialogue process, and having a design basis on which particular models of information state and semantics processing can be built in a tractable manner. We have developed Daisie to help in addressing such tasks, and though the work can only be a basis for intelligent spoken dialogue interface development, we believe Daisie is a useful contribution to the language understanding and dialogue systems community.

Acknowledgement

We gratefully acknowledge the support of the Deutsche Forschungsgemeinschaft (DFG) through the Collaborative Research Center SFB/TR 8 Spatial Cognition - Subproject I5-[DiaSpace].

References

1. Traum, D., Larsson, S.: The Information State Approach to Dialogue Management. In: Smith, R., van Kuppevelt, J. (eds.) Current and New Directions in Discourse and Dialogue, pp. 325–353. Kluwer Academic Publishers, Dordrecht (2003)
2. McTear, M.F.: Spoken dialogue technology: Enabling the conversational user interface. ACM Computing Surveys (CSUR) 34, 90–169 (2002)

3. Bos, J., Klein, E., Lemon, O., Oka, T.: DIPPER: Description and Formalisation of an Information-State Update Dialogue System Architecture. In: 4th SIGdial Workshop on Discourse and Dialogue (2003)
4. Ginzburg, J.: Dynamics and the semantics of dialogue. In: Seligman, J. (ed.) Language, logic and computation, CSLI Stanford. CSLI Lecture Notes, vol. 1, pp. 221–237 (1996)
5. Martin, D., Cheyer, A., Moran, D.: The Open Agent Architecture: a framework for building distributed software systems. Applied Artificial Intelligence 13, 91–128 (1999)
6. Baldridge, J., Kruijff, G.J.: Coupling CCG and Hybrid Logic Dependency Semantics. In: Proceedings of 40th Annual Meeting of the Association for Computational Linguistics, Philadelphia, Pennsylvania, pp. 319–326 (2002)
7. Bateman, J.A.: Enabling technology for multilingual natural language generation: the KPML development environment. Journal of Natural Language Engineering 3, 15–55 (1997)
8. Ross, R.J.: Situated Dialogue Systems: Agency & Spatial Meaning in Task-Oriented Dialogue (Forthcoming)
9. Ross, R.J., Bateman, J.: Agency & Information State in Situated Dialogues: Analysis & Computational Modelling. In: Proceedings of the 13th International Workshop on the Semantics & Pragmatics of Dialogue (DiaHolmia), Stockholm, Sweden (2009)
10. Allen, J., Ferguson, G., Stent, A.: An architecture for more realistic conversational systems. In: Proceedings of the 2001 International Conference on Intelligent User Interfaces, pp. 1–8 (2001)
11. Lemon, O., Gruenstein, A., Peters, S.: Collaborative Activities and Multi-tasking in Dialogue Systems. Traitement Automatique des Langues (TAL) 43, 131–154 (2002)

Linguistic Models Construction and Analysis
for Satisfaction Estimation

Natalia Ponomareva[1] and Angels Catena[2]

[1] University of Wolverhampton, UK
nata.ponomareva@wlv.ac.uk
[2] Universidad Autonoma de Barcelona, Spain
angels.catena@uab.cat

Abstract. Automatic analysis of customer conversations would be beneficial for service companies to improve service quality. In this case, such customer characteristics as satisfaction or competence are of the special interest. Unfortunately, their manual estimation is very laborious and it has a high level of subjectivity. In this work, we aim at parameterization of dialogues for formal (automatic) assessment of customer satisfaction. We elaborate a set of linguistic indicators represented both by lexico-syntactic patterns and rules and introduce their classification by kind, location and sign. We propose several linear regression models for satisfaction estimation and check them and their parameters on statistical significance. The best of the models demonstrates rather high level of concordance between automatic and manual assessments.

1 Introduction

1.1 Problem Setting

Nowadays, dialogue processing is widely used for constructing automatic dialogue systems and for improving service quality. By the word "dialogue" we mean a conversation between a customer and a service center, and by the word "processing" we mean a classification of customers. Politeness, competence, satisfaction, etc. are very important characteristics for customer classification but its formal estimation is quite difficult due to a high level of subjectivity. Thus, these characteristics usually are not taken into account or they are estimated manually [1].

In this work, we aim at constructing an empirical formula to evaluate a customer satisfaction. We believe that this is one the principal customer characteristics the service company is interested to estimate. Our formula is based on two components:

(i) linguistic indicators of satisfaction extracted from dialogues;
(ii) subjective expert assessments of dialogues.

Linguistic indicators are represented by lexico-syntactic patterns related to a characteristic of satisfaction. Their selection depending mostly on an expert experience is a key of the precise evaluation.

Subjective expert assessment may be obtained by means of manual evaluation of a set of dialogues. For this, a fixed scale is taken and each dialogue is evaluated in a framework of this scale. Usually symmetric normalized scale [-1,1] or positive normalized scale [0,1] are considered.

V. Matoušek and P. Mautner (Eds.): TSD 2009, LNAI 5729, pp. 387–394, 2009.

1.2   Data

The data for our research consists of 85 written dialogues between passengers and Barcelona station enquiry service. Although the language of dialogues is Spanish and the analysis has been carried out in Spanish as well, we will translate all dialogue phrases and lexico-syntactic patterns from Spanish into English for better understanding by all readers. Table 1 shows the distribution of the customer satisfaction level within the corpus. As it can be seen the majority of customers either satisfied or neutral (34.1% and 44.7% respectively).

Table 1. Distribution of customer satisfaction value within the corpus

Level of satisfation	Number of dialogues	Relative frequency,%
very satisfied	7	8.2
satisfied	29	34.1
neutral	38	44.7
dissatisfied	9	10.6
very dissatisfied	2	2.4

1.3   Main Difficulties

During our research we encounter with some difficulties of satisfaction assessment. First, we would like to claim that "customer satisfaction" is an ambiguous term since it can be understood in two possible ways: the first one refers to the customer service, and it is closely related to politeness evaluation; the second one concerns the analysis of customer needs focusing on how the service satisfies customer demands. Since such dialogues are simple telephone calls requesting detailed information (mostly about train schedules, train fares and train routes), it seems especially important to draw attention to certain dialogues, during which a customer books or buys a ticket. These types of conversations make it possible to conclude whether customer needs have been satisfied.

Secondly, our corpus consists of written dialogues, and it is reasonable to think that working with spoken corpora would be more beneficial since linguistic indicators would include not only lexico-syntactic marks but also prosodic realization. In fact, prosodic features allow for accurate prediction of customer satisfaction level because emotional content helps to reveal a real meaning of lexico-syntactic structures. On the other hand, the lack of prosodic information in written dialogues may hinder the interpretation of some lexical marks.

Finally, the customer behaviour is determined by their personality and education and, therefore, well-educated customers tend to be reserved and do not show their discontent very evidently. This implies development of finer satisfaction models allowing for extraction of hidden information.

The paper is organized as follows. In Section 2, we introduce linguistic indicators of satisfaction classifying them by different characteristics as kind, location and sign and suggest formulae for their numerical expressions. In Section 3, we present a set of satisfaction models and introduce the scale for manual assessment of satisfaction. Section 4 is devoted to checking statistical significance of satisfaction models. Conclusions and future work are drawn in Section 5.

2 Linguistic Indicators of Satisfaction

Despite the mentioned in Section 1.3 problems with satisfaction assessment the fulfilment of what a customer wants and needs might be partially predicted using some lexical marks. In this section, we present linguistic markers or indicators based on lexico-syntactic patterns we find useful for detection and evaluation of customer satisfaction.

2.1 Classification of Indicators

We propose the following classification of linguistic indicators of satisfation:

1. **By kind: user feedback or question-answer.** We distinguish two different situations: user feedback (a customer just expresses his/her reaction on obtained information) and question-answer (a customer wants to receive more details on an interesting topic). A simple example of user feedback gives the following conversation where "I" stands for informer and "C" for customer:

> I: The train leaves at 9 a.m.
> C: Ok
> I: and arrives to Valencia at 12 p.m.
> C: very well, perfect!

Question-answer markers (**QA**), for its turn, join a rather diverse situations encountered in a dialogue. In this paper, we mostly work with two important groups of QA indicators:

- Negative informer answer. Answering to a query, the informer may use some negative pronouns or adverbs like "none" or "never". In general, it implies that some customer needs can not be completed. In case there is an apology from an informer accompanying these answers (e.g. "None, I am sorry") or in response to a yes-no question, it clearly indicates significant regret for not satisfying customer needs. This is also valid for user queries ending by question tags. For example:

> C: This train stops in Barcelona, doesn't it?
> I: No, I am sorry

- User questions or negative sentences as a reaction of discontent. Some user's questions involving restrictive adverbs (e.g. Is there **only** one train?; Is it the **unique** way to arrive?), contrastive determiners (e.g. Don't you have **other** trains?) or some negative structures (e.g. Don't tell me that it doesn't stop in Barcelona!) indicate dissatisfaction about previous information.

2. **By sign: positive or negative markers.** For example, such user exclamations as: "Perfect!", "Wonderful" are evidently positive and express satisfaction, while "Oh, my God!", "Damn!" are negative and convey displeasure.

3. **By location: body or end of a dialogue.** We believe that customer satisfaction level strongly correlates with a position of linguistic indicators in a dialogue. Indeed,

we can presume that the global level of satisfaction is likely to appear in the farewell phrases. Therefore, we distinguish indicators in the body of a dialogue and at the end of it. In fact, it would be more precise to introduce some function, which assigns different weights to indicators depending on their location in a dialogue. In this work, we restrict ourselves considering only two dialogue positions, the use of more complicated dependencies will be the subject of our further research.

2.2 Numerical Expressions for Satisfaction Indicators

According to introduced classification we suggest the following list of linguistic indicators of satisfaction:

1. **Positive user feedback in a dialogue body (PF).** Table 2 presents some of the lexical marks corresponded to **PF** indicator. We must underline that we assign different weights to distinct lexical marks. For example, such words as "Perfect", "Very well" obviously convey a higher level of satisfaction than "Ok" or "Yes".

Table 2. Some lexical marks of positive user feedback

weight	1	2	3
	well;	thank you	very well;
	OK;	very much;	perfect;
	all right;	very kind	wonderful;
	correct;	of you	splendid;
	thank you		that's enough

2. **Negative user feedback in a dialogue body (NF).** We should notice that it is much more difficult to meet negative marks of user feedback in a dialogue than positive ones since the majority of customers does not express emotions in the case of displeasure and usually starts asking questions for finding other solutions. Such questions correspond to **QA** indicator of satisfaction. Some of the most frequent lexical marks related to negative user reaction are presented below:

Uf, damn, gosh, hell, too expensive, too late

3. **Farewell (F).** We have found no cases expressing negative character of **F** indicator. A customer although being unsatisfied does not tend to say rude things to an informer, however, we expect him to be rather reserved in farewell words. The lexical marks related to **F** indicator are mostly the same as for **PF** indicator (Table 2).

4. **Question-answers.** QA indicators can be both positive and negative and are always expected to occur in the body of a dialogue. In this paper, we work only with negative type of **QA** indicator, although we analyze the possibility to reveal its positive cases as well. For example, affirmative answers of an informer to questions involving the modal verb "can" might be considered as fulfillments of customer requirements, e.g.:

C: Could I buy it later?
I: Of course!

In order to represent numerically introduced above indicators we found upon the following assumptions:

(i) A weight of linguistic indicators should be proportional to the number of their occurrences in a dialogue.

(ii) A weight of linguistic indicators should not depend upon the dialogue's length. Due to the fact that the probability to find more lexical marks grows as the length of conversation increases, we normalize an indicator's weight on the dialog's length. Evidently, this rule is valid for all types of linguistic indicators of satisfaction except for the F indicator.

Therefore, denoting by symbol **I** any type of linguistic indicators except for **F** indicator we can propose the following numerical expressions for indicators of satisfaction:

$$\mathbf{I} = \frac{1 + \sum_{i=1}^{N_I} w_i^I}{L}, \quad \mathbf{F} = \sum_{i=1}^{N_F} w_i^F, \tag{1}$$

where w_i^I, w_i^F are the weights and N_I, N_F are the number of lexical marks of corresponding indicators and L is the dialogue's length.

3 Emperical Formulae for Satisfaction

Logically, each linguistic indicator has a different influence on a result level of satisfaction. The kind of dependency between an indicator and a manual estimation can be approximated using different classes of models. In our previous work, where we constructed empirical formulae for characteristic of politeness a class of polynomial models was taken to approximate the customer level of politeness [2]. We used Inductive Method of Model Self-Organization [3] in order to select the model of optimal complexity where optimality was regulated by external criteria. In this work, we restrict consideration by only linear models. In other words, we suppose that all indicators are linear with respect to the value of satisfaction. Such a simplification allows to reflect the principal tendency in satisfaction assessment, but it can not take into account more detail aspects.

We want to experiment with different models of satisfaction in order to analyze which classification of indicators (either by kind, or by sign or by location) makes the greatest impact into the level of satisfaction. Therefore, our first model has two predictors: one of them joins all indicators of user feedback (**UF** indicator) and the other stands for **QA** indicator[1]. The second model, respectively, distinguishes indicators by sign: we denote by **P** positive indicator and by **N** - negative indicator. Further, in the third model one parameter counts lexical marks in the body of a dialogue (**B** indicator) and the other at its end (**F** indicator). Finally, the last model consists of all significant satisfaction indicators we describe in Section 2.2:

---

[1] Hereinafter, we will name linguistic indicators as predictors or model parameters and manual estimation as response variable.

$$\text{Model 1: } F_1 = A_0 + A_1 \mathbf{UF} + A_2 \mathbf{QA}$$
$$\text{Model 2: } F_2 = A_0 + A_1 \mathbf{P} + A_2 \mathbf{N}$$
$$\text{Model 3: } F_3 = A_0 + A_1 \mathbf{B} + A_2 \mathbf{F} \tag{2}$$
$$\text{Model 4: } F_4 = A_0 + A_1 \mathbf{PF} + A_2 \mathbf{NF} + A_3 \mathbf{F} + A_4 \mathbf{QA},$$

where $A_0, A_1, ..., A_4$ are unknown coefficients.

Evidently, the following relations between indicators must be completed:

$$UF = PF - NF + F$$
$$P = PF + F$$
$$N = NF + QA$$
$$B = PF - NF - QA$$

Manual estimations are used as a real data that models from (2) approximate and whose values they must achieve in the ideal case. In our work, an expert estimates all dialogues giving them one of the following values: 1 - very satisfied, 0.5 - satisfied, 0 - neutral, -0.5 - dissatisfied and -1 - very dissatisfied. For more objective evaluation it is better to work with several independent experts.

Therefore, we have a linear system of equations for each type of the models (2):

$$F_i = e_j, \qquad j = 1, ..., n, \tag{3}$$

where e_j is a manual estimation of the j-th dialogue and n is a number of dialogues. This system can be solved by the least square method, and the solution gives an empirical formula for automatic prediction of customer satisfaction.

4 Analysis of Statistical Significance of Satisfaction Models

This section is devoted to verifying statistical significance of proposed satisfaction models and their parameters. For that, we use standard procedures of analysis of variance and parameters estimation [4].

4.1 Global Test

Global test (F-test) is used to verify the statistical significance of a regression model with respect to the whole set of predictors. It is based on the coefficient of determination R^2 which is equal to the square of correlation coefficient between all modelled variables and the response variable. This coefficient says how well a regression model approximates a given data set. It ranges between 0 and 1, where a value of 1 means perfectness of the model. R^2 has a drawback: it slightly increases when growing the number of predictors that disfigures the results of model comparison with different number of predictors. For this reason so-called adjusted coefficient of determination, which takes into account this circumstance, is used.

Table 3 represents the results of checking the global validity of the models described in Section 3. Although all constructed models reject the null hypothesis slight alterations of the coefficient of determination demonstrates a minor advantage of the classification by sign.

As it can be seen in Table 3 the best results give the Model 4:

Table 3. Global test on satisfaction models significance

Model	R^2	Adjusted R^2	F-value	p-value
1	0.47	0.45	36.08	« 0.0001
2	0.49	0.47	39.59	« 0.0001
3	0.48	0.46	37.71	« 0.0001
4	0.57	0.55	26.83	« 0.0001

$$F_4 = 0.12\mathbf{PF} - 0.68\mathbf{NF} + 0.07\mathbf{F} - 0.91\mathbf{QA}$$

It exploits all types of linguistic indicators and achieves rather good concordance with real data ($R^2 = 0.57$) although 43 % of variation still remains unexplained. Therefore, there is a need for improving our model either by adjusting existing indicators or by introducing new ones.

4.2 Test on Individual Parameters

Rejection of above-stated hypothesis does not contradict the fact that some indicators have null regression coefficients. For checking statistical significance of regression coefficients t-test is used. We calculate t-statistic for coefficients of all constructed models and prove them to be statistically significant with very high reliability (more than 99.9%). In Table 4 we give t-values and p-values for coefficients of Model 4.

Table 4. t and p-values corresponded to t-test for regression coefficients (Model 4)

	PF	NF	F	QA
t-value	3.64	-3.62	4.95	-4.14
p-value	0.0005	0.0005	< 0.0001	< 0.0001

Besides, it seems interesting to analyze the strength of relation between predictors and response variable. In Table 5 corresponding correlation coefficients together with their p-values are presented. All indicators show sufficient correlation with the level of satisfaction, especially farewell phrases and question-answers, whose absolute values of the correlation coefficient exceed the level of 0.5.

Table 5. Correlation coefficients of satisfaction indicators

Satisfaction indicator	Correlation coefficient R	p-value
PF	0.44	< 0.0001
NF	-0.34	0.001
F	0.52	< 0.0001
QA	-0.5	< 0.0001

5   Conclusions and Future Work

In this paper, we study such important customer characteristic as satisfaction and construct the empirical formula of satisfaction founded upon linguistic indicators and manual assessments. We give several classifications of indicators based on their kind, sign and location. In order to evaluate which classification of indicators contributes more into the result level of satisfaction different satisfaction models are considered and estimated. All proposed models are proved to be statistically significant with rather good coefficient of determination that achieves a value of 0.57 for our best model.

In future, we plan a) to check the validity of the model on a control set of dialogues; b) to introduce a weight function depending on the position of lexical marks in a dialogue; c) to extend a list of lexical marks adding more complex cases.

References

1. Alexandrov, M., Sanchis, E., Rosso, P.: Cluster analysis of railway directory inquire dialogs. In: Matoušek, V., Mautner, P., Pavelka, T. (eds.) TSD 2005. LNCS (LNAI), vol. 3658, pp. 385–392. Springer, Heidelberg (2005)
2. Alexandrov, M., Blanco, X., Ponomareva, N., Rosso, P.: Constructing empirical models for automatic dialog parameterization. In: Matoušek, V., Mautner, P. (eds.) TSD 2007. LNCS (LNAI), vol. 4629, pp. 455–463. Springer, Heidelberg (2007)
3. Ivahnenko, A.: Inductive method of model self-organization of complex systems. Tehnika Publ. (1982) (in Russian)
4. Cramer, H.: Mathematical methods of statistics. Mir (1975) (in Russian)

Shallow Features for Differentiating Disease-Treatment Relations Using Supervised Learning
A Pilot Study

Dimitrios Kokkinakis

Department of Swedish Language, Språkdata, University of Gothenburg,
Box 200, 40530 Gothenburg, Sweden
dimitrios.kokkinakis@svenska.gu.se

Abstract. Clinical narratives provide an information rich, nearly unexplored corpus of evidential knowledge that is considered as a challenge for practitioners in the language technology field, particularly because of the nature of the texts (excessive use of terminology, abbreviations, orthographic term variation), the significant opportunities for clinical research that such material can provide and the potentially broad impact that clinical findings may have in every day life. It is therefore recognized that the capability to automatically extract key concepts and their relationships from such data will allow systems to properly understand the content and knowledge embedded in the free text which can be of great value for applications such as information extraction and question & answering. This paper gives a brief presentation of such textual data and its semantic annotation, and discusses the set of semantic relations that can be observed between *diseases* and *treatments* in the sample. The problem is then designed as a supervised machine learning task in which the relations are tried to be learned using pre-annotated data. The challenges designing the problem and empirical results are presented.

1 Introduction

Medical discharge summaries and the narrative part of clinical notes provide an information rich, nearly unexplored corpus of evidential knowledge that is considered as a goldmine for practitioners in the language technology field and a great challenge for data and text mining research. The capability to extract the key concepts (biomedical, clinical) and their relationships will allow a system to properly understand the content and knowledge embedded in the free text and enhance the searching potential by easily finding diagnoses and treatments for patients who have either had positive or negative effect and for which keyword-based techniques *cannot* easily differentiate. E.g., knowing the relationship that prevails between a medication and a disease should be useful for searching the free text and easier finding answers to questions such as: *"What is the effect of treatment X to disease Y?"*, as in the example: '*steroids have not be proven to prevent the genesis of osteoporosis*'. Thus, in our work we are trying to address the following general question, given a training set of previously annotated relations between recognized entity mentions of interest is

V. Matoušek and P. Mautner (Eds.): TSD 2009, LNAI 5729, pp. 395–402, 2009.
© Springer-Verlag Berlin Heidelberg 2009

it possible to use supervised learning to determine the most likely relation between a new set of instances and what performance can we expect?

The paper starts by providing some notes on related work (Section 2). Section 3 gives a brief presentation of the textual material used for learning the semantic relationships and discusses the pre-processing of the texts and the set of relationships that capture the semantic distinctions between two concepts of interest, namely diseases (including symptoms) and treatments (chemical and drugs). The manual annotation is also discussed, which enables the creation of a benchmark data set to allow the evaluation of different data mining algorithms on the semantic relationship prediction. Since the problem is designed as a supervised machine learning task, Section 4 presents the core of the investigation, including a description of the features extracted from the annotated texts. The different learning algorithms applied to the data, results and the comparison of the different approaches are presented and discussed in Section 5, while conclusions and future plans end the paper.

2 Related Work

Deeper semantic analyses, such as automatic recognition of relations between words in free texts (e.g. fact extraction [1]) have gained a renewed interest during the last years, particularly in the context of SEMEVAL competition ([2]). However, very few attempts have been done in order to distinguish the relationships that can occur between domain specific semantic entities. Rosario & Hearst ([3]) examined the problem of distinguishing among eight relation types that may occur between treatment and disease in bioscience texts. After manual annotation of 40 abstracts from Medline, the authors used both graphical models and a neural network achieving an f-score of 96.6% for the recognition of the appropriate relation. Vintar *et al.* [4] used manually constructed lists of pairs of MeSH-classes that represent specific relations (e.g. Nucleic Acids [D13]–Pathological Conditions, [C23]) and a corpus of annotated abstracts for exploring the contextual features of relations. Using hierarchical clustering they could identify a small set of typical context words that characterize the relationships. Roberts *et al.* [5] used supervised learning for the identification of relationships between clinically important entities from patient narratives. Using a corpus of various sizes and eleven sets of shallow features (such as distance between arguments) achieved an f-score of 69%. In general, feature analysis experiments in the task of relation extraction (general corpora) conducted by [6] indicated that full parsing is not as useful as one might have thought since most relation elements are very close and large-distance relations are blurred by parsing errors. They found that 70% of the relations determined by two entity mentions are separated by at most one word. The authors suggest that most of the useful information for relation extraction is shallow and can be captured by chunking. Sibanda [7] applies semantic relation classification between 20 relationship types, such as *test reveals disease*. Using Support Vector Machines she obtained a micro-average of 84.5% and a macro-average of 66.7%. Our work has similarities to [7] and our starting point has been the re-use of some of the relationships defined by her.

3 Material

Although ethical issues might be a barrier to directly accessing patient data, without first de-identifying its content and get the appropriate permits from ethical committees, there are other means that can circumvent this barrier, for instance by looking at similar data but with fewer restrictions upon their content. Thus, in order to avoid such ethical and administrative problems associated with this type of texts we explore texts given to medical students in the form of written examination papers, that directly mirror the "reality" the students are suppose to meet as they start their professional career, particularly in the clinical setting. Parts of these examinations are often in the form of "fake" discharge summaries and case reports, where the students have to read and comprehend in order to evaluate the clinical problems and solutions proposed therein, or answer questions related to the decisions and actions that have to be taken based on symptoms, laboratory findings, history of the patient etc. Apart from gathering such novel text documents from the medical faculties across Sweden, we also included a number of questions/answers from a large database of evaluated drug information called *Drugline*. Drugline contains responses from clinical pharmacologists to specific patient cases, produced by the department of Clinical Pharmacology, Karolinska University Hospital in cooperation with the regional drug information centres. The database contains complex clinical questions posed by physicians (e.g. on side effects, drug interactions and pharmacy).

3.1 Data Pre-processing

Prior to the extraction of suitable features, the following pre-processing steps have been applied to the sampled data. Initially, each relevant document was tokenised and automatically annotated with the 2006 versions of the Swedish and English MeSH thesauri ([8]). From the annotated corpus we selected sentences having both a medication and/or therapeutic as well as a symptom and/or disease MeSH-tags. In MeSH, these tags belong to the categories D (Chemicals and Drugs), E (Analytical, Diagnostic and Therapeutic Techniques), C01-C23 (Diseases & Pathological Conditions, Signs and Symptoms). We collected a random small sample of sentences (188) that fulfilled the above criteria from the corpus in order to perform an in-depth analysis of the structure of such contexts and get valuable insights into the type of linguistic processing required in order to extract suitable features and in the future to refine and to be able to apply the methodology in a much larger scale. For gaining the maximum performance in terms of quality of the results by the automatic classification process that follows, we started by pre-processing the texts with a number of language specific, linguistically-aware annotation tools. These tools recognize multiword linguistic expressions, namely: (i) phrasal verbs, e.g. sätta ut 'to prescribe'; (ii) multiword prepositions, e.g. p g a 'because of', and (iii) multiword adverbs, e.g. t ex 'for example'. Moreover, we merged variants of abbreviated adverbs and prepositions to single variants, counting them as tokens of length '1'. We also applied compound segmentation (in Swedish compounds are written without intervening whitespace), e.g. *virussjukdomar* 'virus diseases'. The identification, segmentation and analysis of compounds, are challenging tasks that poses serious problems for tools that will automatically process Swedish texts. The compound segmentation and the use of the heads of the compounds we apply, is closely correlated to the use of heads of noun phrases as a feature; [6]. We believe that this is an important step for achieving better results during model building.

The texts are also annotated semantically with a rich set of generic named entities, part-of-speech information and they are finally lemmatised. For the semantic processing, we apply an entity recogniser for Swedish described in [9] which includes a large set of numerical subtypes, such as dosage, percent and age, which are relevant for the domain (medical texts contain a plethora of numerical data). Named entities are also important since they can be multiword and can be safely replaced with their label. For instance, in the sample we could identify and annotate multiword entities of various types, such as name of studies, which corresponds to the category *WORK* in the entity tagger, *<Collaborative Atorvastatin Diabetes Study>*; organizations which corresponds to the category *ORG* in the entity tagger, *<Svensk Förening för Bröstkirurgi>* 'Swedish Society of Breast Surgery' and groups of persons which corresponds to the category *PRS* in the tagger, *<kritiskt sjuka intensivvårdspatienter>* 'critically/extremely sick intensive care patients'. All recognised entities were then replaced by their XML-labels, such as *<WORK>, <ORG>* and *<PRS>*.

4 Semantic Relations

As a starting point we began by looking at the set of the five semantic relationship labels that can occur between a *treatment* and a *disease*, presented by [7], namely: *treatment administered-for disease, treatment causes disease, treatment cures disease, treatment does-not-cure/worsens disease*, and *treatment discontinued-in-response-to disease*. In our study *Disease* or *Symptom* are conflated into a single category. We believe that this set can be extended with the addition of a new set of semantic relationships that probably better cover for our available data set. New relationships that we intend to explore in the future are for instance: *treatment1 more-effective-than treatment2 for disease*. Nevertheless, in this study we apply only four rather generic semantic relations namely, *treatment administered-for disease, treatment has-positive-effect-for disease, treatment has-negative-effect-for disease* and finally, when none of the above relationship cannot be assigned between two MeSH-concepts then the context is marked as *irrelevant/unclear/unknown/no-effect*. Table (1) shows the relationship set with frequent verbal predicate occurrences. The Treatment-Disease dataset contains a total of 188 sentences and four relations, which are probably not exhaustive, but which we intend to increase in future experiments, as previously discussed, to at least nine. There are between 74 and 19 annotated contexts per relation, with the *unclear/unknown/no effect* having the lowest number of instances, 19 (Table 1). In a few other contexts a relationship could be clearly distinguished between a *disease* and a *symptom* (in the sense that the first cause the second, rather than a relation between a *disease/symptom* and a *treatment*) but such relations were not considered in the set we examine in the current study. For instance: *<disease> KOL</disease> är en långsamt progredierande sjukdom som kan leda till <symptom>terminal andningssvikt</symptom>, och vissa patienter kommer att ha mycket begränsad nytta av <treatment>intensivvården</treatment>* (i.e. 'Chronic obstructive lung disease /.../ lead to terminal respiratory insufficiency /.../'), implies that a *disease-cause-symptom* relation is relevant for the domain. In some examples it was unclear by the context whether the result of a treatment had a positive or negative effect, so we refrained from choosing a relation. For instance, *<chemical>TNF-a</chemical> vid <disease> ulcerös kolit</disease> har däremot ifrågasatts, och de initiala terapistudierna gav motstridiga resultat* (i.e. 'the role of TNF-alpha in ulcerative colitis is questionable /.../'). Finally, there was one ambiguous case (homography) between a MeSH-term and common vocabulary (*övervikt*, 'advantage'/'overweight'), such as /.../ resultaten gav en *<disease>övervikt</disease>* av /.../ (i.e. '/.../results showed an advantage/.../'). This example was removed from the study, but illustrates that sense disambiguation might be a necessary component for distinguishing overlapping terms with common words.

Table 1. Candidate relationships between *treatments* (T) and *diseases* (D). Occurrences (#).

Relationship	Frequent predicates for the relation	#
T *negative-for* D	*rapportera* 'report', *orsaka* 'cause', *utveckla* 'develop',	74
T *administ.-for* D	*använda* 'use', *få* 'receive', *behandla* 'treat'	61
T *positive-for* D	*ha* 'have', *lindra* 'alleviate', *visa* 'show'	34
T *no-effect-for* D	*ge* 'give', *vara* 'be', *sakna* 'lack'	19

4.1 Feature Extraction

After the manual annotation[1] the next step was to identify and extract local, lexical and syntactic features that are informative for the identification of the relation between the terms. It has been shown that a relatively thorough representation of local context can be more effective then syntactically-based features ([6]). The following set of lexical and syntactic features have been extracted:

• The linear text ordering of the Treatment and Disease entities. Thus, either Treatment precedes Disease or Treatment follows the Disease. The binary feature 0, for the first case, and 1, for the second, are used.
• The number of tokens between Treatment and Disease. In general the closest the two entities the more likely that there is a relation. A numeric feature between 0-n is used. Note that multiword linguistic expressions and named entities, as described in Section 3.1, are reduced (the named entities to their category type) and thus always count as 1.
• Negation plays an important role particularly when the candidate relation is between a *Positive vs.* a *Negative Effect*. For instance, *inte, icke, ej* 'not' as well as the negative prefixes *o-* and *anti-*.
• The first noun (if any) between the annotations, otherwise the one to the left of the first annotation or to the left of the right annotation.
• The verbal main predicate(s) between the Treatment and Disease. This is relevant if the entities are in the same clause, since verbs are usually strong indicators of

---

[1] Because of the limited amount of data used in this pilot, no proper inter-annotator agreement calculation has been performed at this stage. Annotation was made by graduate students. Proper evaluation will be conducted in future studies.

revealing the relation between the two. If no main verb could be found between the entities then we chose the first participle. If even no participle could be found, then the first main verb or participle to the left of the first annotation was used, if once again none could be found, then the first main verb or participle to the right of the second annotation was chosen. In Swedish, participles are commonly used instead of a verbal predicate particularly in passive constructions, e.g. /.../ *<disease>stomatit </disease> orsakad av <treament>cytostatikabehandling </treatment>* /.../ 'stomatitis caused by cytostatic treatment'.

- The first token-trigram to the left of the first annotation, e.g. Hos_1 *<PRS>*$_2$ med_3 *<disease>avancerad bröstcancer</disease>* /.../ and the first token-trigram to the right of the first annotation, e.g. /.../ *<treatment>bildterapi</treatment>* $utgör_1$ ett_2 $värdefullt_3$ /.../

- The first token-trigram to the left of the second annotation, e.g. /.../ $brukar_1$ ses_2 $i\_samband\_med_3$ *<treatment>NSAID-behandling </treatment>*.and the first token-trigram to the right of the second annotation, e.g. *<disease>hyper-ammonemi</disease>* $på\_grund\_av_1$ en_2 $medfödd_3$ /.../.

5   Experimental Setup

For the experiments we used the WEKA platform ([10]). The statistical models we apply are WEKA's implementation of Support Vector Machines learner (SVM), called the Sequential Minimal Optimization algorithm (SMO). SMO is an algorithm for training a support vector classifier and the use of WEKA's metaclassifier mechanism is applied for handling multi-class datasets with 2-class classifiers (standard SVM learning constructs a binary classifier so we require to apply a multi-class classifier). SMO is a pairwise method which constructs a multi-class classifier by combining many binary SMOs. We decided to start with SVM since it has a widespread use in many Natural Language Processing (NLP) tasks shown both robustness and efficiency even in relation extraction ([6] [12] [13]). We also applied a simple Naive Bayes classifier. (NB); BayesNet (BN); a decision table majority classifier (DT); the same algorithm but by using the k-nearest neighbours classifier; instead of the majority class (DTk); the WEKA's implementation of the C4.5 algorithm (J48); with and without binary splits (J48a; J48b); and the OneR classifier which uses the minimum-error attribute for prediction.

5.1   Evaluation

As a baseline scheme we apply WEKA's *ZeroR* classifier which predicts the test data's majority class if this class is nominal as in our sample or the average value if that value happens to be numeric. ZeroR correctly classified 39.36% of the instances failing at the 60.64% of the rest. Thus, for each disease-treatment relationship the baseline is *treatment-has-negative-effect-for disease*. Since there is no gold-standard to use we have evaluated the learners using 10-fold cross validation and for the case of SMO also experimented using the one-against-one evaluation approach. Table 2 presents the number of correct classified instances by the different algorithms. All classifiers significantly outperform the simple baseline (ZeroR) which picks the most common relation in the set (significance level 0.05). The best results were

Table 2. Correctly classified instances by each applied algorithm

Algorithm	Correct classified	Algorithm	Correct classified
ZeroR	39.36% (Baseline)	NB	65.47%
J48a	57.50%	SMO	66.36%
OnerR	61.06%	DTk	67.83%
BN	65.26%	DT	70.16%
J48b	65.33%		

achieved by the decision table classifier using the k-nearest neighbours' algorithm (70.16%).

Verbs are usually considered as an important characteristic of semantic relationships between syntactic elements and syntactic dependencies often composed of subject-verb-object triples *cf.* [14]. We applied WEKA's information gain ranking filter which evaluates the worth of an attribute by measuring the gain ratio with respect to the class. Not surprisingly, the *predicate* attribute was the one highly ranked, followed by the *trigram* to the left of a *disease* or *treatment* annotations. Lowest ranked were the *order* and the *distance* between the disease or treatment annotations which probably have to do with the nature of the relations (we only looked at intra-sentential ones). Moreover, short-distance relations dominate which can be attributed to the fact of the pre-processing of the texts (recognition of multiwords and generic NEs). The average distance between two relations (i.e. the number of tokens between) was 4.7 tokens.

6 Discussion

The goal of this work was to try and learn a set of important relations among medical concepts (MeSH-annotations) at the sentence level by testing a number of supervised learning algorithms. We apply a novel data acquisition approach, in lack of authentic patient data, by looking at similar textual data with fewer restrictions. By learning context models of medical (clinical) semantic relations, new unlabelled instances can be classified and identified in texts, for instance for document retrieval. If such relations can be determined, then the potential of obtaining more accurate results for systems, such as information retrieval and Q&A increases since searching to mere co-occurrence of terms is unfocused and does not by any means guarantee that there can be a relation between the identified terms of interest. Despite the fact that in databases, such as PubMed, the data can be splitted into subsets, e.g. therapeutic use & adverse effects, this still does not imply that we can easily, albeit automatically, identify the semantics of the relations between concepts. Are some relations harder to classify? Since our test data in this pilot study is not very large it is hard to draw any conclusions on whether a relation is harder than another. In the future we intend to improve the whole process in many respects, both by using more data, and also try to incorporate new set of features and relations, not only between treatments and diseases but also between diseases and symptoms. For instance, new features will comprise the use of not only the heads of compounds but also the modifiers which are not currently used. Despite the fact that the importance of syntactic information seems to be controversial and have a limited effect in learning performance (in some studies) while positive effect in some related tasks, particularly when systems combine parsers with different language models for extracting good features ([15] [16]), we intend to also add syntactic information as features as well as explore the effect that word order might have in the assignment of an appropriate relation. We have also observed that valency information and collocations might have a significant roll to play in deciding the right semantic relation, e.g. *ge upphov till* 'cause' is a frequent indicator for the *treatment negative-effect-for disease* relation. Finally, we plan to increase the number of relations and also consider anaphoric expressions which can be the arguments of a relation. Thus, establishing a link between references to discourse entities can enhance the quality and quantity of the learned relations.

References

1. de Bruijn, B., Martin, J.: Literature mining in molecular biology. In: Baud, R., Ruch, P. (eds.) EFMI Workshop on NLP in Biomedical Applications, Nicosia, Cyprus, pp. 1–5 (2002)
2. Girju, R., Nakov, P., Nastase, V., Szpakowicz, S., Turney, P., Yuret, D.: SemEval-2007 Task 04: Classification of Semantic Relations between Nominals (2007)
3. Rosario, B., Hearst, M.A.: Classifying Semantic relations in Bioscience Texts. In: Proceedings of the 42nd Annual Meeting on ACL, Barcelona (2004)
4. Vintar, S., Buitelaar, P., Volk, M.: Semantic relations in concept-based cross-language medical information retrieval. In: Adaptive Text Extraction&Mining Workshop, Croatia (2003)
5. Roberts, A., Gaizauskas, R., Hepple, M.: Extracting Clinical Relationships from Patient Narratives. In: BioNLP 2008, Ohio, USA, pp. 10–18 (2008)
6. Zhou, G., Su, J., Zhang, J., Zhang, M.: Exploring Various Knowledge in Relation Extraction. In: Proc. of the 43rd Annual Meeting of the ACL, Michigan, pp. 427–434 (2005)
7. Sibanda, T.C.: Was the Patient Cured? Understanding Semantic Categories and Their Relationships in Patient Records. Master Thesis. Electrical Engineering & CS. MIT (2006)
8. Kokkinakis, D., Thurin, A.: Applying MeSH® to the (Swedish) Clinical Domain - Evaluation and Lessons learned. 6th Scand. Health Info. Conf. Kalmar, Sweden (2008)
9. Kokkinakis, D.: Reducing the Effect of Name Explosion. LREC Workshop: Beyond Named Entity Recognition Semantic Labeling for NLP tasks. Portugal (2004)
10. Witten, I.H., Frank, E.: Data Mining: Practical Machine Learning Tools and Techniques, 2nd edn. Morgan Kaufmann, San Francisco (2005)
11. Girju, R.: Support vector machines applied to the classification of semantic relations in nominalized noun phrases. In: HLT-NAACL W'hop on Lexical Semantics, Boston. US (2004)
12. Wang, T., Li, Y., Bontcheva, K., Cunningham, H., Wang, J.: Automatic extraction of hierarchical relations from text. In: Sure, Y., Domingue, J. (eds.) ESWC 2006. LNCS, vol. 4011, pp. 215–229. Springer, Heidelberg (2006)
13. Giles, C.B., Wren, J.D.: Large-scale directional relationship extraction and resolution. BMC Bioinformatics 9(Suppl. 9), S11 (2008)
14. Pustejovsky, J., Castaño, J., Zhang, J.: Robust Relational Parsing over Biomedical literature: Extracting Inhibit Relations. In: Proc. 7th Biocomputing Symposium (2002)
15. Mustafaraj, E., Hoof, M., Freisleben, B.: Mining Diagnostic Text Reports by Learning to Annotate Knowledge Roles. In: Kao, A., Poteet, S. (eds.) NLP&TM, pp. 46–67. Springer, Heidelberg (2007)
16. Sætre, R., Sagae, K., Tsujii, J.: Syntactic features for protein-protein interaction extraction. In: 2nd International Symposium on Languages in Biology and Medicine, Singapore (2007)

Extended Hidden Vector State Parser

Jan Švec[1] and Filip Jurčíček[2]

[1] Center of Applied Cybernetics, Department of Cybernetics,
Faculty of Applied Sciences, University of West Bohemia,
Pilsen, 306 14, Czech Republic
honzas@kky.zcu.cz
[2] Cambridge University Engineering Department
Cambridge CB21PZ, United Kingdom
fj228@cam.ac.uk

Abstract. The key component of a spoken dialogue system is a spoken under-standing module. There are many approaches to the understanding module design and one of the most perspective is a statistical based semantic parsing. This paper presents a combination of a set of modifications of the hidden vector state (HVS) parser which is a very popular method for the statistical semantic parsing. This paper describes the combination of three modifications of the basic HVS parser and proves that these changes are almost independent. The proposed changes to the HVS parser form the extended hidden vector state parser (EHVS). The per-formance of the parser increases from 47.7% to 63.1% under the exact match between the reference and the hypothesis semantic trees evaluated using Human-Human Train Timetable corpus. In spite of increased performance, the complex-ity of the EHVS parser increases only linearly. Therefore the EHVS parser pre-serves simplicity and robustness of the baseline HVS parser.

1 Introduction

The goal of this paper is to briefly describe the set of modifications of the hidden vec-tor state (HVS) parser and to show that these changes are almost independent. Every described modification used alone significantly improves the parsing performance. The idea is to incorporate these modifications into a single statistical model. We suppose that the combined model yields even better results. The HVS parser consists of two statisti-cal models - the semantic and the lexical model (see bellow). In the following sections we describe three techniques to improve the performance of the parser by modifying these models.

First, we use a data-driven initialization of the lexical model of the HVS parser based on the use of negative examples which are collected automatically from the semantic corpus.

Second, we deal with the inability of the HVS parser to process left-branching language structures. The baseline HVS parser uses a implicit pushing of concepts during a state transitions and this limits the class of generated semantic trees to be right-branching only. To overcome this constraint we introduce an explicit push operation into the semantic model and we extend the class of parseable trees to the left-branching trees, the right-branching trees and their combinations.

V. Matoušek and P. Mautner (Eds.): TSD 2009, LNAI 5729, pp. 403–410, 2009.

Finally, we extend the lexical model to process a sequence of feature vectors instead of a sequence of words only.

2 Hidden Vector State Parser

The HVS parser is a statistical parser, which implements the search process over the sequence of vector states $S = c_1, c_2, \ldots, c_T$ that maximizes the aposterior probability $P(S|W)$ for the given word sequence $W = w_1, w_2, \ldots, w_T$. The search can be described as

$$S^* = \underset{S}{\mathrm{argmax}}\, P(S|W) = \underset{S}{\mathrm{argmax}}\, P(W|S)P(S) \tag{1}$$

where $P(S)$ is called the *semantic model* and $P(W|S)$ is called the *lexical model*.

The HVS parser is an approximation of a pushdown automaton. The *vector state* in the HVS parser represents a stack of a pushdown automaton. It keeps the semantic concepts assigned to several words during the parsing.

The transitions between vector states are modeled by three stack operations: popping from zero to four concepts out of the stack, pushing a new concept onto the stack, and generating a word. The first two operations modelled by the semantic model which is given by:

$$P(S) = \prod_{t=1}^{T} P(pop_t|c_{t-1}[1,\ldots 4])P(c_t[1]|c_t[2,\ldots 4]) \tag{2}$$

where pop_t is the vector stack shift operation and takes values in the range $0, 1, \ldots, 4$. The variable c_t represents the vector state consisting of four variables - the stored concepts, i.e. $c_t = [c_t[1], c_t[2], c_t[3], c_t[4]]$ (shortly $c_t[1, \ldots 4]$), where $c_t[1]$ is a preterminal concept dominating the word w_t and $c_t[4]$ is a root concept.

The lexical model performs the last operation - a generation of a word. The lexical model is given by:

$$P(W|S) = \prod_{t=1}^{T} P(w_t|c_t[1,\ldots 4]) \tag{3}$$

where $P(w_t|c_t[1,\ldots 4])$ is the conditional probability of observing the word w_t given the state $c_t[1,\ldots 4]$. For more details about the HVS parser see [1].

3 Negative Examples

In this section we briefly describe the first used modification of the HVS parser. It is based on so called *negative examples* which are automatically collected from the semantically annotated corpus. The negative examples are then used to initialize the lexical model of an HVS parser. First, we define a positive example. Then we negate the meaning of the positive example and we get the definition of the negative example. Finally we describe the use of negative examples during the initialization of an HVS parser.

In this paper the positive example is a pair (w, c) of a word and a semantic concept and it says: the word w can be observed with a vector state containing the concept c. For the utterance *jede nějaký spěšný vlak do Prahy kolem čvrté odpoledne* (Lit.: *does any express train go to Prague around four p.m.*) with the semantic annotation DEPARTURE(TO(STATION), TIME), one of many positive examples is the pair (*Prahy*, STATION). Another positive example for the word *Prahy* (Lit.: *Prague, capital city of Czech Republic*) is the pair (*Prahy*, TIME).

The negative example, similarly to the positive example, is a pair of a word w and a semantic concept c. However, the negative example says: the word w *is not* observed together with a vector state containing the concept c. In other words, the negative example is the pair of a word and a concept that do not appear together in any utterance in a training corpus. For example the utterance *jede nějaký spěšný vlak do Prahy kolem čvrté odpoledne* (Lit.: *does any express train go at four p.m.*) with semantic annotation DEPARTURE(TIME). We can see that the word *jede* (Lit.: *does go*) is not generated by the vector state containing the concept STATION. Therefore, the pair (*jede*, STATION) is the negative example.

We analyzed concepts defined in our semantic corpus (see Section 6), and we found four concepts suitable for the negative examples extraction: STATION, TRAIN_TYPE, AMOUNT, LENGTH, and NUMBER. These concepts are selected because they are strong related to their word realization. In other words, the set of all possible words with meaning STATION is finite and well-defined.

Not all utterances are ideal for the extraction of negative examples. For instance, if we use the utterance *dnes je příjemné počasí v Praze* (Lit.: *weather is pleasant in Prague today.*) with the semantic annotation OTHER_INFO for the negative examples extraction, we have to conclude that the word *Praze* is not generated by the vector state [STATION, ...] because the semantics does not contain the concept STATION. However the word *Praze* (Lit.: *in Prague*) is related to the concept STATION[1].

To select the proper utterances, we use only the utterances containing the following top-level concepts: ACCEPT, ARRIVAL, DELAY, DEPARTURE, DISTANCE, DURATION, PLATFORM, PRICE, and REJECT because only these concepts can be parents of suitable leaf concepts. More details on extraction of the negative examples can be found in [2].

The negative examples give us much less information than the positive ones. We have to collect several negative examples to gain the information equal to one positive example. However, using such information brings significant performance improvement.

To utilize negative examples, we modify the initialization phase of the lexical model. We still initialize the lexical model uniformly; however, at the same time, we penalize the probability of observing the word w given the vector stack c_t according to the collected negative examples:

$$p(w, c[1, \ldots 4]) = \begin{cases} \epsilon & \text{if } (w, c[1]) \text{ is a negative example,} \\ 1/|V| & \text{otherwise} \end{cases}$$

---

[1] In this example, the concept STATION does not have to appear in the semantics because the semantics OTHER_INFO is very general and it covers many meanings e.g. STATION as well.

$$P(w|c[1,\ldots 4]) = \frac{p(w, c[1,\ldots 4])}{\sum_{\overline{w} \in V} p(\overline{w}, c[1,\ldots 4])} \quad \forall w \in V \tag{4}$$

where ϵ is a reasonably small positive value and V is a word lexicon. We found that it is better to use some non-zero value for ϵ because the extraction of the negative examples is not errorless and the parser training algorithm (a kind of EM) can deal with such errors.

4 Left Branching

In this section, we describe the semantic model modification which enables the HVS parser to generate not only the right-branching parse trees but also left-branching parse trees and their combinations. The resulting model is called the left-right-branching HVS (LRB-HVS) parser [3].

The left-branching parse trees are generated by pushing more than one concept onto the stack at the same time (the baseline HVS parser pushes only one concept at a time). We analyzed errors of the baseline HVS model and we did not find any error caused by the inability to push more than two concepts. Therefore, we limited the number of concepts inserted onto the stack at the same time to two, but in general it is straightforward to extend the number of pushed concepts to more than two.

To control the number of concepts pushed onto the stack at the same time, we introduced a new hidden variable $push$ into the HVS parser:

$$P(S) = \prod_{t=1}^{T} P(pop_t|c_{t-1}[1,\ldots 4])P(push_t|c_{t-1}[1,\ldots 4]) \cdot$$

$$\begin{cases} 1 & \text{if } push_t = 0 \\ P(c_t[1]|c_t[2,\ldots 4]) & \text{if } push_t = 1 \\ P(c_t[1]|c_t[2,\ldots 4])P(c_t[2]|c_t[3,4]) & \text{if } push_t = 2 \end{cases} \tag{5}$$

In the case of inserting two concepts onto the stack ($push_t = 2$), we approximate the probability $P(c_t[1,2]|c_t[3,4])$ by $P(c_t[1]|c_t[2,\ldots 4])P(c_t[2]|c_t[3,4])$ in order to obtain more robust semantic model $P(S)$.

To illustrate the difference between right-branching and left-right-branching, we can use the utterance *dneska večer to jede v šestnáct třicet* (Lit.: *today, in the evening, it*

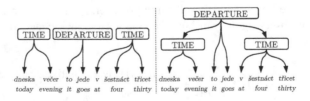

Fig. 1. Incorrect (left) and correct (right) parse trees of left-branching language structure. Input: *dneska večer to jede v šestnáct třicet* (Lit.: *today, in the evening, it goes at four thirty p.m.*).

goes at four thirty p.m.). The incorrect parse tree (Figure 1 left) is represented by the semantic annotation TIME, DEPARTURE, TIME. Such parse tree would be an output of the HVS parser, which allows to generate right-branching parse trees only; the parser is not able to push more than one concept onto the stack. However, the correct parse tree (Figure 1 right) is represented by the semantic annotation DEPARTURE(TIME, ..., TIME). The correct parse tree would be an output of LRB-HVS parser because it is able to push two concepts DEPARTURE and TIME onto the stack at the same time so that the first word *dneska* (Lit.: *today*) can be labeled with the hidden vector state [TIME, DEPARTURE].

5 Input Feature Vector

The input parameterization extends the HVS parser into a more general HVS parser with the input feature vector (HVS-IFV parser) [4]. This parser uses a sequence of feature vectors $F = (f_1, \ldots, f_T)$ instead of a sequence of words W. The feature vector is defined as $f_t = (f_t[1], f_t[2], \ldots, f_t[N])$. Every word w_t has assigned a fixed set of N features. If we use the feature vector f_t instead of the word w_t in Eq. 3, the lexical model changes as follows:

$$P(F|S) = \prod_{t=1}^{T} P(f_t \mid c_t) = \prod_{t=1}^{T} P(f_t[1], f_t[2], \ldots f_t[N] \mid c_t) \qquad (6)$$

To avoid the data sparsity problem, we used the assumption of conditional independence of features $f_t[i]$ and $f_t[j]$, $i \neq j$ given the vector stack c_t. This kind of assumption is also used for example in the naive Bayes classifier. The lexical model of the HVS-IFV parser is then given by:

$$P(F|S) = \prod_{t=1}^{T} \prod_{i=1}^{N} P(f_t[i] \mid c_t) \qquad (7)$$

Because the conditional independence assumption is hardly expected to be always true, we modified the search process defined in Eq. 1. Let's assume that we have the sequence of the feature vectors $F = (f_t[1], f_t[2])_{t=1}^{T}$ where $f_t[1] = f_t[2]$ for every time step t. Then the lexical model is given by $P(F|S) = \prod_{t=1}^{T} [P(f_t[1]|c_t)]^2$. As we can see the probability $P(F|S)$ is exponentially scaled with the factor 2 and it causes the imbalance between the lexical and the semantic model. Therefore, we use the scaling factor λ to compensate the error caused by the assumption of conditional independence:

$$S^* = \arg\max_{S} P(F|S) P^{\lambda}(S) \qquad (8)$$

Then the HVS-IFV parser is defined by equations 5, 7, and 8. The optimal value of λ was found by maximizing the concept accuracy measure defined in Section 6.1 on a development data. In our experiments the feature set consists of two linguistic features - a lemma and a morphological tag assigned to the original word.

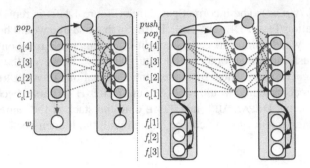

Fig. 2. Graphical models of a transition between two consecutive vector states for the original HVS model (left) and the HVS-IFV model (right)

6 Experiments

The semantic parsers described in this paper were trained and tested on the Czech human-human train timetable (HHTT) dialog corpus [5]. The HHTT corpus consists of 1,109 dialogs completely annotated with semantic annotations. Both operators and users have been annotated. It has 17,900 utterances in total. The vocabulary size is 2,872 words. There are 35 semantic concepts in the HHTT corpus. The dialogs were divided into training data (798 dialogs - 12972 segments, 72%), development data (88 dialogs - 1,418 segments, 8%), and test data (223 dialogs - 3,510 segments, 20%). Each segment has assigned exactly one abstract semantic annotation.

The training of the semantic and the lexical models of HVS parser is divided into three parts: (1) initialization, (2) estimation, (3) smoothing. All probabilities are initialized uniformly; however, the negative examples are used to alter the probabilities in the lexical model. To estimate the parameters of the models, it is necessary to use the expectation-maximization (EM) algorithm because the abstract semantic annotations do not provide full parse trees. We use a simple back-off model to smooth probabilities. To build the semantic parser, we use the Graphical modeling toolkit (GMTK, see Figure 2) [6].

We evaluate our experiments using two measures: semantic accuracy and concept accuracy. These measures compare the reference tree (human annotated) with the hypothesis tree (parser output).

6.1 Performance Measures

When computing the semantic accuracy, the reference and the hypothesis annotations are considered equal only if they exactly match each other. The *semantic accuracy* of a model output is defined as $SAcc = \frac{E}{N} \cdot 100\%$, where N is the number of evaluated semantics, E is the number of hypothesis semantics which exactly match the reference.

The exact match is a very tough standard. It does not measure a fine differences between similar semantics. Therefore we introduced the *concept accuracy*. Similarity scores between the reference and the hypothesis semantics can be computed by a tree edit distance algorithm [7]. The tree edit distance algorithm uses a dynamic programming to find the minimum number of substitutions (S), deletions (D), and insertions (I) required to transform one semantic tree into another one. The operations act on nodes and modify the tree by changing parent/child relationships of given trees. The *concept accuracy* of a model output is defined as $CAcc = \frac{N-S-D-I}{N} \cdot 100\%$, where N is the total number of concepts in the corresponding reference semantics.

6.2 Results

Table 1 and Table 2 show the results achieved by different modifications of the baseline HVS parser on development and test data. To measure the statistical significance we use the paired t-test. The p-value < 0.01 of this test indicates significant difference.

The baseline HVS parser corresponds to the implementation of He and Young. The "negative examples" method for the HVS parser initialization used the negative examples extracted for the concepts AMOUNT, LENGTH, NUMBER, STATION, and TRAIN_TYPE. The LRB-HVS parser combines the extension which allows the generation of left-right-branching parse trees and the "negative examples" method. The HVS-IFV parser then adds the ability to parse the sequence of feature vectors instead of single sequence of words. According to [4] the feature vector consists of two features - the original word and its corresponding lemma. Finally the EHVS parser corresponds to the developed extended hidden vector state parser which yields the better results.

Table 1. Performance of parsers evaluated on the development data

Parser type	Development data		
	SAcc	CAcc	p-value
HVS (baseline)	50.7	64.3	
HVS with neg. examples	52.8	67.0	< 0.01
LRB-HVS	60.1	70.6	< 0.01
HVS-IFV	58.2	73.1	< 0.01
EHVS	**65.4**	**75.7**	< 0.01

Table 2. Performance of parsers and estimates of performance lower- and upper-bound evaluated on the test data

Parser type	Test data		
	SAcc	CAcc	p-value
HVS (baseline)	47.9	63.2	
HVS with neg. examples	50.4	64.9	< 0.01
LRB-HVS	58.3	69.3	< 0.01
HVS-IFV	57.0	69.4	< 0.01
EHVS	**63.1**	**73.8**	< 0.01

7 Conclusions and Future Work

In this paper we presented the combination of three well-known methods of improving the performance of the HVS parser. We used the data-driven initialization of the lexical model using the negative examples. We also modified the semantic model to support a wider class of semantic trees. The modified parser allows to generate left branching, right branching and left-right branching parse trees. The last modification extends the lexical model to be able to parse the sequence of feature vectors instead of the sequence of single words. We used a lemma and a morphological tags of the original words as features. All these modifications alone significantly improve the performance of the baseline HVS parser. The key result of this paper shows that these modifications (negative examples, left-right branching and the input feature vector) are almost independent. The performance gain of the resulting *extended hidden vector state parser* (EHVS) is composed of the performance gains of the partial modifications.

All in all, we improved the performance of the parser from 47.9 % to 63.1 % in SAcc and from 63.2 % to 73.8 % in CAcc measured on the test data. The absolute improvement achieved by the suggested modifications of the original HVS parser was about 15 % in SAcc and more than 10 % in CAcc.

Acknowledgment

This work was supported by the Ministry of Education of the Czech Republic under project No. 1M0567 (CAK).

References

1. He, Y., Young, S.: Semantic processing using the hidden vector state model. Computer Speech and Language 19(1), 85–106 (2005)
2. Jurčíček, F.: Statistical approach to the semantic analysis of spoken dialogues. Ph.D. thesis, University of West Bohemia (2007)
3. Jurčíček, F., Švec, J., Müller, L.: Extension of HVS Semantic Parser by Allowing Left-Right Branching. In: Proc. IEEE ICASSP (2008)
4. Švec, J., Jurčíček, F., Müller, L.: Input Parameterization of the HVS Semantic Parser. In: Proceedings of TSD, Pilsen, The Czech Republic (2007)
5. Jurčíček, F., Zahradil, J., Jelínek, L.: A Human-Human Train Timetable Dialogue Corpus. In: Proceedings of Interspeech, Lisboa, Portugal (2005)
6. Bilmes, J., Zweig, G.: The graphical models toolkit: An open source software system for speech and time-series processing. In: Proc. IEEE ICASSP (2002)
7. Klein, P.: Computing the edit-distance between unrooted ordered trees. In: Proceedings of the 6th Annual European Symposium, Venice, Italy. Springer, Berlin (1998)

Semantic Annotation of City Transportation Information Dialogues Using CRF Method

Agnieszka Mykowiecka[1,2] and Jakub Waszczuk[1]

[1] Institute of Computer Science, Polish Academy of Sciences,
J. K. Ordona 21, 01-237 Warsaw, Poland
{agn,Jakub.Waszczuk}@ipipan.waw.pl
[2] Polish-Japanese Institute of Information Techniques
Koszykowa 86, 02-008 Warsaw, Poland

Abstract. The article presents results of an experiment consisting in automatic concept annotation of the transliterated spontaneous human-human dialogues in the city transportation domain. The data source was a corpus of dialogues collected at a Warsaw call center and annotated with about 200 concepts' types. The machine learning technique we used is the linear-chain Conditional Random Fields (CRF) sequence labeling approach. The model based on word lemmas in a window of length 5 gave results of concept recognition with an F-measure equal to 0.85.

1 Introduction

One of the most difficult areas of NLP (Natural Language Processing) is language understanding. Apart from many efforts devoted to different areas of linguistic engineering, programs which could communicate with users in natural language are still rare and inflexible. The lack of satisfying results in this field even caused a lowering of interest in NLP applications for several years. Nowadays, with rapid development of machine learning methods, and their utilization in almost all NLP domains, a lot of new solutions of many problems are being proposed.

Although universal methods of solving the natural language understanding task do not already exist, practically usable applications can be built e.g. by limiting ourselves to recognizing a restricted set of semantic concepts defined for a given domain. However, assigning concept names to fragments of text is not trivial – it either requires manual work and expert knowledge or data previously annotated with the same set of labels (manually or by rule based methods).

In this paper we present the results of an experiment consisting in assigning semantic concept names to real human-human dialogues (collected in a city transportation call center) by the system using a model trained on previously annotated data. The prospective aim of the work was to test the possibility of building a system which aids call center personnel in answering queries in a way similar to the real operator-customer dialogue. The first stage of the project defined an ontology of the transportation domain and a set of rules for assigning concepts from this ontology to phrases occurring within the recorded dialogues. The set of rules was applied to the entire corpus and the results were manually corrected. Here, we present the experiment consisting in using

V. Matoušek and P. Mautner (Eds.): TSD 2009, LNAI 5729, pp. 411–418, 2009.

this manually corrected semantic annotation as a data source for the machine learning approach. The method we chose for the experiment is linear-chain Conditional Random Fields ([12]) which is a modification of an original CRF model introduced in [4]. It is a discriminative modeling method that maximizes a conditional probability of a label sequence given a word sequence. It became very popular and was used (in many variations) for many NLP tasks due to very good results. Among many tested applications we can list biomedical named entities recognition (e.g. [8], [10]), shallow parsing ([11]), information extraction ([12]), word sense disambiguation by assigning WordNet synsets ([2]).

In the rest of the paper, we first describe the domain of the dialogues and the annotated corpus we used for performing the experiments. Then, we present the assumptions made while building a learning model, and the results we achieved.

2   Data Description

The data set which we operated on is a corpus of Polish dialogues collected in 2007, at the Warsaw City Transportation information center where people can obtain various kinds of data concerning transport lines, e.g. which line to use to reach a given point in the city, what are departure times of buses of a given line, how long does a trip last, where one should get off to reach a particular place, and what the fare reduction rules are. The process of data collection and their transliteration, as well as the rules and format of annotation at the morphosyntactic and semantic levels were described in [6]. The first level of semantic annotation consisted in assigning names of concepts and their values to sequences of words. In contrast to the annotation on the morphosyntactic levels, semantic annotation had to be defined from scratch – there was no existing domain model which could be used as the source of the concept names. A domain model which was defined within the project has the form of an OWL ontology covering all relevant knowledge areas. The ontology evolved during the project and now it consists of about two hundreds concepts organized in 4 main groups representing notions related to person characteristics (only features important form the transport domain point of view), time, city transport network, and dialogues about city transportation. Some details about the types and number of concepts are presented in Table 1.

The entire corpus consists of 500 dialogues divided into 5 groups: transportation routes, itinerary, schedule, stops, and reduced and free-fares. For the experiment the first four groups of 399 dialogues were chosen (the fifth group contains dialogues concerning different subjects). Their quantitative characteristics are given in Table 2. All dialogues are annotated at the morphological level – for every wordform we know the lemma, part of speech and values of the morphological categories (number, gender, case). As an additional source of knowledge, a lexicon with all domain related proper names and their types was also prepared, [5]. The next level of anotation – semantic annotation by concept names – is much more complicated than annotation on morphosyntactic level and is still very often done manually. For example, French MEDIA corpus, [1], was annotated manually by about 80 basic concept. Later, HMM models were trained on that data, [9]. Our semantic annotation was done using a set of rules and then manually corrected. The process of rule based semantic annotation was incremental. The rules were corrected on the basis of the results achieved on a small set of

Table 1. The most important concept groups

Group name	Most important concepts	Concepts nb
Person_Feature	Educational_Status, Age Person_Occupation	8
Transport_Dialogues	Question, Conv_Form Exclamation, Reaction	54
Time	Date, Repetitive_Moments, Time_Desc_Partial, Time_Point_Desc Time_Span, Time_WeekPeriod, Time_PeriodYear	40
Transport_Domain	Connection_Quality, Discount_Title, Fare Location, Place, Road_Disturbance, Route, Route_Comparison, Route_Feature, Transport_Line, Trip, Trip_Feature Transport_Mean, Transp_Line_Feature	157

Table 2. Quantitative characteristics of the selected groups of dialogues

Category	Number of calls	Average number of user's turns per call	Average number of user's words per call	Total number of wordforms types	Total number of wordforms
Transportation routes	93	14	98	1975	17019
Itinerary	140	16	96	2562	29038
Schedule	111	11	61	1339	13313
Stops	55	11	86	1332	9061
Total	399	14	85	4130	68431

dialogues. After three rounds of rules corrections the entire corpus was annotated – the concept/value error rate of this process was about 20%, [7]. After final manual verification, in the selected subcorpus there were 24732 occurrences of 188 concepts. The most common concepts are listed in Table 3.

Table 3. The most common concept types

concept	nb	concept	nb
REACTION	4694	Action	3612
BUS	1626	CONV_FORM	1316
LOCATION_REL	1202	STOP_DESCRIPTION	916
AT_HOUR	777	GOAL_STR	610
STREET	471	LOCATION_STR	458
TRAM	419	PERSON_NAME	410
Q_CONF	408	SOURCE_STR	401

3   Model Creation

For the purpose of the experiment described here, a new implementation of the linear-chain CRF method was performed on the basis of the description given in [12]. As CRF model allows for the use of various additional features describing the particular words, the choice of these features is crucial for the model itself and influence the results achieved.

In our approach, features have forms of triples (y, y/, x) or pairs (y, x), where x is an observation concerning the given word, y – a concept which is assigned to that word, and y/ is a concept assigned to the previous word. Models were trained on the corpus described in section 2. The data used for the training procedure was converted into a text file. Every line contains one word from the dialog annotated with values of the selected "observations". In (1), we cite a fragment of a training file showing the format of the data (for third and fourth words from the sequence *dostać się do Konstancina z* 'to get to Konstancin from'). Every line (in the example, the second line had to be split in two) contains the lemma of a word, two lemmas of the two preceding words, the lemma of the next word, and the part of speech the current word belongs to. For each noun we added the value of its case, and if it is a (part of) proper noun – type of the object it refers to (e.g. *Konstancin* is a town near Warsaw). Finally, every line ends with a name of a concept which was attributed to the fragment of the text this word belongs to (a special null concept for non annotated words is added).

(1)   do L_do PL1_dostać PL2_się NL1_konstancin PREP GOAL_TOWN
        konstancina L_konstancin PL1_do PL2_dostać NL1_z NP GEN TOWN
                                                                    GOAL_TOWN

The linear-chain CRF model is a discriminative sequence classifier which maximizes the conditional probability of label sequance $\mathbf{y}$ given the sentence $\mathbf{x}$, see (2a), where T is the sentence length, K,L - number of f_k and g_l feature functions, I - number of observations about word $\mathbf{x}_t$ and $\mathbf{Y}$ is the set of all label sequences of length T). Functions f_k and g_l take the value of 1 or 0, $x_{t,i}$ are singular observations. If we assume that every f_k is equal to 1 for only one triple (y, y', x) and every g_l is equal to 1 for only one pair (y, x), we get an equation (2b). Assuming that $\theta(y, y', x) = \lambda_k + \delta_l$ for $f_k(y, y', x) = 1$ and $g_l(y, x) = 1$ (we set $\delta_l = 0$ if such (y, x) does not exists, similarly for λ_k), we get the equation given in (2c).

(2)   a. $p(\mathbf{y}|\mathbf{x}) = \frac{1}{Z(\mathbf{x})} exp\{\Sigma_{t=1}^{T}(\Sigma_{l=1}^{L}\delta_l g_l(y_t, \mathbf{x}_t) + \Sigma_{k=1}^{K}\lambda_k f_k(y_{t-1}, y_t, \mathbf{x}_t))\}$, where
        $Z(\mathbf{x}) = \Sigma_{\mathbf{y}\in\mathbf{Y}} exp\{\Sigma_{t=1}^{T}(\Sigma_{l=1}^{L}\delta_l g_l(y_t, \mathbf{x}_t) + \Sigma_{k=1}^{K}\lambda_k f_k(y_{t-1}, y_t, \mathbf{x}_t))\}$

        b. $p(\mathbf{y}|\mathbf{x}) = \frac{1}{Z(\mathbf{x})} exp\{\Sigma_{t=1}^{T}\Sigma_{i=1}^{I}(\Sigma_{l=1}^{L}\delta_l g_l(y_t, x_{t,i}) + \Sigma_{k=1}^{K}\lambda_k f_k(y_{t-1}, y_t, x_{t,i}))\}$

        c. $p(\mathbf{y}|\mathbf{x}) = \frac{1}{Z(\mathbf{x})} exp\{\Sigma_{t=1}^{T}\Sigma_{i=1}^{I}\theta(y_{t-1}, y_t, x_{t,i})\}$

To use the above equation, we have to estimate λ_k, δ_l (θ) parameters. This is done on the basis of training data $\{\mathbf{x}^{(i)}, \mathbf{y}^{(i)}\}_{i=1}^{N}$ (N is the number of sentences in the training set). In this process we maximize the log likelihood function given in (3a) with regularization added to avoid overfitting (where $\frac{1}{2\sigma^2}$ is the *regularization parameter*) (3b).

(3)   a. $l(\theta) = \Sigma_{i=1}^{N} log(p(\mathbf{y}^{(i)}|\mathbf{x}^{(i)}))$
        b. $l(\theta) = \Sigma_{i=1}^{N} log(p(\mathbf{y}^{(i)}|\mathbf{x}^{(i)})) - \Sigma_{k=1}^{K}\frac{\lambda_k^2}{2\sigma^2} - \Sigma_{l=1}^{L}\frac{\delta_l^2}{2\sigma^2}$

The standard algorithm used for computing l partial derivatives (which are needed for optimization) has time complexity $\Theta(k|Y|^2 M)$ (see [12]), where k – number of observations for each word, Y - set of concepts, and $M = \Sigma_{i=1}^{N} len(\mathbf{x}^i)$. For corpus of this size ($M \approx 70000$) with a big number of concepts ($|Y| \approx 200$) this complexity is unacceptable. An algorithm used in our implementation takes advantage of data sparseness (specifically, of the small number of features $(y, y\prime, x)$ for given observation x in the training set). This allows to compute l partial derivatives in $\Theta(k\overline{Y}M)$ time, where $\overline{Y}$ is an average number of pairs $(y, y\prime)$ per word $\mathbf{x}$.

A program for model creation was written in Python with some fragments written in C. For calculating the maximum of the l function we used Python SciPy library module which implements L-BFGS (limited memory Broyden-Fletcher-Goldfarb-Shannon) method, [13].

4 Results

CRF allows for the representation of many features, so is a natural candidate to deal with annotated linguistic data.

Table 4 shows the results obtained for 32 models trained on 4 different randomly selected subsets of the corpus (about 9/10 of the data each time), while the rest of the corpus has been used for evaluation which was performed using standard F_1-measure. To calculate it we compared the assigned label with the correct label for every word separately. Special null concepts, assigned to semantically non-annotated words have not been taken into account during F-measure calculation, because otherwise results would be far too optimistic.

Table 4. Results

	Set 1	Set 2	Set 3	Set 4	Average
Words1	80.0	82.7	80.0	81.6	81.1
Words2	78.7	82.7	80.8	81.2	80.9
WordsPN	83.2	85.0	82.5	84.0	83.7
WordsMorphPN	83.8	86.6	84.8	85.2	85.1
WordsPref4PN	84.6	86.4	84.9	84.6	85.1
WordPref5MorphPN	85.6	87.7	85.7	86.9	86.4
WordPref5PN	84.4	86.5	84.4	84.4	84.9
WordPref5Prev5PN	84.4	85.8	84.3	84.3	84.7

The second dimension of model creation was the choice of features. For this evaluation, the following features have been taken into account: WORD – a particular word-form, LEMMA – a base form, PREF – a word prefix, SUF – a word suffix, POS – a part of speech name and CASE – one of seven Polish cases defined for nouns and adjectives (the tagset used is described in [6]). Additionally, we used a type of a proper name of which a given word is a part (PROP_NAME). Forms or lemmas of words from the context can also be addressed as well as their prefixes and suffixes. The maximal context analyzed contained two words before and one word after the one analyzed. For the experiments performed, the following combinations of features were chosen:

Words1   — WORD[-1:1]
Words2   — WORD[-2:1]
WordsPN   — WORD[-2:1] PROP_NAME
WordsMorphPN   — WORD LEMMA[-2:1] POS CASE PROP_NAME
WordsPref4PN   — WORDS [-2:1] SUF[3:4] PREF[3:4] PROP_NAME
WordPref5MorphPN   — WORDS[-1:0], LEMMAS[-2:1], PREF[2:5], SUF[2:5],
                           POS, CASE, PREV_POS PREV_CASE, PROP_NAME
WordPref5PN   — WORD[-2:1], PREF[2:5], SUF[2:5], PROP_NAME
WordPref5Prev5PN   — WORD[-2:1], PREF[2:5], SUF[2:5], PREV_PREF[2:5],
                           PREV_SUF[2:5], PROP_NAME

The results achieved for four test sets are given in Table 4. The best result was achieved for the model with both prefix/postfix and morphological information.

The results of the process of concept recognition have different quality for different concepts. This is caused partially by the fact that some concepts are assigned only to very few types of phrases while others have a lot of different possible values (e.g. street names, line numbers or time descriptions). In this particular case, an added difficulty arises from the very detailed differentiation of names describing places. Not only did we divide them into several types (buildings, street, areas, stops ...) but separate concepts are introduced for differentiating places being location of something or someone, source of the trip, its goal, its overall direction, or an element of the path. These resulted in about 76 concepts which describe places and their role in the dialog context. As the word context is not always sufficient and one needs more pragmatic knowledge to interpret a particular phrase, these concepts are frequently recognized incorrectly.

Table 5. Best and worst recognized numerous concept types

phrase meaning	concept name	F	prec	recall	nb
conventional formula	CONV_FORM	0.99	0.99	0.99	344
name of person	PERSON_NAME	0.99	0.98	1.00	53
from a stop	SOURCE_STP	0.97	0.95	1.00	56
bus number	BUS	0.97	0.96	0.98	395
reaction	REACTION	0.97	0.97	0.97	462
to a railway station	GOAL_RS	0.96	1.00	0.92	24
from a railway station	SOURCE_RS	0.95	0.95	0.95	44
'precise time description	AT_HOUR	0.95	0.94	0.96	268
from a street	SOURCE_STR	0.94	0.96	0.92	106
...					
question about departures	Q_DEPARTURE	0.47	0.44	0.50	14
street number	STREET_NUMB	0.38	0.38	0.38	13
fragment of bus number	NUM_BUS_PART	0.33	0.40	0.29	7
town district	TOWN_DISTRICT	0.31	0.33	0.29	7
number	NUMB	0.14	0.14	0.14	7

5 Conclusions

The semantic annotation of natural language text is very difficult as it requires preparing a detailed model of the domain, and then a method of assigning its elements to identified fragments of data. An annotation process is very laborious – it has to be either done manually or by rules whose preparation is also time consuming. When such annotated data are already constructed, they can be used to train a model which might be even better in annotating new data sets than rule based procedures. In the experiment presented here we tried to test such an approach in the relatively complicated domain represented by about 200 not easily separable concepts. The results of the experiment show that machine learning techniques are powerful enough to deal with such complicated domains. At the same time they show the benefits of basic linguistic processing. Taking into account lemmas and values of morphosyntactic categories of analyzed wordforms, improved the results.

For the most complex models, the obtained results are even better than those achieved by the rule based method. This gives good perspectives for achieving the initial goal of building a more effective concept recognition system after further improvements of the model. CRF method which allows for easy introduction of various features describing source elements, seems to be very suitable for dealing with linguistic data. This is especially important for highly inflectional languages, where a lot of information is carried by the values of morphological categories of wordforms.

To complete the understanding module, we plan to connect the procedure of assigning concept names with rule based concept values assignment. To test if there are possibilities of further model improvement, we plan to test a semi-Markov CRF hybrid model ([3]).

Acknowledgment

This work was supported by LUNA – STREP project in the EU's 6th Framework Programme (IST 033549) which started in 2006.

References

1. Bonneau-Maynard, H., Rosset, S.: A semantic representation for spoken dialog. In: Eurospeech, Geneva (2003)
2. Deschat, K., Moens, M.: Efficient Hierarchical Entity Classifier Using Conditional Random Fields. In: Proceedings of the 2nd Workshop on Ontology Learning and Populations, Sydney Australia, pp. 33–40 (2006)
3. Galen, A.: A Hybrid/Semi-Markov Conditional Random Field for Sequence Segmentation, In. In: Proceedings of the 2006 Conference on Empirical Methods in Natural Language Processing, ACL, Sydney (2006)
4. Lafferty, J., McCallum, A., Pereira, P.: Conditional Random Fields: Probabilistic Models for Segmenting and Labeling Sequence Data. In: Proceedings of 18th International Conf. on Machine Learning, pp. 282–289. Morgan Kaufmann, San Francisco (2001)
5. Marciniak, M., Rabiega-Wiśniewska, J., Mykowiecka, A.: Proper Names in Dialogs from Warsaw Transportation Call Ceter. In: Intelligent Information Systems XVI, Zakopane, EXIT, Warsaw (2008)

6. Mykowiecka, A., Marasek, K., Marciniak, M., Gubrynowicz, R.J.: Rabiega-Wiániewska: Annotation of Polish spoken dialogs in LUNA project. In: Proceedings of the Language and Technology Conference, Poznan (2007)
7. Mykowiecka, A., Marciniak, M., Głowińska, K.: Automatic semantic annotation of polish dialogue corpus. In: Sojka, P., Horák, A., Kopeček, I., Pala, K. (eds.) TSD 2008. LNCS (LNAI), vol. 5246, pp. 625–632. Springer, Heidelberg (2008)
8. Ponomareva, N., Rosso, P., Pla, F., Molina, A.: Conditional Random Fields vs. Hidden-Markov Models in a Biomedical Named Entity Recognition Task. In: Proceedings of the RANLP 2007 conference, Bulgaria, Borovets (2007)
9. Raymond, C., Bechet, F., De Mori, R., Damnati, G.: On the use of finite state transducers for semantic interpretation. Speech Communication 48, 288–304 (2006)
10. Settles, B.: Biomedical Named Entitiy Recognition Using Conditional Random Fields and Rich Feature Sets. In: Proceedings of the COLING 2004 International Joint Workshop on Natural Language Processing in Biomedicine and its Applications (NLPBA), Geneva, Switzerland (2004)
11. Sha, F., Pereira, F.: Shallow Parsing with Conditional Random Fields. In: Proceedings of Human Language Technology-NAACL 2003, Edmonton, Canada (2003)
12. Sutton, C., McCallum, A.: An Introduction to Conditional Random Fields for Relational Learning. In: Getoor, L., Taskar, B. (eds.) Introduction to Statistical Relational Learning. MIT Press, Cambridge (2006)
13. Zhu, C., Byrd, R.H., Nocedal, J.: L-BFGS-B: Algorithm 778: L-BFGS-B, FORTRAN routines for large scale bound constrained optimization. ACM Transactions on Mathematical Software 23(4), 550–560 (1997)

Towards Flexible Dialogue Management Using Frames

Tomáš Nestorovič

University of West Bohemia in Pilsen
Univerzitní 22, 30614 Pilsen, Czech Republic
nestorov@kiv.zcu.cz

Abstract. This article is focused on our approach to a dialogue management using frames. As we show, even when dealing with this simple technique, the manager is able to provide a complex behaviour, as for example maintenance of context causality. Our research goal is to create a domain-independent dialogue manager accompanied with an easy-to-use dialogue flow editor. At the end of this paper, future work is outlined, as the manager is still under development.

1 Introduction

Dialogue management focuses on finding the best machine response to a user's (spoken) input. Many approaches to this issue emerged. They are based on different backgrounds ranging from finite state machines to intelligent agents. However, we decided to follow a way of using frames. Our aim is to implement and test a generic manager equipped with (relatively) uncommon way of keeping track of coherence of changing circumstances within a dialogue. This approach is based on application of a journaling system to the construction of frames. Last but not least, we also would like to offer the manager as a free software on our website[1] as the development progresses promisingly towards the specified goal.

In the rest of the paper, we first describe the term frame and try to summarize it in short. Next, we move to our particular approaches and describe manager's context and history modules. The paper is concluded with the planned future work.

2 Frame-Based Dialogue Management in Brief

Frame-based management attempts to reflect most of the disadvantages exhibited by the state-based approach (inflexibility above all). Here the basic construction asset is a *frame* (also referred to as an entity, topic, template, etc.) consisting of a set of *slots*.

To control the dialogue flow the system needs to select one of unacceptably filled slots. To inform the user about which slot was chosen, an appropriate prompt needs to be uttered by the system. The prompt is usually attached to a slot and invoked as a reaction to the "value-needed" event. Traditionally, additional event handlers are assigned as well, defining actions to be carried out when these events arise. However,

---

[1] http://liks.fav.zcu.cz

V. Matoušek and P. Mautner (Eds.): TSD 2009, LNAI 5729, pp. 419–426, 2009.
© Springer-Verlag Berlin Heidelberg 2009

the main purpose of a frame still remains to accumulating information gathered from the user. A variety of frame types evolved during research – [1] provides an overview.

Under frame-based management, a dialogue gets more flexible – a possibility to exhibit initiative during the discussion is granted not only to the system itself, but instead it is distributed between both partners [2] (the so-called *mixed initiative*). The scenario is always the same: at the beginning, the user provides an incomplete demand (due to his/her unfamiliarity with the system or speech recognition errors). To complete the demand, the system takes the initiative over and elicits additional information. Therefore, the frame-based management is mainly involved in information retrieval systems (traveling, weather or timetable services) [7].

3   An Example of Frame-Based Technique

Due to the fact that our dialogue management approach employs hierarchical extension to basic flat frames, many of algorithms solving particular issues are needed. They will be described in the next sections, however, now on to a top-level description.

The manager is divided into four collaborative modules (Fig. 1). The *Context* module maintains information about the current dialogue (active frames and relations between them are stored here). The *History* module serves as a source of historical data – it provides a basis for dereferencing/disambiguating user's utterances (for example "*the previous train*"). The *Core* module controls the behaviour of both modules – it interprets the current state of the dialogue and coordinates information stream flows. Additionally, the Core produces CTS (Concept-to-Speech) utterance descriptions and feeds them into the *Prompt planer* module. Here, we will add natural speech paradigms into descriptions – however, this module still remains unimplemented.

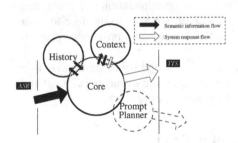   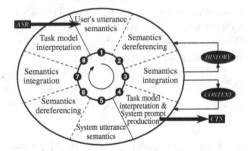

Fig. 1. Dialogue manager modules topology   **Fig. 2.** Action loop performed by the manager

From the top-level point of view, the manager loops in a cycle "system prompt – user's answer." All related actions are depicted in Fig. 2: the semantic information received from the ASR (Automatic Speech Recognition) module (1) needs to be disambiguated based on given dialogue history (2). Next, it is integrated into the current task context (3), and finally, the new context is interpreted and system prompt produced (4). Note that the system utterances follow the same way of processing as the user's ones do (5-8). The reason is that even the system may introduce new information that needs to be anchored within given context and recorded to history ("The next train leaves at 15:00"). Additionally, the manager deals with several key situations arising during the conversation:

a)   introduction of a new concept by the user,
b)   corrections (of both current concepts and relations between them),
c)   confirmations (of context fragments), and
d)   recalling information from the History module.

Solutions to these issues are the following. Every incoming semantics fragment is a priori supposed to either refer to historical data (d), or to introduce new information (a). Situations (b) and (c) are perceived as very similar – in particular, we deal with confirmation as with a special case of correction. Hence, the input semantics model get more simple as it is possible to represent both of them using the same element (Fig. 3).

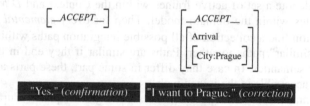

Fig. 3. Semantic information for confirmation (on the left) and correction

3.1   Frames and Relations between Them

As mentioned above, both frames and relations are parts of the Context module. Our notion of a frame is quite "concept-like" since it may hold single domain information at most. Hence, we design a frame to handle a specific concept type (for example *Time* concept). A frame is additionally equipped with a message queue containing demands for actions to be performed on this frame, and a journal for a roll-back operation.

Relations express how active frames are bound to each other. Templates for possible relations are defined within the manager's editor environment, and in run-time they are constructed in accordance with these templates. The Context module contains two types of relations – *standard relations* (to maintain relevant bindings) and *disambiguation relations* (to express a detailed description of a frame). An example of a proposed domain structure may be seen in Fig. 4.

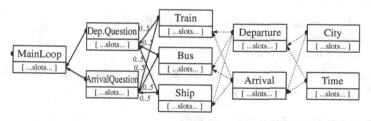

Fig. 4. Proposed structure for a timetable domain. Solid lines represent standard relation templates, whereas the dotted ones are disambiguation relation templates.

Let us stress that the Context is built upon *relations only*, i.e., every frame is within the Context registered using a *registration relation* (it is a special case of the standard relation). We found this approach of Context useful as operations with historical entities get simpler; see below.

3.2 Integrating the Semantic Information into the Context

Let us stick to the Context module description and skip the process of semantics dereferencing (see Fig. 2), we shall return to it later. Suppose that the input semantics went through dereferencing and is about to be integrated into the current context model. Here, the basic idea is a production and evaluation of all possible unification trees. Only the best evaluated one is then used as a template for the semantics integration. The complete algorithm is described in the following four steps.

1. Let F denote a set of active frames within the Context and D a set of frame templates within the domain model. Then for every *elemental* semantic information find a collection of all possible integration paths within $F \times D$.
2. Join "similar" paths together. Paths are similar if they end in the same elemental semantics. In case they differ in some part, these parts are made parallelly accessible in the joined path.
3. Build and evaluate all possible trees upon joined paths. Currently in our manager, there are six criteria for evaluation, as for example, whether a particular relation does exist or not, or whether a particular frame is on the path to the one interpreted as last.
4. Select the best evaluated tree and process (interpret) it as a LISP program structure. The basic interpretation may be affected by processing *system semantics* (for correction, for instance).

3.3 Dialogue Stack

The manager maintains currently discussed "topics" in a form of a stack. This approach found an inspiration in Grosz and Sidner's framework [6]. However, in comparison to it, our stack topics are frames themselves, not abstract descriptors. There is another difference: an absence of interruption detection, i.e., absence of a capability to detect a discussion topic change – to make a change, user is supposed to utter an explicit correction demand, for example *"No, I want to get there by ship."* Therefore in our approach, the stack plays a role of a purely passive component of the manager, designed to collect:

■ newly emerged concepts (i.e., frames) in the discussion,
■ frames with an updated content, and
■ currently discussed frames.

Frames are stored in the stack as long as they meet at least one of the conditions above, otherwise they are popped out.

3.4 Corrections Made from the User Side

The ability to accept corrections of a current context must be an essential part of every manager due to ASR errors arising during an interaction (the ASR module serves as the weakest part of every dialogue system [7]). However, sentences similar to "*I don't want Y, but X instead*" provide semantics distinguishable by the ASR only, but say nothing about user's intentions. It is the manager's task to guess them.

Our approach to this issue is a restriction to a last manipulation with a particular frame. We distinguish between two types of manipulations – *construction of frame*, and *its use as a super-frame* for another one (Train is a super-frame for Arrival). Hence, if Y has not been used as any super-frame until now, user's correction "*I don't want Y*" is perceived as a rejection of Y. Similarly, if the last manipulation was making Y a super-frame for Z, then by uttering the same sentence user is rejecting the relation between Y and Z, not the existence of Y itself.

Our current approach is best suited for system's prompts informing a user about recognized frames immediately. For example, instead of "*Which time do you want to leave?*" a production of "*Which time do you want to leave by train from Prague?*" offers a possibility to make instant changes of transportation means or departure city.

The manager is able to infer an invalidation of related parts of the context on a basis of one particular change. We call this mechanism a *causality consistence mechanism* (its description follows). Using it, information dependent on changed fragment disappears from the context and the system is forced to re-elicit it. However, we would like to extend the current "correction/causality" mechanisms with the possibility of recovering last confirmed fragments. An open question remains whether this introduces rather more confusion than help.

3.5 Causality Consistence Mechanism

As mentioned, the mechanism helps to keep the context in causal consistence. We decided for a *distributed approach*, i.e., every part of the context (frame and relation) maintains its own agenda of what operations it was involved in. Compared to a centralized approach, the distributed one offers more flexibility regarding a roll-back.

We distinguish between two types of operations: information *reading* and *interpretation* of a frame. Entries of these operations are inserted into particular frame journals, and in a case of reading, also into journals of relations the operation covers. Information *changing* causes a roll-back of journals. For an illustration of a roll-back, let us stick to the timetable domain and consider a context fragment depicted in Fig. 5. Here, the system uttered a particular transportation means in Q/3 (DepartureQuestion frame, slot *3*; "*The next ship from Delft to London leaves at 10:15*"), and an additional back-end reading was performed in S/1 (Ship). Frames' journal contents are depicted in Fig. 6. The figure also serves as a trace of the interpretation algorithm as time is involved. Consider the user changed the City of departure from Delft to Oslo. Now, neither of the previous readings R_1 and R_4 is valid and the journal of the City of departure will be rolled-back up to R_4. However, the rolled-back fragment still remains stored in a REDO part of the journal. The same must be done with both readers (Ship and DepartureQuestion), temporarily losing the Arrival branch (during the reinterpretation it is recovered utilizing the REDO). As for the Departure, it remains unaffected. To keep track of what parts of the context were modified, notification messages with D/0, S/0 and Q/0 are sent to Departure, Ship and DepartureQuestion, respectively. By this, the model reaches the consistency and a new interpretation may begin. (Example continues.)

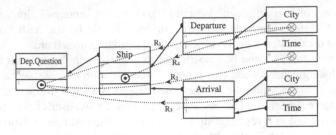

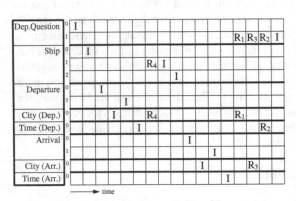

Fig. 5. Readings within given context fragment are dotted

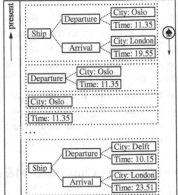

Fig. 6. Contents of frame journals in time; *R* represents entry with a reading, *I* with an interpretation

Fig. 7. History module content

3.6 Context Interpretation

An essential goal of the interpretation is to search for newly emerged, updated or missing fragments of the context, and invoke their integration, validation (proving they are true) or elicitation, respectively. As mentioned above, the behaviour of frames is modeled by message passing. To reach not only a flexible, but a collaborative interpretation environment as well, we additionally employ:

- *the dialogue stack* (only the message queue of a frame on its top is processed as long as it is not empty), and
- *an interpretation token* (a frame which holds it may pass it over to one of its sub-frames; the interpretation token is realized as a standard message).

In combination of both of these components the dialogue flow is managed on one side quite strictly, however, on the other hand it is still easily adaptable to new circumstances. For example, although the user provides new information (new frames are pushed onto the stack), from the manager's point of view, the information may be *currently* irrelevant (until the frames do not hold the interpretation token, the manager *mostly* ignores them; mostly = except for necessary operations like disambiguation). If the information is currently really irrelevant, it will disappear from the stack (however, not from the context) and the dialogue will continue in accordance with manager's original plan. Note that this plan may be affected by user's corrections.

Let us continue and finish the "roll-back" example above. The interpretation starts with obtaining the messages, and hence, reinterpreting the City of departure. Next, it continues reevaluating S/1 and moves to S/2. Here, REDO part of the journal will be employed and the formerly lost branch recovered. Finally, the interpretation reaches Q/1 and a new prompt is generated – *"The next ship from Oslo to London leaves at 11:35."* Note that the new time was obtained by searching in database, initiated by the DepartureQuestion frame.

3.7 History Module

Now, as the Context module is described, let us return to the semantics dereferencing depicted in Fig. 2. The structure of the History module consists of a series of previously used entities, similarly as proposed in [4]. We define an *entity* to be a set of relations (i.e., a fragment of context) which meet the following conditions.

- All information held in frames is confirmed.
- All standard relations are confirmed.
- Every frame content is acceptable (i.e., it does not need further disambiguation).

The history is built automatically after semantics has been integrated. If a context fragment meets the three conditions listed above, a set of entities based on this fragment is created. The process of generation starts with an entity containing the most concrete information and ends with the most general one – Fig. 7 demonstrates.

The inverse operation, reading the history, is initiated implicitly, i.e., every incoming semantic unit is perceived as a reference to historical data. The process of dereferencing tries to take as big fragment of semantics as possible and match it against the most general historical entity found closest to the "present." If a match is found, the entity is transformed into a semantics replacing the original fragment in the input. However, the reading is a complex issue. For example in the reference *"the previous ship,"* first an entity expressing a ship must be found (in Fig. 7 the above one), and once found, the *"previous <entity>"* must be dereferenced (in Fig. 7 the below one). We approach this by introducing a *stack of pointers* to the history time line where successful dereferences were realized. Therefore, once the inner reference is resolved, the outer starts searching from the point the inner was satisfied either back or forward in the history (in our case marked with ♠ sign in Fig. 7).

4 Future Work

In this paper, we have omitted to describe the production of system utterances. We currently use a XML-based sentence description to first express the content itself, and second to mark distinguished fragments. For example, *"<q><concept ship> <TheShipFrom> <concept departure><r _parent.#1/></concept><Leave/>...</concept> </q>"*, is our current (simplified) description of *"The ship from Oslo leaves...".* Hopefully, this approach leads to a CTS output fed into the Prompt planner (Fig. 1) which is not realized yet.

We also omitted to mention a disambiguation process – currently, it is in progress. We found an inspiration in the McGlashan's approach [5] consisting of lists of related topics (frames) which need to be discussed prior to a database query is performed.

Also the manager lacks a confirmation process. However, because finding entities in the context depends on it, we simulate it on-the-fly. As a real solution, we propose an introduction of a special type of slot which question will be built with respect to information to be confirmed. This would enable the manager to automatically detect fragments of context which should be considered as believable after a user's positive answer is obtained.

5 Conclusion

We are on a development of a domain-independent dialogue manager. It utilizes frames to represent context knowledge. We adjusted well known approaches to fit our purposes of creating a manager with complex behaviour. In [1] we found a motivation for nested frames technique, [4] served us as a basis for historical entities processing we augmented with stack of pointers – another stack besides the one (partially) adopted from [6]; we are inspired of the disambiguation process in [5], however, we would like to extend it to work "reversibly" as well (enabling relaxation). Last but not least, we presented here our journaling system for keeping the context in causal and coherent state. Once the manager is finished, it will be applied in car navigation and timetable domains to thoroughly test its management skills. Its previous version [3] was applied to car navigation domain only. We expect to obtain far better results in this domain since the previous version employed flat frames only (extended with another features).

References

1. Cenek, P.: Hybrid dialogue management in frame-based dialogue system exploiting VoiceXML. Ph.D. thesis proposal, Masaryk University, Brno (2004)
2. McTear, M.F.: Modeling spoken dialogues with state transition diagrams: experiences with the CSLU toolkit. In: Proc. of ICSLP, paper 0545 (1998)
3. Nestorovič, T.: Navigation System: An Experiment. In: Proc. of NAG-DAGA, paper 470, Rotterdam (2009)
4. Zahradil, J., Müller, L., Jurčíček, F.: Model světa hlasového dialogového systému. In: Proc. of Znalosti, Ostrava, pp. 404–409 (2003)
5. McGlashan, S.: Towards Multimodal Dialogue Management. In: Proc. of Twente Workshop on Language Technology, vol. 11, pp. 1–10
6. Grosz, B.J., Sidner, C.L.: Attention, intention and the structure of discourse. Computational Linguistics 12(3), 175–204
7. Gustafson, J.: Developing Multimodal Spoken Dialogue Systems - Empirical Studies of Spoken Human-Computer Interaction. Ph.D. thesis, KTH, Department of Speech, Music and Hearing (2002)

Author Index